MEASURING AND AUDITING BROILER WELFARE

MEASURING AND AUDITING BROILER WELFARE

Edited by

C. Weeks and A. Butterworth

School of Veterinary Science
University of Bristol, UK

CABI Publishing

CABI Publishing is a division of CAB International

CABI Publishing
CAB International
Wallingford
Oxfordshire OX10 8DE
UK

Tel: +44 (0)1491 832111
Fax: +44 (0)1491 833508
E-mail: cabi@cabi.org
Web site: www.cabi-publishing.org

CABI Publishing
875 Massachusetts Avenue
7th Floor
Cambridge, MA 02139
USA

Tel: +1 617 395 4056
Fax: +1 617 354 6875
E-mail: cabi-nao@cabi.org

A catalogue record for this book is available from the British Library, London, UK.

Library of Congress Cataloging-in-Publication Data

Measuring and auditing broiler welfare / edited by C. Weeks and A.
Butterworth.
    p. cm.
Based on papers from a conference held in Bristol, UK, May 2003; with
additional material.
Includes bibliographical references (p.).
  ISBN 0-85199-805-4 (alk. paper)
  1. Broilers (Poultry)--Health.   2. Broilers (Poultry)--Quality.   3.
Poultry products--Quality.   I. Weeks, C. (Claire)   II. Butterworth, A.
(Andrew)   III. Series.
SF498.7.M43 2004
636.5′083--dc22

                                                    2003027109

ISBN 0 85199 805 4

Typeset by AMA DataSet Ltd, UK.
Printed and bound in the UK by Cromwell Press, Trowbridge.

Contents

Contributors*

Algers, B., Department of Animal Environment and Health, Faculty of Veterinary Medicine, Swedish University of Agricultural Sciences, P.O. Box 234, SE-532 23 Skara, Sweden.

Appleby, M.C., The Humane Society of the United States, 2100 L Street, NW Washington, DC 20037, USA.

Barnett, J.L., Animal Welfare Centre, Department of Primary Industries, Victorian Institute of Animal Science, 600 Sneydes Road, Werribee, Victoria 3030, Australia.

Berg, C., Department of Animal Environment and Health, Faculty of Veterinary Medicine, Swedish University of Agricultural Sciences, P.O. Box 234, SE-532 23 Skara, Sweden.

Bessei, W., Applied Farm Animal Ethology and Small Animal Production, Institute for Animal Breeding and Husbandry, University of Hohenheim, D 70 593, Stuttgart, Germany.

Butterworth, A., School of Veterinary Science, University of Bristol, Langford, Bristol BS40 5DU, UK.

Charles, D.R., DC R&D, Loughborough, Leicestershire.

Coleman, G.J., Department of Psychology, Monash University, Caulfield Campus, Victoria 3145, Australia.

Cook, P., RL Consulting Ltd, The Field Station, Wytham, Oxford OX2 8QJ, UK.

Glatz, P.C., South Australian Research and Development Institute, Pig and Poultry Production Institute, Roseworthy, SA 5371, Australia.

Haslam, S.M., School of Veterinary Science, University of Bristol, Langford, Bristol BS40 5DU, UK.

---

* This list excludes authors of poster papers.

Hemsworth, P.H., Animal Welfare Centre, University of Melbourne and Department of Primary Industries, Victoria Institute of Animal Science, Werribee, Victoria 3030, Australia.

Hocking, P.M., Roslin Institute (Edinburgh), Roslin, Midlothian EH25 9PS, UK.

Gordon, S.H., ADAS Gleadthorpe, Meden Vale, Mansfield, Nottinghamshire NG20 9PF, UK.

Jansen, R.G., Noldus Information Technology b.v., P.O. Box 268, 6700 AG Wageningen, The Netherlands.

Julian, R.J., Department of Pathobiology, Ontario Veterinary College, University of Guelph, Guelph, Ontario N1G 2W1, Canada.

Kestin, S.C., School of Veterinary Science, University of Bristol, Langford, Bristol BS40 5DU, UK.

Kettlewell, P.J., Silsoe Research Institute, Wrest Park, Silsoe, Bedfordshire MK45 4HS, UK.

Kristensen, H.H., Royal Veterinary and Agricultural University, Grønnegaardsvej 8, DK-1870 Frederiksberg C, Denmark.

Lymbery, P., World Society for the Protection of Animals (WSPA), 14th Floor, 89 Albert Embankment, London SE1 7TP, UK.

Main, D.C.J., School of Veterinary Science, Division of Farm Animal Science, University of Bristol, Langford, Bristol BS40 7DU, UK.

Mench, J., Department of Animal Science, One Shields Avenue, University of California, Davis, CA 95616, USA.

Mitchell, M.A., Roslin Institute (Edinburgh), Roslin, Midlothian EH25 9PS, UK.

Morton, D.B., Centre for Biomedical Ethics, Division of Primary Care, Public and Occupational Health, University of Birmingham, Edgbaston, Birmingham B15 2TT, UK.

Noldus, L.P.J.J., Noldus Information Technology b.v., P.O. Box 268, 6700 AG Wageningen, The Netherlands.

Pattison, M., Rossmoyne, Barbon, Carnforth, Cumbria LA6 2LS, UK.

Prescott, N.B., Silsoe Research Institute, Wrest Park, Silsoe, Bedfordshire MK45 4HS, UK.

Sanotra, G.S., Department of Animal Science and Animal Health, Division of Ethology and Health, The Royal Veterinary and Agricultural University, Grønnegaardsvej 8, DK-1870 Frederiksberg C, Denmark.

Walker, A.W., ADAS Gleadthorpe, Meden Vale, Mansfield, Nottinghamshire NG20 9PF, UK.

Wathes, C.M., Silsoe Research Institute, Wrest Park, Silsoe, Bedfordshire MK45 4HS, UK.

Weeks, C.A., School of Veterinary Science, University of Bristol, Langford, Bristol BS40 5DU, UK.

Whay, H.R., School of Veterinary Science, Division of Farm Animal Science, University of Bristol, Langford, Bristol BS40 7DU, UK.

Wilkins, L.J., School of Veterinary Science, University of Bristol, Langford, Bristol BS40 5DU, UK.

Wotton, S., School of Veterinary Science, University of Bristol, Langford, Bristol BS40 5DU, UK.

Foreword

Forty years ago, when questions about the welfare of animals in modern production systems started to be raised, inspired by Ruth Harrison's seminal book *Animal Machines* (Harrison, 1964), the main targets were battery cages for laying hens, stalls for gestating sows, and crates for veal calves. Criticisms were also levelled at certain invasive practices, such as castration, tail docking and beak trimming, and at some management procedures, such as transport and preslaughter handling. The plight of chickens kept for meat production was hardly considered. After all, they appeared to have much more behavioural freedom than did caged hens, and, unlike hens, their beaks were not trimmed. Moreover, breeding for fast growth meant that surgical castration was no longer practised and chemical castration was rapidly being phased out. Compared with many other classes of intensively kept livestock, broilers seemed to have a good life. Of course, there was a dark side, but the catching, transport and slaughter of broilers were swift procedures that took place at night or behind closed doors and were largely unavailable for public scrutiny. The argument was made that death is never pleasant in any case, and premature death is a necessary part of all meat production.

How things have changed! Whereas the welfare of most other farm livestock has been slowly improving in recent years – for example, by the introduction of alternative husbandry systems to replace battery cages, sow stalls and veal crates, and by the adoption of more humane handling, transport and preslaughter management techniques – the lot of broilers has not shown a parallel improvement. There can be little doubt that the average broiler today has poorer welfare than its predecessor of 40 years ago. The poultry industry's incessant drive for efficiency and reduction of costs has largely been to blame. Admittedly, there are some encouraging signs. First, the development of controlled-atmosphere stunning will undoubtedly lead to huge welfare benefits for broilers when eventually the method is adopted world-wide. Second, there have been some very promising innovations in the catching and transport of broilers. Third, the setting of welfare standards for suppliers of poultry by restaurant chains and fast food companies, such as McDonalds, Burger King and KFC, will eventually result in welfare gains, particularly in North America, where welfare progress has been slow. However, the most encouraging sign is the publication of books such as this, where welfare problems are openly discussed and solutions suggested.

It used to be thought that all farm animal welfare problems could be solved by correct environmental design. Experience with modern broilers and their parent stock, broiler breeders, has cast doubt on this assumption. They have been bred so intensively for fast growth that it is becoming increasingly difficult to provide them with an environment that will satisfy their needs. I would maintain that we have already reached the point where it is impossible to keep modern broiler breeders humanely; either we keep them so restricted of food that they suffer hugely from hunger, or we allow them to satisfy their appetite and they suffer from diseases of obesity. Already, broilers cannot be grown at altitude without encountering an incidence of ascites that is unacceptable from both a production and a humane stand-point. Litter that might be acceptable as a foraging substrate becomes unacceptable when the birds are so changed physically and behaviourally that they cannot or will not adopt the normal resting posture of perching, and instead sit or lie in the litter, thus promoting hock burn and breast blisters. An atmosphere polluted with ammonia and dust becomes much more of a hazard to broilers which are so unfit that the slightest exertion results in open-mouth breathing, thus bypassing the filtration mechanisms in the upper respiratory tract. So, to a large extent, the welfare problems described in this book will not be solved by environmental manipulations. It is the bird that must be changed, and the long-term solution is in the hands of the primary breeding companies. It is hoped that they will take note and consider new breeding strategies that will reverse the current trend. Of course, this book will be invaluable to a much wider audience than just the primary breeders; everyone in any way connected to the poultry industry can gain enormously from the wealth of information it contains. In addition, I hope that it might serve an even broader function, and that is to warn other sectors of the animal production industry that there is a cost to be paid for a blinkered drive towards ever-cheaper animal products – and that cost is paid by the animals.

Ian J.H. Duncan
Professor of Poultry Ethology,
Chair in Animal Welfare,
University of Guelph,
Guelph, Ontario, Canada

Introduction

CLAIRE WEEKS

School of Veterinary Science, University of Bristol, Langford, Bristol, UK

The rearing of chickens for meat is one of the few areas of agriculture that continues to expand. From its conception only half a century ago, current global production has increased rapidly and is now approaching the 50 billion mark. The explosion in poultry output has been particularly marked in South America, where production in the last 5 years alone has increased by 30%, and Asia, which has seen a 22% increase. With these sorts of numbers, most chickens cannot be farmed under traditional husbandry in a sustainable way with minimal environmental impact. In Chapter 22, Philip Lymbery discusses some of the issues concerned with globalization and large-scale production. Not least is the startling fact that poultry consume a sixth of world cereal production (Qureshi, 2001). The major broiler-producing countries are the USA, with almost 9 billion, followed by China, with more than 7 billion, and then Brazil, with an output of some 5 billion birds annually. Mexico, Indonesia and Thailand all produce over a billion broilers each year. The global trade in chickens is significant, but in most markets home consumption predominates.

Broilers, through intensive genetic selection, are among the fastest-growing farmed species. At the moment of hatch, a chick weighs about 50 g and grows to a slaughter weight of nearly 2 kg at 37 days of age. Modern strains of broilers have an average weight gain of 50–55 g daily. The intensity of genetic selection is such that the age of slaughter decreases by 1 day per year under optimal conditions. However, great emphasis on the selection of animals for high economic production efficiency is undoubtedly related to the risk of behavioural, physiological and immunological problems. Growth rate is so extremely rapid that viability becomes affected, and it is a challenge to keep breeding stock fit for a sufficiently long time for them to reproduce. The modern broiler breeder represents an animal whose welfare needs are impossible to meet: either the animal suffers hunger (which is generally the lesser of two evils) or it becomes obese and suffers health problems and reproductive failure. It would be impossible to exclude the welfare of breeders from a consideration of broiler welfare, and Paul Hocking outlines the main issues in Chapter 2.

With the scaling up of numbers, language and attitudes alter. The farmer becomes a producer; with up to 60,000 birds in one shed, the individual chicken becomes a broiler or a unit of production, and the whole is geared up into an industry that, more often than not, is vertically integrated and increasingly multinational, with a few very large players. In Chapter 5, veterinary surgeon Andrew Butterworth emphasizes the importance of considering ill health in terms of its consequences for the individual bird affected rather than for the bottom line. It is easy also to forget the sentience of animals that are factory-farmed, to use a term coined by Ruth Harrison in her book _Animal Machines_, which so shocked the British public when it came out in 1964 that the government set up a committee of enquiry chaired by Professor F.W. Rogers Brambell. Much of the UK farm animal welfare legislation has grown from the recommendations of this committee (Brambell, 1966), although basic anti-cruelty laws have been in place for over 100 years. David Morton outlines the philosophy of concern about animals in Chapter 20.

Public concern about animal welfare has led to laws and codes of recommendation in many countries, and in Switzerland and Sweden, for example, this has limited intensification (see Chapter 18). The public increasingly demands that human food is not only safe but is also humanely produced. Michael Appleby outlines some of the current expectations in Chapter 21. Undoubtedly, the various food safety scares have focused attention on the way in which food is produced, and many Europeans are deeply suspicious of genetic modification. With food safety as the major issue, a plethora of farm assurance schemes has developed in recent years. With them has come the need for auditing to deliver credibility, and Paul Cook gives a practical view of the auditing of broiler production in Chapter 16. The scope of assurance has, in most cases, expanded to embrace animal welfare and environmental issues. David Main and Helen Whay assess whether such schemes assure good welfare in Chapter 17.

The production of broiler chickens is one of the most intensive, but also the most uniform forms of animal production in terms of genotypes, feed, housing and handling, and thus it follows that the auditing of broiler welfare is possible on a global scale. Most multiple retailers and restaurant chains are developing, or have in place, global standards that include animal welfare as well as quality issues that will be audited. There are numerous challenges in meeting cultural and religious beliefs (for example, kosher slaughter) and inevitable differences of opinion, not only in what constitutes welfare but also in how to measure it. We hope that this book will provide some guidance. Our aim has been to provide the rationale and science needed to identify the major concerns of broiler welfare, such as lameness (this is considered by Joy Mench in Chapter 1), and also, where possible, to identify what is practicable to measure or audit. In a few cases it has been appropriate to identify how to measure or audit a particular aspect. However, on the whole this has not been done, because there will be such a variety of priorities for those setting up an auditing scheme.

This is a new, developing area. Auditing of broiler welfare is in the expansion phase, and rationalization is likely to take place sooner rather than later. Among the practical issues are the numbers of potential inspectors and auditors needed for broiler units, threats to biosecurity, the time required and the administration involved. It is quite possible for a single farm to have visits from auditors representing government, animal welfare organizations, farm assurance schemes, commercial firms, multiple retailers and restaurant chains, to name but a few!

This book considers, in Part 1, the various ways in which the chickens themselves can indicate that their welfare is not optimal. These are often termed 'outcomes' and may be considered as the consequences for the animals of not getting the inputs right. Part 2 then examines some of the more important aspects of the chickens' environment and feed that may affect their welfare. These include the conditions at the end of their short life, when they are caught and transported (Chapter 12) and then slaughtered (Chapter 13). Quite properly, these inputs are often considered in terms of the consequences for the birds, and so outcomes may be the most effective measure of welfare, as Werner Bessei suggests for stocking density (Chapter 11), for example. Increasingly it is recognized that defining the resources does not ensure welfare. Not only management and husbandry but also – especially – stockmanship has a vital part to play. Paul Hemsworth and Grahame Coleman outline the influence of humans in determining welfare in Chapter 15. Clearly, all those who have responsibility for the care of sentient animals ought to have training in welfare and many assurance schemes are now specifying this.

Part 3 of the book also considers, sometimes critically, some practical issues concerning the measurement and auditing of broiler welfare, with examples from current systems. The broiler industry is so large that further developments in automation are inevitable. Automated catching is increasing and automatic shackling is quite feasible. We can expect more automatic measuring systems that will assist with the auditing and assurance of compliance. Some commercial plants are automatically measuring and scoring footpad lesions, for example. The final chapter speculates about which aspects of behaviour it might be possible to measure automatically on-farm in the future.

Acknowledgements

Thanks are due to Safeway plc for sponsorship of the conference in Bristol, UK, in May 2003, from which this book has developed. I have included some material from Philip Lymbery and G. Singh Sanotra in this chapter, for which I thank them.

References

Brambell, F.W.R. (1965) Report of the technical committee to enquire into the welfare of animals kept under intensive livestock husbandry systems. HMSO, London.

Harrison, R. (1964) *Animal Machines. The New Factory Farming Industry*. Vincent Stuart, London.

Qureshi, A.A. (2001) Disparities in world poultry production in 1999. *World Poultry* 17 (2), 22–23.

1 Outcomes (Animal-based)

1 Lameness

J. MENCH

Department of Animal Science, University of California, Davis, California, USA

Introduction

One of the most serious welfare problems in broiler production is the high incidence of skeletal disorders, particularly those that lead to impaired mobility or lameness (European Commission, 2000). The development of many of these disorders is related to selection and management for rapid growth, since they are rare in slow-growing meat strains and laying strains of poultry but common in modern commercial broilers, broiler breeders, ducks and turkeys (Wise and Nott, 1975; Duff and Hocking, 1986; Havenstein *et al.*, 1994; Hester, 1994; Kestin *et al.*, 2001). Kestin *et al.* (1992) evaluated more than 2000 broilers reared under commercial conditions and found that 90% had detectable gait abnormalities at market age while 26% had gait abnormalities that could be considered severe, causing an impairment of function. Despite the initiation of programmes by primary breeders to select for leg soundness, more recent commercial surveys of flocks in Scandinavia found similar rates of severe gait disorders, although rates depended on the genetic stock of the birds and showed wide between-farm variation (Sanotra *et al.*, 2001; Sanotra and Berg, 2003). Skeletal problems are not just a welfare issue – they are also costly to the industry. Bennett *et al.* (1999) calculated the direct costs associated with various endemic farm animal diseases in the UK, and found that skeletal problems were by far the most costly diseases for poultry producers in terms of output loss, resource wastage, and treatment and prevention costs. In the USA, it is estimated that leg problems are responsible for 1.1% of broiler mortality and 2.1% of carcass condemnations and downgrades annually, and cost the poultry industry billions of dollars each year (Morris, 1993).

Although mobility problems in poultry are often generically referred to as 'leg weakness' or 'leg problems', mobility can actually be affected by a range of different bone and soft-tissue disorders, and is also influenced by the conformation (e.g. breast muscle development) of the birds. Determining the relative prevalences of specific problems on commercial farms, the factors affecting their occurrence, and assessing their effects on mobility are critical to the development of selection and management programmes to reduce the incidence and severity of these problems. This chapter will provide an overview of some of the more common skeletal disorders, describe the assessment methods currently available, and explore the relationships between skeletal disorders, gait and welfare issues.

Skeletal Disorders

There have been a number of recent reviews of the aetiology of skeletal disorders in meat-type birds (Hester, 1994; Thorp, 1994; Julian, 1998; Butterworth, 1999; European Commission, 2000; McNamee and Smyth, 2000; Whitehead *et al.*, 2003; Chapter 4 in this book). Both infectious and non-infectious skeletal conditions are seen in commercial broiler flocks. Among the most common non-infectious skeletal disorders are long-bone deformities such as valgus–varus deformities and 'twisted legs'. Birds with long-bone deformities usually have bent or twisted tibiae and tarsometatarsi (Fig. 1.1), frequently accompanied by slippage of the gastrocnemius tendon. The femur can also be affected. Long-bone deformities can cause severe lameness. Broilers with long-bone deformities may become emaciated, and the muscle of the affected leg may atrophy (Julian, 1998).

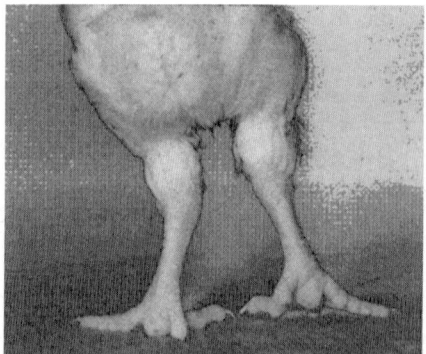 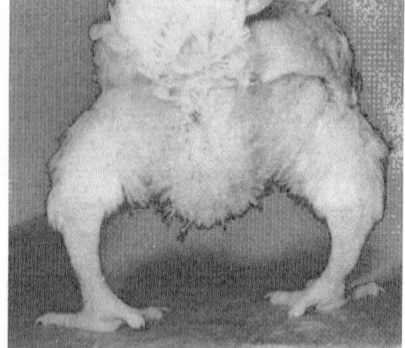

Fig. 1.1.   Broilers with valgus (knock knees), varus (bowed legs) and twisted leg deformities. The primary contributors to long-bone deformities like these are body weight and growth rate. (Photographs courtesy of Richard Julian.)

Another common leg problem is tibial dyschondroplasia (TD), a heritable disorder that can cause lameness. TD results from inadequate vascularization and ossification of the growth plate, and is characterized by the presence of an opaque cartilage mass located below the epiphyseal plate that extends into the metaphysis (Fig. 1.2). If the area of cartilage is large, the angle of the tibial plateau can be altered sufficiently to cause angular and rotational deformities (Lynch *et al.*, 1992). Severe lesions cause abnormal biomechanical forces, and therefore gait alteration, additional bone abnormalities (Farquharson and Jeffries, 2000), and even fractures (Julian, 1988). TD is seen in meat-type poultry but is rare or absent in other birds, suggesting that it is related to rapid growth. However, the expression and severity of TD may or may not be correlated with body weight (Cook *et al.*, 1984; Wong-Valee *et al.*, 1993; Kuhlers and McDaniel, 1996; Sanotra *et al.*, 2001a; Sanotra and Berg, 2003).

Other common non-infectious skeletal disorders in broilers include spondylolisthesis ('kinky back'), which is due to dislocation of the thoracic vertebrae, and rickets. Older birds, such as broiler breeders, also manifest degenerative changes in the joints and rupture of the ligaments and tendons (Duff and Hocking, 1986). Infectious disorders causing leg problems in broilers include arthritis/tenosynovitis, infectious stunting syndrome, and bacterial chondronecrosis with osteomyelitis (BCO), sometimes called 'femoral head necrosis' or 'proximal femoral head degeneration'. BCO is an infection of the proximal part of the femur or tibia, usually with *Staphylococcus*, although other infectious organisms may also be present (Butterworth, 1999; McNamee and Smyth, 2000).

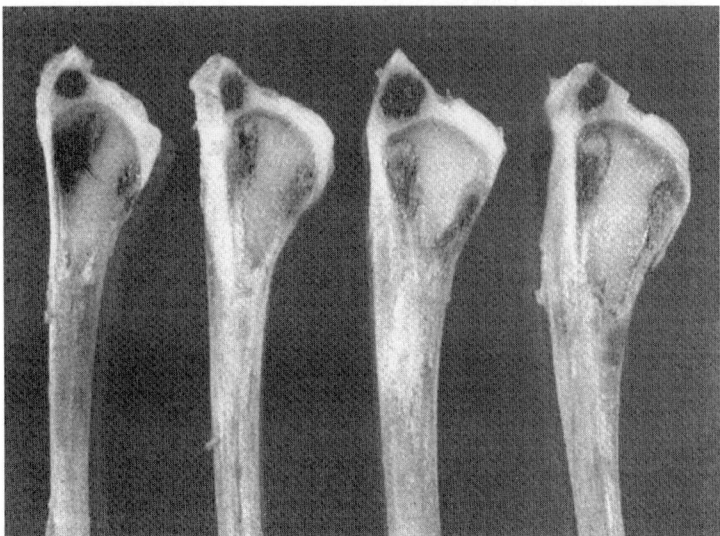

Fig. 1.2. Cross-sections through tibiae showing tibial dyschondroplasia lesions. Tibial dyschondroplasia is an overgrowth of cartilage in the bone growth plate, and is common in poultry selected for rapid growth. The lesions shown here are severe and have caused varying degrees of bending of the tibiae. (Photograph courtesy of Richard Julian.)

Some of the additional problems that can cause lameness in broilers include epiphyseal separation, ruptured gastrocnemius tendons, and other traumatic disorders, such as dislocation during catching, which can result from or be exacerbated by other skeletal disorders. While osteoporosis has not been considered a significant problem in broilers, there is evidence that selection for rapid growth in broiler strains is associated with decreased bone mineralization and increased porosity of cortical bone, and hence reduced bone quality (Williams *et al.*, 2000). Heavier broilers and turkeys have poorer bone strength than lighter birds (Yalcin *et al.*, 1998; Crespo *et al.*, 2000) and heavier turkeys are more likely to sustain femoral fractures during growth (Crespo *et al.*, 2000).

Many commercial broilers have crooked toes (Sanotra *et al.*, 2001a), which is probably caused by uneven tension in the tendons rather than being a skeletal disorder *per se* (Nairn and Watson, 1972). However, crooked toes do contribute to gait abnormalities (Sanotra *et al.*, 2001), as does pododermatitis (footpad dermatitis or 'ammonia burns'), a disorder that typically occurs as a result of the birds' feet and legs being in contact with poorly maintained litter (Su *et al.*, 2000; Chapter 3 in this book).

The reported incidence of particular skeletal disorders varies widely from one flock to another. For example, recent surveys of commercial flocks in Scandinavia (Sanotra *et al.*, 2001a; Sanotra and Berg, 2003) showed an incidence of mild to severe TD ranging from 32% to nearly 90% in different flocks; on average, 17.4% of birds in Swedish flocks had moderate to severe TD lesions. The incidence of BCO in Swedish flocks ranged from 0 to 24% (average 10.4%), while the incidence of angular deformity (specifically valgus–varus) ranged from 5 to 74% (on average 50% in Swedish flocks and 37% in Danish flocks). However, these numbers do not take into account birds culled for lameness. In a smaller-scale study, McNamee *et al.* (1998) performed histological examinations on legs from 44 lame and 22 non-lame birds culled from commercial flocks, and found that the most common primary cause of lameness was BCO, followed by angular deformities and spondylolisthesis; however, 61.3% of lame broilers and 23% of non-lame broilers also had TD lesions, and the birds could have multiple disorders. BCO was also found to be the most common cause of lameness in cull birds in broiler flocks in Northern Ireland, being present in more than 17% of lame birds examined (McNamee and Smyth, 2000). In contrast, Julian (Chapter 4 in this book) estimates that valgus–varus, TD and spondylolisthesis together cause 65–80% of the leg deformities and lameness seen in North American broiler flocks, but again the frequency and severity of these problems in birds that are culled or condemned in commercial flocks varies widely (e.g. Riddell and Springer, 1984).

The causes of the different skeletal disorders are complex and multifactorial (Sullivan, 1994; Thorp, 1996), which no doubt accounts for the reported variation. The incidence of infectious skeletal disorders in a particular flock, for example, is influenced by any factor that affects rates of infection generally; for instance, environmental bacterial loads and the immune status of the flock (Butterworth, 1999). In addition, pre-existing pathologies and injuries (for example, cuts and abrasions) that provide possible entry points for bacteria probably play a role (Thorp *et al.*, 1993; McNamee and Smyth, 2000). Slower-growing birds have a lower incidence of BCO (McNamee and Smyth, 2000), so selection and management for growth are contributing factors. Hatcheries and breeder flocks seem to be important sources of

Staphylococcus, so good hygiene in these areas is important in reducing the risk of infection in growing birds (McNamee and Smyth, 2000).

The causation of non-infectious disorders is similarly complex, but as Sorensen (1992, page 221) states: 'there is no doubt that the rapid growth rate of birds used for meat production is the fundamental cause of [these] leg disorder problems.' High body weight and rapid growth cause the production of bone, cartilage and support-ing tissue of poor structural quality, and abnormal and uneven loads on developing bone, which can lead to abnormal bone growth (Rowland, 1989; Rath *et al.*, 2000). Rapid growth rate is, of course, influenced by genetics, and the incidence and sever-ity of gait and non-infectious skeletal disorders vary significantly from one genetic stock of commercial broiler to another (Kestin *et al.*, 1992, 1999; Sanotra and Berg, 2003). These disorders have heritabilities ranging from 0.1 to 0.4%, a range similar to those of other traits used as selection criteria in poultry (Mercer and Hill, 1984; Sorensen, 1992; Bihan-Duval *et al.*, 1996, 1997). While poor gait and high body weight are highly genetically correlated, the genetic correlations between body weight and specific disorders are generally low, raising the possibility that careful multitrait selection could improve leg health with little effect on growth rate (White-head *et al.*, 2003). Body weight and growth rate both contribute to non-infectious skeletal problems and impaired gait, however, and each may affect different types of problems (Bihan-Duval *et al.*, 1997; Kestin *et al.*, 1999). For this reason, aspects of management that increase body weight and growth rate, such as feeding high-density diets *ad libitum* and providing nearly constant light stimulation, which encourages high rates of feeding behaviour and does not allow a resting period, are important contributors to lameness and leg problems. Accordingly, restricting growth using qualitative or quantitative feed restriction (Yu and Robinson, 1992) and/or provid-ing a daily period of darkness either continuously or intermittently (Hester, 1994) helps to decrease non-infectious disorders. Birds kept at high stocking densities also have poorer gait (European Commission, 2000; Sanotra *et al.*, 2001b), possibly partly because activity is impeded (see below). Infectious agents such as mycoplasma can also play a role in the aetiology of non-infectious disorders (Butterworth, 1999), as can nutritional excesses and deficiencies (Edwards, 1992).

Assessing Leg Problems

As is obvious from the foregoing discussion, one method of assessing leg problems is the examination of affected birds in the flock. The probable cause of some disorders can be determined from the appearance and behaviour of the birds. For example, broilers with spondylolisthesis have abnormal posture, sit on their hocks, and make creeping movements backwards using their wingtips for balance (Sullivan, 1994). Birds with BCO lesions in the proximal end of the femur typically use one or both wing-tips for support during locomotion and vocalize loudly when pressure is applied to the affected area (Thorp *et al.*, 1993). Angular deformities that result in severe bowing or twisting of the legs can also be diagnosed in live birds. In most cases, however (including cases of spondylolisthesis), diagnoses need to be made or confirmed by gross (see Chapter 4) and sometimes histological post-mortem examination. For example, BCO lesions are not visible macroscopically in most birds

(McNamee *et al.*, 1998; McNamee and Smyth, 2000), and so accurate diagnosis requires histological examination.

Leg problems can be assessed non-invasively using radiography. A lixiscope, which is a portable, hand-held imaging device (similar to a fluoroscope) that allows manipulation of the limbs of live birds to evaluate leg pathology, can be used to produce a real-time image of the legs (Bartels *et al.*, 1989). Although problems such as necrosis, fractures and septic arthritis can be seen using this method, it is most useful for examining TD lesions, and primary breeders have used this method in their selection programmes against TD. Real-time imaging is expensive and the images can be interpreted only by highly trained and experienced individuals, so this technique is unlikely to be widely used for the on-farm evaluation of leg problems.

The most common method used for the evaluation of leg problems on-farm is probably gait scoring, using the Bristol system developed by Kestin *et al.* (1992). Small groups of birds are surrounded with a catching pen and are then observed as they walk out of the pen, either freely or when they are tapped on the back. Birds with a normal gait are scored 0, those with a slight gait impairment 1, and so on, with completely lame birds that cannot walk being given a score of 5. This system has been criticized because there is an element of subjectivity involved in assigning scores (European Commission, 2000), which can result in poor reliability between observers (Kestin *et al.*, 1992). Various attempts have been made to change the Bristol system to improve reliability; for example, by requiring multiple observers to reach consensus on scores (Classen *et al.*, 1994; Martrenchar *et al.*, 1999) or by collapsing categories (Yalcin *et al.*, 1998). We recently modified this system (Garner *et al.*, 2000) to include more objective inclusion and exclusion criteria by setting time limits for how long birds take to stand or remain standing or walking after being touched or prodded. This modified system proved to have significantly better test–retest and inter-rater reliabilities than the Bristol system when used by naive observers, although reliabilities for the Bristol system were also acceptably high, suggesting that reliability problems with gait scoring have perhaps been overemphasized. Nevertheless, even with the modified system there is an element of subjectivity. Gait scoring is also time-consuming and may require two people, one of whom must be trained to observe and score the birds properly.

Another method recently developed for assessing lameness is the 'latency to lie' (LTL) test (Weeks *et al.*, 2002). For this test, broilers are placed in a waterproof test pen; after a period of acclimatization of 15 min the pen is flooded with a shallow layer (30 mm) of tepid water. Chickens prefer not to sit in water, so flooding the pen motivates all but the most severely lame birds to stand. The time taken for each bird to lie down is then recorded. Latencies to lie are highly inversely related to gait scores (Weeks *et al.*, 2002; Berg and Sanotra, 2003), and can be measured more objectively. However, the original LTL method was more time-consuming than gait scoring, and for commercial use this meant that a special pen had to be set up in the broiler house. Berg and Sanotra (2003) modified this test to make it simpler to use in a commercial setting. For their test, birds are placed individually in water-filled tubs (Fig. 1.3); there is no need for special pens and the acclimatization period is eliminated. Testing the birds individually also ensures that birds do not influence one another's behaviour. Using this method, 12–15 birds can be evaluated on a commercial farm in about 30 min by one person. The LTL test appears to be robust for use in between-flock

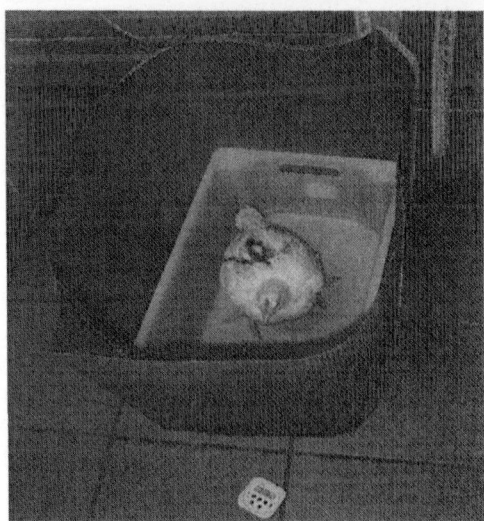

Fig. 1.3. The latency to lie method for measuring gait. This test (Weeks *et al.*, 2002) involves placing groups of broilers in a waterproof pen and then flooding it with shallow tepid water; the amount of time the birds take to lie down is then recorded as a measure of gait. Below is the modified version proposed by Berg and Sanotra (2003), in which birds are instead tested individually in a small container. (Photographs courtesy of Claire Weeks and Charlotta Berg.)

comparisons, in that it is not influenced by the genotype of the birds, the type of litter on which the birds have been raised, or the specific housing and management environment of a particular flock (Weeks *et al.*, 2002; Berg and Sanotra, 2003).

A force-measuring plate has also been developed (Sandilands *et al.*, 2003; J. Savory, R. McGovern and V. Sandilands, unpublished) which has the potential to partially automate the collection of gait information in commercial houses. Broilers are encouraged to walk across the plate (for example, by directing a puff of air towards them), which allows various aspects of the birds' gait to be recorded and analysed using a software program. Preliminary studies showed that gait was correctly classified (i.e. corresponded to the Bristol gait score) using only two measures (step length and standard deviation of vertical force during foot placement) in about 90% of the birds that crossed the plate once or twice. However, when the method was tried in a commercial house, a large percentage (25%) of the birds, especially the lamer birds, could not be motivated to cross the plate; of the birds that did cross the plate, only 73% of the runs produced analysable data, meaning that the overall success rate of this method was less than 50%. The method therefore requires further development before it can be implemented on a wide scale.

Leg Problems, Lameness and Welfare

Measuring lameness (by scoring walking ability or willingness of the birds to stand) is therefore currently the most practical method for the on-farm evaluation of skeletal problems. But is there a correspondence between lameness and specific leg problems? Julian (1998) notes that many heavy, older broilers walk as if they are in pain and prefer to sit, although no specific cause of their abnormal gait can be determined by gross or histological examination post-mortem. Some percentage of impaired walking ability may be due to the conformation of fast-growing birds rather than to pain associated with skeletal disorders *per se*. These birds develop large amounts of breast muscle, which moves their centre of gravity forward, and they have short legs and large thigh muscle masses (Corr *et al.*, 2003a). They also show a gait pattern different from that of slower-growing birds, walking slowly with their toes pointed outwards, and taking short, wide steps. While this increases their stability during walking, this gait is inefficient and probably tiring (Corr *et al.*, 2003b), and possibly affects the behaviour of the birds.

Studies of activity patterns in broilers all indicate that the amount of locomotor activity declines with age, while the amount of time spent lying increases (Newberry *et al.*, 1988; Preston and Murphy, 1989; Lewis and Hurnik, 1990; Newberry and Hall, 1990; Gordon and Tucker, 1993; Weeks *et al.*, 1994; Estevez *et al.*, 1997). A decrease in the level of activity is pronounced during the last few weeks of growth, but differences in activity level between fast- and slow-growing strains are seen at even earlier ages (Bizeray *et al.*, 2000). Although several factors, including high stocking densities (Lewis and Hurnik, 1990; Estevez *et al.*, 1997), could affect the ability of broilers to move around, the most likely reason for these changes is an increase in the severity of skeletal problems and/or progressively decreased walking ability with age. Gait scores worsen with age (Fig. 1.4). Weeks *et al.* (2000) found a direct relationship between gait impairment and reduced activity. They recorded behaviours in broilers, aged 39–49 days, that had varying gait scores, and found that lamer birds (gait scores 2 and 3) spent more time lying and less time standing and walking than birds with better walking ability (gait scores 0 and 1), and also reduced the number of visits they made to the feeder. Lamer birds (gait score > 1) also dustbathe less with age than sound birds (Vestergaard and Sanotra, 1999).

As mentioned, these altered behaviour patterns could be attributed to fatigue or other problems associated with weight or conformation-related gait alterations.

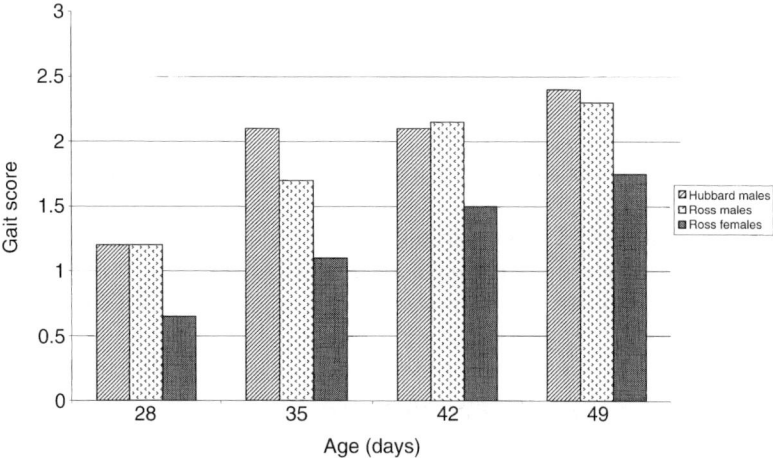

Fig. 1.4. Gait scores as a function of age. Gait scores worsen similarly with age in different stocks (Hubbard and Ross) of male broilers, particularly between days 28 and 35. They also worsen in females (Ross), but overall scores for females are better than those for males. Data for Hubbard males are from unpublished work (J.A. Mench, C.J. Falcone and J.P. Garner), and data for Ross birds are estimated from data provided by Sorensen *et al.* (2000) for broilers housed at three stocking densities.

Bokkers and Koene (2000) found that older fast-growing broilers sat more and walked less than slow-growing broilers, and while they were also found post mortem to have more tendon degeneration, tibial rotation and scoliosis than slower-growing birds, there was no relationship between these disorders and their behaviour. However, they only included birds with gait scores of 0–2 in their study. There is evidence, some suggestive and some direct, that more severe gait impairment is associated with bone and joint pathologies that cause pain. First, changes in gait and behaviour with age parallel increases in the incidence of specific skeletal disorders in flocks. BCO, for example, can be seen in broilers as young as 14 days of age, but most cases occur when the birds are around 35 days old (McNamee and Smyth, 2000). The number of birds needing to be culled due to angular deformities and spondylolisthesis also increases with age; in one study conducted on commercial farms (Riddell and Springer, 1984), there was an approximately sixfold increase in the number of birds culled for valgus deformities from week 1 to week 3 of age, and another large increase from week 3 to week 5. It is also at around 35–42 days of age that the absolute growth rate, and absolute growth in breast muscle mass, of male broilers is greatest (Goliomytis *et al.*, 2003), of course, so conformation could also contribute to a change in walking ability at these ages.

Nevertheless, there does seem to be a correspondence between lameness and particular leg disorders, at least in birds with gait scores of 3 or more. As mentioned previously, cull birds with scores of 3 or more were found to have a variety of pathologies not seen in sound birds, particularly BCO and severe TD lesions (McNamee *et al.*, 1998). Similarly, Kestin *et al.* (1994) conducted post-mortem examinations of 300 commercial broilers that had varying gait scores, and found that a large percentage (69%) of birds with gait score 4 or 5 had osteomyelitis, a disorder that was absent in birds with lower gait scores. The most common disorder seen in birds with a gait score of 3 was TD; 25% of these birds had TD compared with 0–5.9% of birds with lower gait scores, and there was correspondingly greater tibial angulation in birds with score 3. In commercial flocks in Scandinavia, gait score has been found to be highly positively correlated with severe TD lesions, valgus–varus deformities and macroscopically observable BCO (Sanotra *et al.*, 2001; Sanotra and Berg, 2003).

It is obvious from their clinical manifestations that many gait disorders must be painful, since they involve inflammation, spinal cord damage, tension on the joints or rupture of tendons. Exceptions are TD and other disorders affecting the cartilage. Nerves for pain perception are limited to the periosteum, bone and marrow, and cartilaginous lesions are probably not painful *per se* until the lesions are severe and other tissues are involved (Rowland, 1989; McNamee *et al.*, 1998). This probably explains why TD lesions are often (e.g. Sanotra *et al.*, 2001a; Sanotra and Berg, 2003), but not always (e.g. Sorenson *et al.*, 1999; Garner *et al.*, 2002), found to be associated with impaired gait.

There is also experimental evidence for a relationship between lameness and pain. Broilers with gait scores of 3 or more tend to self-select more feed containing an analgesic and anti-inflammatory drug, carprofen, than broilers with no gait impairment, and the amount ingested increases with the severity of lameness (Danbury *et al.*, 2000). Broilers with scores of 3 also take more than three times longer to negotiate an obstacle course to reach food than broilers with scores of 0–2, but their times decrease to the levels of the sounder broilers 90 min after they are injected with carprofen (McGeown *et al.*, 1999). Ingestion of carprofen improves gait (Danbury *et al.*, 2000), strongly suggesting that altered gait is associated with pain. In contrast, Hocking (1994) found that broiler breeder males with musculoskeletal lesions (cartilage degeneration and ruptured ligaments and tendons) responded equivocally on several measures of behavioural activity to administration of the anti-inflammatory drug betamethasone, and it was suggested that pain associated with the lesions may have been alleviated by endogenous opioids. However, the nature of the lesions may also have had an effect: as mentioned previously, cartilage lesions are not themselves likely to cause pain, and tendons and ligaments are likely to be most painful at the time that they rupture (see Chapter 4).

Regardless of whether or not impaired mobility is associated with pain, it has welfare consequences for the birds. While there is a positive relationship between body weight and the presence of moderate gait abnormalities, birds with more severe abnormalities tend to weigh less, possibly because they have difficulty competing with more mobile birds for food and water (Kestin *et al.*, 1992). Because birds with leg problems are likely to spend more time lying down, they are also probably more prone to developing potentially painful skin problems such as breast blisters and footpad and hock burns (Kestin *et al.*, 1999; Chapter 3 in this book). Birds that are lying down are often stepped upon by other birds (Murphy and Preston, 1988), which can result in skin scratches (Proudfoot and Hulan, 1985) that can become infected. Broiler breeder males with severe musculoskeletal lesions have reduced fertility and are often culled on that account (Hocking and Duff, 1989).

Paradoxically, the lack of activity seen in broilers may also worsen their skeletal problems. Physical loading in the form of mechanical stresses and strains is essential for normal bone formation (Thomas and Howland, 1964). Insufficient loading due to inactivity can contribute to abnormal bone development. Free-range broilers usually have a lower incidence of leg problems and fewer severe gait abnormalities than intensively reared broilers (Rodenhoff and Dämmrich, 1971; Kestin *et al.*, 1992), providing indirect evidence for a beneficial effect of activity on leg problems. However, this also seems to depend upon genotype, since fast-growing commercial broilers have also been reported to make little use of free-range areas (Weeks *et al.*, 1994), particularly compared with slower-growing strains (Nielsen *et al.*, 2003).

Several experimental studies have shown that increasing activity can decrease skeletal problems. For example, Haye and Simons (1978) increased the distance between the feeder and drinker to encourage exercise in caged broilers. They found a decrease in the incidence of twisted legs in one trial, although not in a subsequent trial. Thorp and Duff (1988) exercised broilers on a treadmill daily for four periods of 15 min beginning at 8 days of age, and found that 43% of unexercised birds had abnormalities of the physis and physeal vasculature compared with only 20% of the exercised birds. Exercise thus promotes the blood flow and perfusion through skeletal tissue that is necessary for the maintenance of ossification, and could lead to a decreased incidence or severity of TD. Mench *et al.* (2001) provided broilers with an opportunity to exercise on an apparatus containing ramps, perches and dustbaths, and found a moderate improvement in gait score. However, other methods for increasing activity designed to be practical in a commercial setting that have been tried experimentally, such as providing perches (Su *et al.*, 2000), a free-range area (Weeks *et al.*, 1994) or light of greater or varying intensity (Newberry *et al.*, 1985, 1986), have proved ineffective in improving gait, so more work is required in this area.

Summary and Conclusions

Skeletal disorders and mobility impairments in meat-type birds are a serious problem from the perspectives of both welfare and production. Some gait impairment may be due simply to conformation, but more serious impairment is generally found to be associated with infectious or non-infectious skeletal disorders. Affected birds may have difficulty reaching food and water, may be in pain, and are more prone to skin conditions because they spend more time lying on the litter and are also stepped on by other birds, causing scratches. The aetiology of skeletal disorders is still not completely understood, and the incidence of specific disorders varies widely from one flock to another. Some progress has apparently been made in decreasing these disorders through genetic selection, but according to recent surveys incidences on commercial farms are still unacceptably high. Some management strategies to reduce problems also exist, including decreasing stocking density, providing a period of darkness, and slowing growth rate by modifying nutrition, but it is not known how widely these are used commercially. Several methods can be used to determine the incidences of problems on commercial farms, including gross and microscopic post-mortem examination and evaluation of lameness by gait-scoring or assessing the reluctance of birds to stand. Gait scoring has proved very valuable for large-scale assessment of commercial flocks, but new and refined methods for evaluating lameness that decrease subjectivity and require less personnel time and training will be very helpful in ensuring consistency across studies. The increased understanding that will come from on-farm assessment of the risk factors for skeletal problems and lameness will be invaluable in the development of effective selection and management strategies to decrease these problems.

Acknowledgements

Preparation of this manuscript was supported by grant USDA-NRI 2001-02498.

References

Bartels, J.E., McDaniel, G.R. and Hoerr, F.J. (1989) Radiographic diagnosis of tibial dyschondroplasia in broilers: a field selection technique. *Avian Diseases* 33, 254–257.

Bennett, R.M., Christiansen, K. and Clifton-Hadley, R.S. (1999) Direct costs of endemic diseases of farm animals in Great Britain. *Veterinary Record* 145, 376–377.

Berg, C. and Sanotra, G.S. (2003) Can a modified LTL test be used to validate gait-scoring results in commercial broiler flocks? *Animal Welfare* 12, 655–659.

Bihan-Duval, E. Le, Beaumont, C. and Colleau, J.J. (1996) Genetic parameters of the twisted legs syndrome in broiler chickens. *Genetics, Selection and Evolution* 28, 177–195.

Bihan-Duval, E. Le, Beaumont, C. and Colleau, J.J. (1997) Estimation of the genetic correlations between twisted legs and growth or conformation traits in broiler chickens. *Journal of Animal Breeding and Genetics* 114, 239–259.

Bizeray, D., Leterrier, C., Constatin, P., Picard, M. and Faure, J.M. (2000) Early locomotor behaviour in genetic stocks of chickens with different growth rates. *Applied Animal Behaviour Science* 68, 231–242.

Bokkers, E.A.M. and Koene, P. (2000) Behaviour of fast- and slow-growing broilers up to 12 weeks of age and the physical consequences. *Applied Animal Behaviour Science* 81, 59–72.

Butterworth, A. (1999) Infectious components of broiler lameness: a review. *World's Poultry Science Journal* 55, 327–352.

Classen, H.L., Riddell, C., Robinson, F.E., Shand, J.P. and McCurdy, A.R. (1994) Effect of lighting treatment on the productivity, health, behaviour and sexual maturity of heavy male turkeys. *British Poultry Science* 35, 215–225.

Cook, M.E., Patterson, P.H. and Sunde, M.L. (1984) Leg deformities: inability to increase severity by increasing body weight of chick and poults. *Poultry Science* 63, 620–627.

Corr, S.A., Gentle, M.J., McCorquodale, C.C. and Bennett, D. (2003a) The effect of morphology on the musculoskeletal system of the modern broiler. *Animal Welfare* 12, 145–157.

Corr, S.A., Gentle, M.J., McCorquodale, C.C. and Bennett, D. (2003b) The effect of morphology on walking ability in the modern broiler: a gait analysis study. *Animal Welfare* 12, 159–171.

Crespo, R., Stover, S.M., Taylor, K.T., Chin, R.P. and Shivasprasad, H.L. (2000) Morphometric and mechanical properties of femora in young adult male turkeys with and without femoral fractures. *Poultry Science* 79, 602–608.

Danbury, T.C., Weeks, C.A., Chambers, J.P., Waterman-Pearson, A.E. and Kestin, S.C. (2000) Self-selection of the analgesic drug carprofen by lame broiler chickens. *Veterinary Record* 146, 307–311.

Duff, S.R.I. and Hocking, P.M. (1986) Chronic orthopaedic disease in adult male breeding fowls. *Research in Veterinary Science* 41, 340–348.

Edwards, H.M. (1992) Nutritional factors and leg disorders. In: Whitehead, C.C. (ed.) *Bone Biology and Skeletal Disorders in Poultry*. Carfax, Abingdon, UK, pp. 167–193.

Estevez, I., Newberry, R.C. and de Reyna, L.A. (1997) Broiler chickens: a tolerant social system? *Etologia* 5, 19–29.

European Commission (2000) The welfare of chickens kept for meat production (broilers). Report of the Scientific Committee on Animal Health and Animal Welfare. European Commission Report B3, R15, 2000. P Unit B3, Directorate B of the European Commission, Brussels, Belgium.

Farquharson, C. and Jeffries, D. (2000) Chondrocytes and longitudinal bone growth: the development of tibial dyschondroplasia. *Poultry Science* 79, 994–1004.

Garner, J.P., Falcone, C., Wakenell, P., Martin, M. and Mench, J.A. (2002) The reliability and validity of a modified gait scoring system and its use in assessing tibial dyschondroplasia in broilers. *British Poultry Science* 43, 355–363.

Goliomytis, M., Panopoulou, E. and Rogdakis, E. (2003) Growth curves for body weight and major component parts, feed consumption, and mortality of male broiler chickens raised to maturity. *Poultry Science* 82, 1061–1068.

Gordon, S.H. and Tucker, S.A. (1993) Broiler walking behavior. In: Savory, C.J. and Hughes, B.O. (eds) *Fourth European Symposium on Poultry Welfare*. Universities Federation for Animal Welfare, Potters Bar, UK, p. 291.

Haye, U. and Simons, P.C.M. (1978) Twisted legs in broilers. *British Poultry Science* 19, 549–557.

Havenstein, G.B., Ferket, P.R., Scheideler, S.E. and Larson, B.T. (1994) Growth, livability, and feed conversion of 1957 vs 1991 broilers when fed 'typical' 1957 and 1991 broiler diets. *Poultry Science* 73, 1785–1794.

Hester, P.Y. (1994) The role of environment and management on leg abnormalities in meat-type fowl. *Poultry Science* 73, 904–915.

Hocking, P.M. (1994) Assessment of the welfare of food restricted male broiler breeder poultry with musculoskeletal disease. *Research in Veterinary Science* 57, 28–34.

Hocking, P.M. and Duff, S.R.I. (1989) Musculo-skeletal lesions in adult male broiler breeder fowls and their relationships with body weight and fertility at 60 weeks of age. *British Poultry Science* 30, 777–784.

Julian, R.J. (1998) Rapid growth problems: ascites and skeletal deformities in broilers. *Poultry Science* 77, 1773–1780.

Kestin, S.C., Knowles, T.G., Tinch, A.E. and Gregory, N.G. (1992) Prevalence of leg weakness in broiler chickens and its relationship with genotype. *Veterinary Record* 131, 190–194.

Kestin, S.C., Adams, S.J.M. and Gregory, N.G. (1994) Leg weakness in broiler chickens, a review of studies using gait scoring. *Proceedings of the 9th European Poultry Conference*. World's Poultry Science Association, Glasgow, UK, pp. 203–206.

Kestin, S.C., Su, G. and Sorensen, P. (1999) Different commercial broiler crosses have different susceptibilities to leg weakness. *Poultry Science* 78, 1085–1090.

Kestin, S.C., Gordon, S., Su, G. and Sorensen, P. (2001) Relationships in broiler chickens between lameness, liveweight, growth rate and age. *Veterinary Record* 148, 195–197.

Kuhlers, D.L. and McDaniel, G.R. (1996) Estimates of heritabilities and genetic correlations between tibial dyschondroplasia expression and body weight at two ages in broilers. *Poultry Science* 75, 959–961.

Lewis, N.J. and Hurnik, J.F. (1990) Locomotion of broiler chickens in floor pens. *Poultry Science* 69, 1087–1093.

Lynch, M., Thorp, B.H. and Whitehead, C.C. (1992) Avian tibial dyschondroplasia as a cause of bone deformity. *Avian Pathology* 21, 275–285.

McGeown, D., Danbury, T.C., Waterman-Pearson, A.E. and Kestin, S.C. (1999) Effect of carprofen on lameness in chickens. *Veterinary Record* 144, 668–671.

McNamee, P.T. and Smyth, J.A. (2000) Bacterial chondronecrosis with osteomyelitis ('femoral head necrosis') of broiler chickens: a review. *Avian Pathology* 29, 253–270.

McNamee, P.T., McCullagh, J.J., Thorp, B.H., Ball, H.J., Graham, D., McCullough, S.J., McConaghy, D. and Smyth, J.A. (1998) Study of leg weakness in two commercial broiler flocks. *Veterinary Record* 143, 131–135.

Martrechenar, A., Huonnic, D., Cotte, J.P., Boilletot, E. and Morisse, J.P. (1999) Influence of stocking density on behavioural, health and productivity traits of turkeys in large flocks. *British Poultry Science* 40, 323–331.

Mench, J.A., Garner, J.P. and Falcone, C. (2001) Behavioral activity and its effects on leg problems in broiler chickens. In: *6th European*

Symposium on Poultry Welfare. World's Poultry Science Association, Zollikofen, Switzerland, pp. 152–156.

Mercer, J.T. and Hill, W.G. (1984) Estimation of genetic parameters for skeletal defects in broiler chickens. *Heredity* 53, 193–203.

Morris, M.P. (1993) National survey of leg problems. *Broiler Industry* May, 20–24.

Murphy, L.B. and Preston, A.P. (1988) Time budgeting in meat chickens grown commercially. *British Poultry Science* 29, 571–580.

Nairn, M.E. and Watson, A.R.A. (1972) Leg weakness of poultry – a clinical and pathological characterization. *Australian Veterinary Journal* 48, 645–656.

Newberry, R.C. and Hall, J.W. (1990) Use of pen space by broiler chickens: effects of age and pen size. *Applied Animal Behaviour Science* 25, 125–136.

Newberry, R.C., Hunt, J.R. and Gardiner, E.C. (1985) Effect of alternating lights and strain on behavior and leg disorders of roaster chickens. *Poultry Science* 64, 1863–1868.

Newberry, R.C., Hunt, J.R. and Gardiner, E.C. (1986) Light intensity effects on performance, activity, leg disorders, and sudden death syndrome of broiler chickens. *Poultry Science* 65, 2232–2238.

Newberry, R.C., Hunt, J.R. and Gardiner, E.C. (1988) Influence of light intensity on behavior and performance of broiler chickens. *Poultry Science* 67, 1020–1025.

Nielsen, B.L., Thomsen, M.G., Sorensen, P. and Young, J.F. (2003) Feed and strain effects on the use of outdoor areas by broilers. *British Poultry Science* 44, 161–169.

Preston, A.P. and Murphy, L.B. (1989) Movement of broiler chickens reared in commercial conditions. *British Poultry Science* 30, 519–532.

Proudfoot, F.G. and Hulan, H.W. (1985) Effects of stocking density on the incidence of scabby hip syndrome among broiler chickens. *Poultry Science* 64, 2001–2003.

Rath, N.C., Huff, G.R., Huff, W.E. and Balog, J.M. (2000) Factors regulating bone maturity and strength in poultry. *Poultry Science* 79, 1024–1032.

Riddell, C. and Springer, R. (1984) An epizootiological study of acute death syndrome and leg weakness in broiler chickens in Western Canada. *Avian Diseases* 29, 90–102.

Rodenhoff, G. and Dämmich, K. (1971) Ber die Beeinflussung des Skeletts der Masthänchen durch Haltung und Auslaf im Freien. *Zentralblatt für Veterinärmedizin Reihe A* 18, 297–309.

Rowland, G.N. (1989) Broiler leg problems: some causes and effects. *Poultry Digest* 48, 54–56.

Sandilands, V., McGovern, R. and Savory, C.J. (2003) Objectively measuring broiler walking style using a force plate. In: *Proceedings of the 37th International Congress of the ISAE.* Fondazione Iniziative Zooprofilattiche e Zootechniche, Brescia, Italy, p. 61.

Sanotra, G.S. and Berg, C. (2003) Investigation of lameness in the commercial production of broiler chickens in Sweden. Sveriges Landtbruks Universitet, Skara, Sweden.

Sanotra, G.S., Lund, J.D., Erksboll, A.K., Petersen, J.S. and Vestergaard, K.S. (2001a) Monitoring leg problems in broilers: a survey of commercial broiler production in Denmark. *World's Poultry Science Journal* 57, 55–69.

Sanotra, G.S., Vestergaard, K.S., Thomsen, M.G. and Lawson, L.G. (2001b) The influence of stocking density on tonic immobility, lameness and tibial dyschondroplasia in broilers. *Journal of Applied Animal Welfare Science* 4, 71–87.

Sorenson, P. (1992) Genetics of leg disorders. In: Whitehead, C.C. (ed.) *Bone Biology and Skeletal Disorders in Poultry.* Carfax, Abingdon, UK, pp. 213–230.

Sorenson, P., Su, G. and Kestin, S.C. (1999) The effect of photoperiod:scotoperiod on leg weakness in broiler chickens. *Poultry Science* 78, 336–342.

Sorenson, P., Su, G. and Kestin, S.C. (2000) Effects of age and stocking density on leg weakness in broiler chickens. *Poultry Science* 79, 864–870.

Su, G., Sorenson, P. and Kestin, S.C. (2000) A note on the effects of perches and litter substrate on leg weakness in broiler chickens. *Poultry Science* 79, 1259–1263.

Sullivan, T.W. (1994) Skeletal problems in poultry: estimated annual cost and descriptions. *Poultry Science* 73, 879–882.

Thorp, B.H. (1994) Skeletal disorders in the fowl: a review. *Avian Pathology* 23, 203–236.

Thorp, B.H. (1996) Diseases of the musculoskeletal system. In: Jordan, F.T.W. and

Pattison, M. (eds) *Poultry Diseases*. W.B. Saunders, London, pp. 290–305.

Thorp, B.H. and Duff, S.R.I. (1988) Effect of exercise on the vascular pattern in the bone extremities of broiler fowl. *Research in Veterinary Science* 45, 72–77.

Thorp, B.H., Whitehead, C.C., Dick, L., Bradbury, J.M., Jones, R.C. and Wood, A. (1993) Proximal femoral degeneration in growing broiler fowl. *Avian Pathology* 22, 325–342.

Vestergaard, K.S. and Sanotra, G.S. (1999) Relationships between leg disorders and changes in the behaviour of broiler chickens. *Veterinary Record* 144, 205–209.

Weeks, C.A., Nicol, C.J., Sherwin, C.M. and Kestin, S.C. (1994) Comparison of the behaviour of broiler chickens in indoor and free-range environments. *Animal Welfare* 3, 179–192.

Weeks, C.A., Danbury, T.D., Davies, H.C., Hunt, P. and Kestin, S.C. (2000) The behaviour of broiler chickens and its modification by lameness. *Applied Animal Behaviour Science* 67, 111–125.

Weeks, C.A., Knowles, T.G., Gordon, R.G., Kerr, A.E., Peyton, S.T. and Tillbrook, N.T. (2002) New method for objectively assessing lameness in broiler chickens. *Veterinary Record* 151, 762–764.

Whitehead, C.C., Fleming, R.H., Julian, R.J. and Sorensen, P. (2003) Skeletal problems associated with selection for increased production. In: Muir, W.M. and Aggrey, S.E. (eds) *Poultry Genetics, Breeding and Biotechnology*. CAB International, Wallingford, UK, pp. 29–52.

Williams, B., Soloman, S., Waddington, D., Thorp, B. and Farquharson, C. (2000) Skeletal development in the meat-type chicken. *British Poultry Science* 41, 141–149.

Wise, D.R. and Nott, H. (1975) Studies on tibial dyschondroplasia in ducks. *Research in Veterinary Science* 18, 193–197.

Wong-Valle, J., McDaniel, R., Kuhlers, D.L. and Bartels, J.E. (1993) Correlated responses to selection for high or low incidence of tibial dyschondroplasia in broilers. *Poultry Science* 72, 1621–1629.

Yalcin, S., Settar, P. and Dicle, O. (1998) Influence of dietary protein and sex on walking ability and bone parameters of broilers. *British Poultry Science* 39, 251–256.

Yu, M.W. and Robinson, F.E. (1992) The application of short-term feed restriction to broiler chicken production: a review. *Journal of Applied Poultry Research* 1, 147–153.

2 Measuring and Auditing the Welfare of Broiler Breeders

P.M. HOCKING

Roslin Institute (Edinburgh), Roslin, Midlothian, UK

Introduction

In most of Europe, commercial broiler parent stocks (broiler breeders) are reared in separate-sex groups on the floor of large, environmentally controlled sheds. At 16–18 weeks of age they are moved to similar conditions, where nest boxes are provided for egg-laying and at least one-half of the floor area is littered with wood shavings or, occasionally, chopped wheat straw. In the most modern sites, up to one-half of the floor area comprises raised slats or perforated plastic flooring that leads to nest boxes fitted with automatic egg collection equipment. During rearing, the birds are fed increasing quantities of a pelleted ration (Fig. 2.1) that is scattered on the floor to encourage uniform feed intake. In the breeding period, males and females are fed different quantities of feed daily from different delivery systems to control body weight gain. In North America and many other parts of the world, feed is provided on alternate days (a practice that is not permitted in the UK and in several other countries in Europe) and the birds may be reared in the same sheds that they will be kept in during the laying period. Water is provided *ad libitum* during laying and usually during rearing, but may also be restricted in order to control litter moisture, and rice hulls are frequently used for litter in east Asian countries. Temperature and ventilation are controlled in temperate regions by the use of electric fans. In hot climates the walls may be open to the outside environment and in very hot conditions evaporative or other cooling systems may be used.

The body weight gains of broiler breeders are limited by feed restriction to control reproduction and mortality. If female broiler breeders are fed *ad libitum*, egg production and hatchability are poor and mortality is high (Table 2.1). Feed restriction controls ovulation rate and restores normal function to the reproductive process (see below). In males, control of body weight by feed restriction maximizes fertility (Table 2.2) and the ability to mate (Hocking, 1990; Hocking *et al.*, 1996). High mortality in broiler breeders fed *ad libitum* is largely related to cardiovascular failure in females and culling for lameness in males.

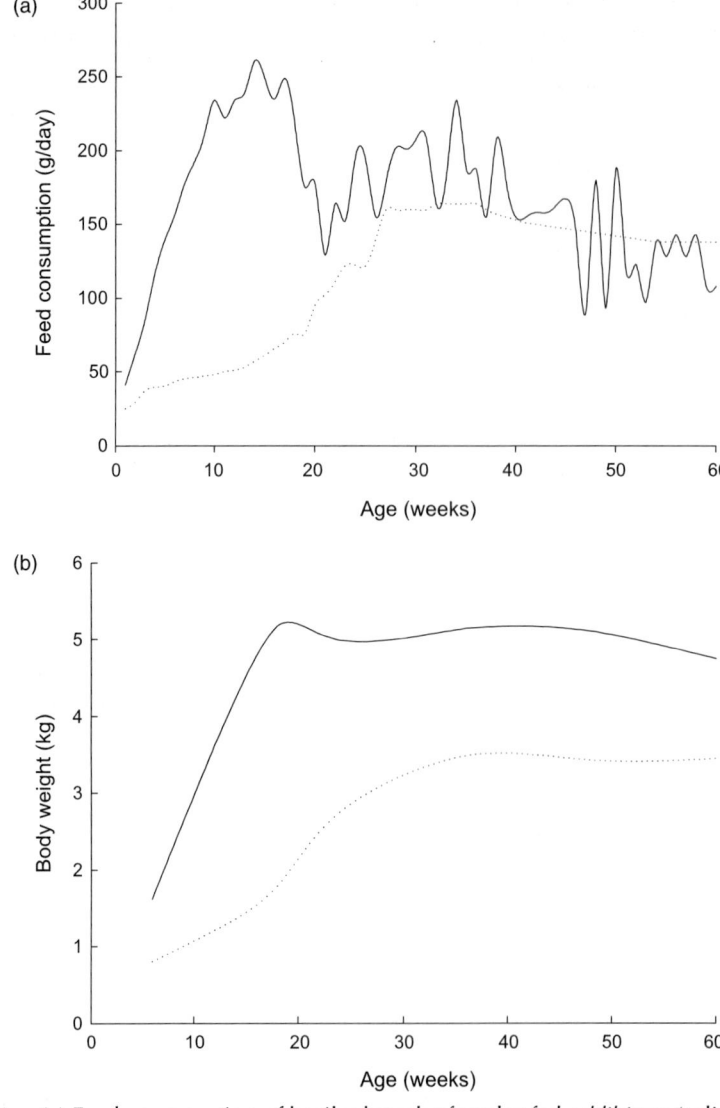

Fig. 2.1. (a) Feed consumption of broiler breeder females fed *ad libitum* (solid line) and feed allocations to birds restricted to achieve target body weight gain (dotted line) from 0 to 60 weeks of age. (b) Actual body weights attained by the same birds. (From Hocking *et al.*, 2002a.)

From a welfare perspective, the housing of commercial broiler breeders compares favourably to that of battery hens, whereas the degree of feed restriction to which they are subjected has raised concerns that they may be chronically stressed. This review will discuss the assessment of the welfare of feed-restricted broiler breeders of both sexes and how it can be audited. Commercial systems of production raise a number of other welfare issues that are largely specific to this group of farm

Table 2.1. Productivity and mortality to 60 weeks of age in feed-restricted and *ad libitum*-fed broiler breeder females mated to feed-restricted males. The feed restriction programme followed the breeder's recommended target body weight schedule. (From Hocking *et al.*, 2002a.)

Trait	Restricted	*Ad libitum*
Final body weight (kg)	3.7	5.3
Mortality (%)	4	46
Eggs (number/hen day)	157	44
Hatching eggs (number/hen day)	140	35
Egg weight (g)	65	65
Fertility (%)	86	87
Hatchability (%)	86	43

Table 2.2. Fertility of boiler breeder males of different body weight at 60 weeks of age. (P.M. Hocking, unpublished data.)

Weight (kg)	Fertility (%)
3	45
4	86
5	91
6	79
7	22

animals, which will be discussed before a brief examination of the welfare of broiler breeders in alternative systems.

Auditing is defined (in *Collins English Dictionary*) as to 'inspect, verify and correct', and I will consider these three aspects in this chapter as they relate to the welfare of broiler breeders.

Feed Restriction in Females

Genetic selection for high body weight is associated with an increase in the incidence of ovulation of two or more ova. These potential egg yolks may be lost into the body cavity, where they are reabsorbed or, if they enter the oviduct, result in the production of a large proportion of double-yolked, soft-shelled, thin-shelled and misshapen eggs that are not suitable for incubation, even if they can be collected. Restricting body weight controls ovulation rate and restores the reproductive system to single ovulations and the production of normal, hatching-quality eggs (Hocking *et al.*, 1987). The results of a series of experiments to examine the factors governing ovarian function are summarized in Box 2.1. The conclusions from these experimental results were important because they suggested that any departure from the optimum feeding schedule published by the various breeding companies would result in lower productivity, and they are consistent with reports that assess the impact of minor changes in target body weight curves (e.g. Fattori *et al.*, 1991; Robinson and

---

Box 2.1.   Experimental results on the control of ovulation rate

---

- Feed restriction must be substantial and last from 14 weeks of age (Hocking *et al.*, 1989)
- Achieving the recommended target body weight at the onset of lay is critical but it does not matter what growth trajectory the birds follow to reach that body weight (Hocking, 1993a)
- Any departure from the recommended target body weight is associated with depressed egg production (Hocking, 1996)
- Feed supplied after the onset of lay should be more generous in order to support the peak of egg production but must still be controlled (Hocking, 1996)
- After the peak of egg production has been reached, feed intake can be gradually decreased to match the requirements for lower egg production and to limit the effects of excessive body weight gain on mortality and productivity (Hocking *et al.*, 2002a)

Robinson, 1991; Robinson *et al.*, 1995). A higher recommended rate of gain in the first two-thirds of the rearing period combined with less rapid growth during the last third (designed to achieve the target body weight at the onset of lay) was associated with lower productivity and higher mortality and did not improve the welfare of the birds (Hocking *et al.*, 2002a). Decreases in the protein concentrations of the rations, while achieving the same body weight at puberty, also resulted in lower egg production and higher mortality and, despite higher feed intakes, no improvement in the welfare of the birds (Hocking *et al.*, 2002a).

In assessing the welfare of commercial broiler breeders, the auditor should therefore expect the flock to be managed in accordance with the breeder's flock management manual with respect to optimum body weight gains and feeding schedules. These should be available from the records of the regular weighings, possibly recorded automatically, that are performed to guide the daily feed allocation during rearing. Because broiler breeder performance (egg weight and numbers, fertility and hatchability) is sensitive to small deviations from the optimum management and feeding of the birds, records of egg weight and egg production during lay are a good source of information about their welfare. High numbers of double-yolked eggs in early lay indicate that the control of body weight or (more likely) of body weight uniformity is not adequate and that the welfare of individual birds in the flock may be compromised. Deviations from the expected performance (egg numbers or average egg weight) may indicate that the feed allocation or water supply has been inadequate in quantity or quality or that the flock has been affected by disease. Many individual birds will be laying at a relatively high rate at the end of lay and feed allocations after peak should not be decreased too far. The auditor should therefore ensure that feed restriction after the peak, and particularly after 50 weeks of age, does not subject these birds to inadequate nutrition. This will be detectable as a sharp decline in egg weight or production and as the presence of birds in poor condition and with low body weight.

The experimental data provide an understanding of the biological basis of the commercial practice of attempting to minimize the variability of body weight during rearing: large birds will have lower egg production because of increased ovulation rates and smaller birds because they do not have enough follicles for a consistent daily ovulation. During rearing, smaller birds may obtain less feed than they require for body maintenance and egg production and their welfare will suffer. Thus, as part of good management, broiler breeders are graded; smaller birds are removed to separate pens, where they are subjected to less competition for feed. Feed allocation may be increased in these pens to ensure that the birds achieve target body weights by 10 weeks of age. Larger birds may also be removed for separate feeding when this is justified. Auditing the potential welfare of individual female broiler breeders should therefore involve an assessment of the uniformity records of the flock and the success of grading and selecting birds for special treatment. The flock should be inspected for the presence of very small or large birds, which indicates that feed is not being provided in such a way that all birds have an equal chance of obtaining their share. In many modern systems the feed is pelleted and scattered over the floor by spin-feeders suspended from the ceiling. The pellets must be hard enough to withstand feed distribution, enable the birds to locate and consume the feed, and to minimize the formation of dust comprised of fine feed particles. During the laying period, feed is presented in meal form via conventional chain-link feeding tracks. It is important that sufficient space (track length) is provided to enable all birds to feed at the same time: typically, 15 cm per female is recommended. It is also essential that the feeding system is capable of distributing the feed as rapidly as possible to ensure that all birds are able to consume their share of the feed allowance. This is of the utmost importance where birds are reared in their laying accommodation, and this consideration may underlie the alternate-day feeding practised in North America and elsewhere.

Beak and oral lesions (including necrosis of the distal part of the tongue) may result if meal feed contains a proportion of very fine particles (Fig. 2.2a). They were more common in males than females, probably because the females were attracted to peck at fine particles of feed on the beaks of the males (Duff *et al.*, 1990). These conditions have also been observed in commercial flocks (P.M. Hocking and R. Bernard, unpublished) and may be detected only by manual handling and inspection of individual birds. If they are discovered, the auditor should ensure that a coarser feed is obtained.

During lay, female birds are generally fed from a conventional feeding track fitted with a grid to exclude males (see next section). The gap should be sufficiently wide to enable the hens to feed without suffering abrasion and facial swelling (Fig 2.2b). In practice, the grid aperture should be at least 4 mm wider than the average female head width or 2 mm wider than the widest head to facilitate ready access to the feed without incurring facial damage (Hocking, 1993b). A major problem in early separate-sex feeding systems was variation in the width of the aperture due to faulty manufacturing or physical damage. This gave access for a proportion of males, allowing them to gain excessive body weight. Modern systems may use a continuous bar that is easier to maintain and cheaper to manufacture, sometimes in conjunction with a grid. If males are not dubbed (removal of the comb at hatch), this double limitation of the aperture will be found to be more effective in excluding the males from the females' feed and will enhance the uniformity (and welfare) of the birds. Nevertheless, inspection of the birds when feed is available should be conducted regularly in order to monitor the system for structural faults. An auditor should include direct observation of feeding behaviour immediately

Fig. 2.2.   (a) Beak necrosis in an adult male associated with feed that was too fine. (b) Physical damage to the head caused by insufficient space for females to feed from their feeding track. (From Duff *et al.*, 1990, 1989. Reproduced with permission from the editors of *Veterinary Record* and the publishers of *Avian Pathology* (http://www.tandf.co.uk/journals), respectively.)

after feeding and later if there is still feed left in the female trough. Similarly, the feeding system should be inspected for evidence of damage and for the presence of sharp edges that may cause physical injury. In general, observation of the behaviour of the birds in a flock over an extended period of time (30–60 min) at regular intervals throughout the day will show whether there are management issues that should be addressed.

Feed Restriction in Males

Early surveys of musculoskeletal disease in breeding males suggested that ligament and tendon ruptures and destructive cartilage loss were common in culled males and at the end of the breeding period (Duff and Hocking, 1986). Many of these birds were feed-restricted during rearing but then fed in open troughs with females and were relatively heavy. Separate-sex feeding systems were introduced in the late 1980s, based on the difference in the width of the heads of males and females. Females were fed from a conventional trough fitted with a grid or bar that presented a gap that was too narrow for the males to gain access to the feed. The males were fed via a separate feeding line that was placed at a height that the smaller females could not easily reach, and competition ensured that they quickly consumed the feed that was allocated to them. The prevalence of musculoskeletal disease in experiments conducted over several years is summarized in Table 2.3. In the most recent experiment there was very little evidence of skeletal disease (P.M. Hocking and R. Bernard, unpublished).

Controlling body weight has undoubtedly been responsible for a large part of the decrease over time in the prevalence of musculoskeletal disease. Genetic selection for improved leg health in growing birds may also have contributed. In practice, lame birds may be attacked if they are unable to defend themselves, and are unlikely to be fertile. Most farm managers will immediately cull lame birds and males that are too heavy or too light, and will cull those that do not appear to be fertile to maintain good fertility and to avoid overmating. The auditor should check that this has been done and that poor males have been removed from the flock, by inspection of culling records and by observing the birds.

Dubbing to prevent excessive bleeding following aggression after sexual maturity and removal of the dew claw to diminish damage to the hen during mating have largely been abandoned since the introduction of separate-sex feeding: the presence of a large comb helps to prevent the male birds feeding from the female trough. Thus, males are lighter and better able to mate without damaging the hen. Traditionally, the spur buds might have been cut off at hatch, but this may be unnecessary in modern systems of production, where the birds are culled at 60–70 weeks of age. The selection of male breeding stock from commercial lines with short spurs obviates the need for despurring and is preferable to the mutilation. Conversely, breeders who supply male lines that have long spurs should select

Table 2.3. Musculoskeletal disease in male broiler breeders at 60 weeks of age in a series of experiments.

Year	Body weight (kg)	Destructive cartilage loss (%)	Ruptured tendons and ligaments (%)	Reference
1989[a]	5.3	43	29	Hocking, 1990
1989	3.6	4	13	Hocking, 1990
1994	4.5	67	0	Hocking and Bernard, 1996
1998	4.8	11	0	Hocking *et al.*, 2002a

[a]Mixed-sex feeding.

breeding stock to decrease the length of the spurs. Dubbing and de-toeing are unnecessary; the auditor should insist that these traditional practices are not performed and that steps are taken to avoid any need for them by changing management and husbandry practices.

Little work has been conducted on the welfare of male broiler breeders and it is assumed that feed restriction has the same effect on the welfare of males as of females. Unpublished research on conventionally restricted males in the author's laboratory is consistent with this assertion. Research on alternate-day feeding showed increased aggression in male broiler breeders on off-feed days (Mench, 1988). It is possible that this could be avoided by changing husbandry practices, such as everyday feeding on the floor via spin-feeders.

As in females, body weight uniformity is important in maintaining male fertility and health, and auditors should assess the flock records and the birds for evidence of sufficient access to the available feed. Adult breeding males should also be fed to permit a small but steady increase in body weight to ensure optimum fertility, fitness and health (Hocking and Bernard, 1997). Large males (over 5 kg) are relatively infertile and may dominate smaller birds, preventing them from mating or obtaining sufficient feed to maintain their health and reproductive fitness (Hocking, 1990).

Aggression in males has been the focus of recent research. In some flocks, aggression towards poultry keepers is a problem and may be related to the use of the Cornish (cock-fighting) breed in the development of some male lines (because of its deeply muscled, broad-breasted conformation). Male broiler breeders were more sexually aggressive (i.e. they attempted to force the female to crouch for mating without prior sexual display) than layer and traditional strains (Millman and Duncan, 2000b; Millman *et al.*, 2000). However, there was no evidence that selection for greater body weight gain had increased the prevalence of aggressive mating or that feed restriction was associated with increased aggression. In fact, feed-restricted males showed less aggression towards both males and females compared with *ad libitum*-fed birds (Millman and Duncan, 2000a). Aggression leading to the death of young hens has been a problem in North America but not in Europe. The most likely explanation is that in North America the females are mated at a younger age, when they are not sexually mature, and do not respond to the (aggressive) sexually mature males (J.A. Mench, 2003, personal communication). Delaying mating should prevent this welfare problem. A clear distinction must be made between these forms of male aggression and 'overmating', which is caused by too many males mating with individual females. Overmating affects hen welfare because the constant treading of the females by the males causes excessive demands on the hen and the removal of back feathers, which may in turn lead to skin tears. In well-managed flocks the mating ratio is progressively deceased from about 10 males to every 100 females when the sexes are first introduced to as few as 6 males per 100 females at 60 weeks. This is accomplished by removing overweight and sexually inactive males at regular intervals (about 5% of the male birds every 5 weeks). This also helps to maintain the welfare of the males by decreasing body weight variation and eliminating subordinate males, which may not consume enough feed to maintain their body weight and sexual condition.

Assessing Hunger in Broiler Breeders

The objective of assessing the subjective psychological feeling of stress associated with hunger in feed-restricted broiler breeders is problematic and has not been resolved. The degree of feed restriction changes with age and reaches a maximum of about 0.75 at 10–12 weeks of age, as measured by the reduction in feed intake as a proportion of birds fed *ad libitum* (Fig. 2.1). Comparisons of the body weight or feed intake of restricted-fed as a proportion of that of *ad libitum*-fed birds are misleading because of the large differences in body weight arising from the disparate allocation of feed. Prediction of the feed intake of restricted broiler breeders as a proportion of the feed consumed by *ad libitum*-fed birds of the same body weight by regression analysis showed that feed intake declined to 0.45 during rearing (Hocking, 1993c). Restricted birds continue to gain weight throughout the rearing period in response to regular increases in feed allocation (Fig. 2.1). They cannot therefore be described as starving and are substantially better fed than many farm animals in traditional farming systems where, historically, maintenance feeding prevailed for many months. Compared with birds fed *ad libitum*, restricted birds rapidly consume their daily feed allocation. However, using the rate of eating as a measure of hunger (Savory *et al.*, 1993b) is subject to the problems of comparing birds of different size, physiological maturity and prior experience, over very short periods of time (several minutes). Birds fed on restricted diets adopt different feeding strategies from those fed *ad libitum*. Large quantities of feed are stored in the crop of restricted birds and competition ensures that they quickly learn to consume feed as rapidly as possible, whereas birds fed *ad libitum* eat regularly throughout the day and night and do not store large quantities of feed in the crop. Attempts have therefore been made in our research to assess how well feed-restricted birds cope with the hunger that is imposed upon them, and whether the normal biological systems for surviving limited feed supplies in natural conditions are compromised in any way.

Compared with *ad libitum* feeding, feed restriction is invariably associated with a change in behaviour: restricted birds spend less time resting and more time foraging (scratching and pecking the litter), drinking and spot-pecking than birds fed *ad libitum* (Kostal *et al.*, 1992; Savory *et al.*, 1992, 1993a,b; Hocking, 1993c, 1996; Savory and Maros, 1993; Hocking *et al.*, 2001). Increases in the heterophil–lymphocyte ratio, the proportion of basophilic cells and plasma corticosterone concentration (recognized indexes of physiological stress) were reported in many but not all experiments (Maxwell *et al.*, 1990, 1992; Hocking *et al.*, 1993, 1996, 2001; Savory and Maros, 1993; Savory *et al.*, 1993a,b). There was little evidence that feed restriction compromised fundamental bodily functions, as indicated by a profile of several enzyme systems, and when differences were detected they indicated better welfare in the feed-restricted birds (Hocking *et al.*, 1993, 1996, 2001). Freeman (1985) argued that many of the responses of birds to perceived stressors were normal adaptations for survival and that those which would leave the animal unprotected, such as impaired immune function, are more satisfactory indices of stress. Indirect measures of immune function indicate that it is at least as good if not better in feed-restricted broiler breeders than in *ad libitum* controls (Katanbaf *et al.*, 1987; O'Sullivan *et al.*, 1991; Hocking *et al.*, 1996). The outcome of this research is summarized in Table 2.4.

Table 2.4.   Changes in welfare criteria of broiler breeders fed *ad libitum* or restricted during rearing (+ = better, − = worse).

Criterion	Restricted	Ad libitum
Production	+	−
Health	+	−
Behaviour: activity	+	−
Behaviour: hunger	−	+
Fear	+	−
Corticosterone	−	+
Haematology	−	+
Organ function	+	−
Immunology	+	−

Hunger and satiety are not discrete opposing states but represent two extremes of a continuous scale that might be measured by determining the strength of the motivation to feed. Research into motivation to feed, by measuring operant responses in feed-restricted broiler breeders, has shown that the strength of motivation is largely dependent on the size of the restricted bird in relation to its potential and, somewhat anomalously, that the motivation to feed is greater after a larger than after a smaller meal (Savory *et al.*, 1993b; Savory and Mann, 1999; Savory and Lariviere, 2000).

If hungry birds are able to conduct foraging activity in response to feed restriction, the psychological feelings of hunger may be dissipated into a normal behavioural response. Experimental evidence by Savory and co-workers suggests that stereotypic behaviour, particularly that of spot-pecking at the walls, has de-arousing properties that may help the birds to cope with the stress of feed restriction (Kostal *et al.*, 1992; Savory *et al.*, 1992, 1993a). Support for this proposal is found in the results of an experiment that compared *ad libitum*-fed and conventionally fed broiler breeders with those in which feed restriction led to body weights that were 0.24, 0.40, 0.55, 0.70 and 0.85 of those of *ad libitum*-fed contemporaries. The sum of oral behaviours (drinking, feeding, foraging and spot-pecking) was similar in all seven groups and comparable with the proportion of time spent foraging in feral fowls (Hocking *et al.*, 1996). Furthermore, spot-pecking occurred at a greater frequency in restricted broiler breeders in experimental pens than in a commercial flock (Fig. 2.3) of several thousand birds with ready access to litter. The key practical application of these results is that restricted broiler breeders should have access to a suitable foraging substrate, and it is worth noting that many studies of stereotypic behaviour in restricted broiler breeders have been conducted on birds kept in cages. A corollary of this is that the auditor should ensure that litter is dry and friable and permits the birds to scratch and peck the litter in an appropriate manner.

It is well known that plasma corticosterone concentrations rise when feed intake is limited and that all of the behavioural and physiological indices of welfare used in our research are affected by increasing plasma corticosterone: interpretation of the results is therefore not easy. We therefore examined the statistical relationships among the various indices and found that each of them helped to explain some of the observed variation. Furthermore, the responses to feed restriction from 0.25 to 0.85

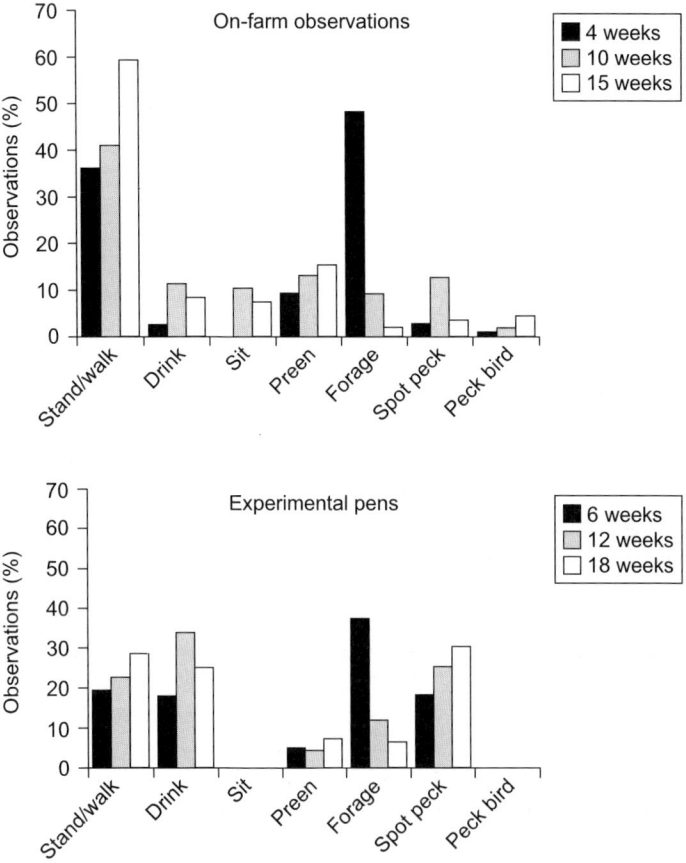

Fig. 2.3. Behavioural time budgets of broiler breeder females in a commercial flock (on-farm observations) and in contemporary birds in six experimental pens of 14 birds at different ages during rearing. (From C.E. Channing and P.M Hocking, unpublished on-farm data and Hocking *et al.*, 2001.)

of *ad libitum* for several physiological traits was curvilinear, suggesting that they were not simply alternative measures of feed intake (Hocking *et al.*, 1996). These results suggest that each trait was measuring a different response to feed restriction and that there was no single criterion of welfare. Taken together, the data on welfare indices, productivity, health and mortality suggest a curvilinear relationship between welfare and body weight (Fig. 2.4). Mortality is high and productivity is low – indicating poor welfare – at both extremes. Physiological indices of welfare are little affected at intermediate levels (perhaps to 60% of *ad libitum* body weight) and increase steeply thereafter in larger or smaller birds (Hocking *et al.*, 1996). (It should be noted that in commercial flocks broiler breeders are fed to target body weights, not feed intakes, and that in any given environment body weight is directly proportional to feed intake.)

Body weight at sexual maturity in control lines typical of the early 1970s averaged 3.2 kg compared with 4.5 to 5.6 kg in contemporary male and female

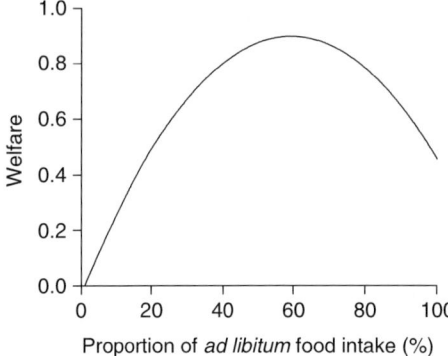

Fig. 2.4. A model of welfare in feed-restricted broiler breeders. The scale of welfare is arbitrary and represents a subjective integration of changes in physiology, immune function, behaviour, health, mortality and productivity in response to the degree of feed restriction based on the responses reported by Hocking *et al.* (1996).

parent lines (Hocking and Robertson, 2000, and unpublished data of same authors). Recommended body weights for commercial broiler breeder females have changed little over the last 30 years; they remain at about 2.8 to 3.0 kg at sexual maturity rising to 3.6 to 3.8 kg at the end of lay (60–64 weeks of age). The net result is that the body weight of restricted birds as a proportion of potential (mature) body weight has increased with time and there is no sign that this trend is diminishing. Selection for greater feed efficiency has been a major objective in genetic improvement programmes for several years and it is possible that modern broiler breeders require less feed than previously to achieve the same body weight gain, thereby increasing hunger. Alternatively, selection on feed efficiency determined in individual cages may have led to indirect selection for less sociable lines of broiler breeders in the same way that selection of laying hens in individual cages may have resulted in an increase in feather-pecking and cannibalism in modern layers (Muir, 1996). Whatever the cause, in the longer term, increasing attention will need to be given to the management of broiler breeders.

It is possible that the recent increase in beak-trimming in commercial flocks may be related to an effective increase in the degree to which the body weights of modern birds are restricted compared with their potential. Debeaking is mutilation and is clearly a welfare issue, and methods of avoiding having to beak-trim should be investigated: this may involve changing the line of broiler breeder. An important management strategy is to insist on a regular increase in feed allocation (at least weekly), which should also allow for changes in environmental temperature (e.g. sudden severe weather). Regular body weight records are essential for this purpose and automatic weighing devices are an important management tool in this respect.

The most promising long-term strategy to improve the welfare of feed-restricted broiler breeders, and to prevent the degree of feed restriction becoming greater with further selection for improved feed efficiency, is genetic selection to decrease the prevalence of multiple ovulation. If broiler breeders could be selected to ovulate a single ovum in each daily cycle, a more generous feed restriction programme that optimizes the welfare of the birds could be adopted. The use of genetic markers for single ovulation is promising in this respect and is a potential solution. Dwarf broiler breeders that inherit a sex-linked gene affecting body weight are about one-third smaller than birds carrying the normal allele, have fewer ovarian follicles (Hocking *et al.*, 1987) and require less severe feed restriction than conventional broiler breeders

(Whitehead *et al.*, 1987). The replacement of standard broiler breeders by dwarf lines is one strategy that may improve the welfare of commercial flocks of broiler breeders in the short term.

There are two current methods that might be used to decrease the severity of feed restriction and the hunger that ensues. These are the use of appetite suppressants and dietary manipulation. Specific appetite suppressants were ineffective or had undesirable attributes that limit their practical application (Hocking and Bernard, 1993; Savory *et al.*, 1996). Manipulating dietary components by using indigestible material to induce satiety resulted in poorer welfare, as indicated by physiological indices, although excessive oral behaviour was abolished, and there was no apparent improvement in the motivation to feed (Savory *et al.*, 1996; Savory and Lariviere, 2000). In another experiment it was shown that, whereas the behaviour of birds fed high concentrations of inert dietary diluents was comparable with that of those fed a conventional ration *ad libitum*, the physiological indices were similar to those of quantitatively restricted birds (Kubikova *et al.*, 2001). Manipulating growth curves and feeding low-protein rations were ineffective in improving welfare and led to relatively poor productivity (Hocking *et al.*, 2001, 2002a,b). Early experiments that used diets very low in protein to limit body weight gain in broiler breeders resulted in greater mortality and signs of nutrient deficiency (Wilson *et al.*, 1971; Wambeke and Okerman, 1976).

General Husbandry

Many aspects of the welfare of broiler breeder flocks also pertain to the husbandry of their progeny; further details will be found in the chapters in this book that discuss air, light, water and litter quality, and will not be considered here. Many criteria that affect the welfare of broiler breeders are covered in welfare codes and breeders' management guides (e.g. stocking densities, light intensities, feeding and drinking space allowances). These are updated regularly and may change as experience evolves, new information is obtained and different circumstances arise. The auditor should review these recommendations regularly and seek to assess the validity of departures from recommendations from the perspective of the welfare of the birds. However, there are two aspects that warrant particular mention.

Good health is important in order to maximize the welfare of both broilers and broiler breeders. Furthermore, maximizing the immunity of broiler breeders during the breeding period is necessary in order to ensure the adequate protection of the young broiler chick against infection with certain disease organisms via maternally derived antibodies. It is therefore essential to inspect the vaccination records of breeder flocks and the results of any tests of acquired immunity, to verify adherence to an agreed programme of effective immune protection. In addition, the biosecurity arrangements (e.g. clean sites, proper rodent control, hand washing, etc.) should be audited by inspection of the environment, records and procedures.

Broiler breeder flocks are kept in the same accommodation for a longer period of time than broilers, and litter management is both more challenging and more important. Feed restriction during rearing tends to increase drinker-directed activity and may lead to the addition of excessive water to the environment, which may result in wet or capped litter. This is undesirable for health and management reasons and may inhibit normal foraging behaviour (scratching and pecking) in response to feed restriction, to the detriment of the welfare of the birds. Wet or dirty litter in the breeding period may result in dirty eggs (a potential source of infection during incubation and hatching) and strenuous efforts must be made to maintain good litter conditions by adequate ventilation, turning the litter and the provision of fresh material whenever needed. The auditor should inspect the broiler breeder shed for signs of inadequate litter management; for example, evidence of the recent addition of fresh litter and the presence of birds with soiled plumage. There are no detectable welfare consequences for the birds if access to water is limited for 3–4 hours during and after feeding to control litter moisture (Hocking, 1993c).

Alternative Systems

In the past, some East European dwarf broiler breeder flocks were kept in conventional cages and artificial insemination was used for producing fertile hatching eggs. Colony cages (similar to large conventional battery cages) have been manufactured and used in The Netherlands for broiler breeders. For practical reasons and because both systems are likely to suffer from the same welfare and public relations problems as battery cages for laying hens, they have not been more widely adopted. Experience of broiler breeders in cages for experimental purposes for even a short period of time does not encourage the belief that the welfare of these birds will be acceptable in commercial conditions. The condition of the feet in particular should give cause for concern (Pearson, 1983).

Alternative systems of production using traditional lines and methods, such as the Label Rouge, have been practised widely in France. The birds are slow-growing, have access to the outdoors for at least part of their lives, and are fed traditional rations containing whole grain. The welfare of the parent stock has not been assessed and there are doubtless issues in these systems that are similar to the well-known problems in free-range laying flocks (e.g. high mortality and the necessity for debeaking to control feather-pecking and cannibalism, exposure to disease organisms carried by wild birds and predation) that auditors should be aware of. The use of modern lines of broiler breeders in free-range habitats may be inadvisable because effective control of body weight will be difficult to achieve in addition to the potential problems listed here.

Coda

In many respects the welfare of well-managed intensively housed broiler breeders is excellent and the purpose of auditing a broiler breeder flock is to verify that the welfare of the birds is optimized. In this context, auditing has functions that pertain to both external and internal criteria. Externally, the purpose of auditing is to verify for the consumer that adherence to legislation about the welfare of broiler breeders has occurred. Internally, auditing may be used by managers to provide an independent check on the operation of a particular farm. The advice presented in this chapter will be as useful to the concerned farm manager as to the independent auditor charged with examining a broiler breeder farm. Both should be familiar with the very excellent publications on the management of broiler breeders issued by the leading broiler breeding companies because, in virtually all respects, the welfare of broiler breeders will be maximized when the management of individual flocks is optimized for maximum health and productivity. Auditors and managers should also be familiar with the statutory codes of welfare pertaining to national legislation, such as the UK statutory codes of recommendation for the welfare of broiler chickens and breeding birds. A summary of auditing criteria is presented in Box 2.2.

Box 2.2. Summary of major auditing criteria for broiler breeder flocks

- Uniformity of body weight during rearing
    - Coefficients of variation of body weight
    - Grading and selection of low weight birds
    - Number of double-yolked eggs in early lay
- Post peak-feed allocation adequate in later lay
- Feed increase and body weight gains continuous during rearing
- Adequate feeding space
- Feeding equipment well maintained and safe
- Litter managed to maintain dry friable surface
- Proper environmental control (ventilation, heat, air quality)
- Absence of mutilations
- Mating ratio managed to avoid overmating
- Mating at correct stage of development
- Effective culling of sick and injured birds
- Mortality and culling within normal limits
- Vaccination programme effective

Acknowledgements

The work at the Roslin Institute on which this paper is based was supported by MAFF (DEFRA) and the breeding companies generously supplied day-old chicks. The Roslin Institute is supported by the BBSRC.

References

Duff, S.R.I. and Hocking, P.M. (1986) Chronic orthopaedic disease in adult male broiler breeding fowls. *Research in Veterinary Science* 41, 340–348.

Duff, S.R.I., Hocking, P.M., Randall, C.J. and Mackenzie, G. (1989) Head swelling of traumatic aetiology in broiler breeder fowl. *Veterinary Record* 126, 133–134.

Duff, S.R.I., Hocking, P.M. and Randall, C.J. (1990) Beak and oral lesions in broiler breeder fowl. *Avian Pathology* 19, 451–456.

Fattori, T.R., Wilson, H.R., Harms, R.H. and Miles, R.D. (1991) Response of broiler breeder females to feed restriction below recommended levels.1. Growth and reproductive-performance. *Poultry Science* 70, 26–36.

Freeman, B.M. (1985) Stress and the domestic fowl: physiological fact or fantasy? *World's Poultry Science Journal* 41, 45–51.

Hocking, P.M. (1990) The relationship between dietary crude protein, bodyweight and fertility in naturally mated broiler breeder males. *British Poultry Science* 31, 743–757.

Hocking, P.M. (1993a) Effects of body weight at sexual maturity and the degree and age of restriction during rearing on the ovarian follicular hierarchy of broiler breeder females. *British Poultry Science* 34, 793–801.

Hocking, P.M. (1993b) Optimum size of feeder grids in relation to the welfare of broiler breeder females fed on a separate sex basis. *British Poultry Science* 34, 849–855.

Hocking, P.M. (1993c) Welfare of broiler breeder and layer females subjected to food and water control during rearing: quantifying the degree of restriction. *British Poultry Science* 34, 53–64.

Hocking, P.M. (1994) Assessment of the welfare of food restricted male broiler breeder poultry with musculoskeletal disease. *Research in Veterinary Science* 57, 28–34.

Hocking, P.M. (1996) The role of body weight and food intake after photostimulation on ovarian function at first egg in broiler breeder females. *British Poultry Science* 37, 841–851.

Hocking, P.M. and Bernard, R. (1993) Evaluation of putative appetite suppressants in the domestic fowl (*Gallus domesticus*). *British Poultry Science* 34, 393–404.

Hocking, P.M. and Bernard, R. (1997) Effects of male body weight, strain and dietary protein content on fertility and musculoskeletal disease in naturally mated broiler breeder males. *British Poultry Science* 38, 29–37.

Hocking, P.M. and Robertson, G.W. (2000) Ovarian follicular dynamics in selected and control (relaxed selection) male- and female-lines of broiler breeders fed *ad libitum* or on restricted allocations of food. *British Poultry Science* 41, 229–234.

Hocking, P.M., Waddington, D., Walker, M.A. and Gilbert, A.B. (1987) Ovarian follicular structure of White Leghorns fed *ad libitum* and dwarf and normal broiler breeders fed *ad libitum* or restricted to point of lay. *British Poultry Science* 28, 493–506.

Hocking, P.M., Waddington, D., Walker, M.A. and Gilbert, A.B. (1989) Control of the development of the ovarian follicular hierarchy in broiler breeder pullets by food restriction during rearing. *British Poultry Science* 30, 161–174.

Hocking, P.M., Maxwell, M.H. and Mitchell, M.A. (1993) Welfare of broiler breeder and layer females subjected to food and water control during rearing. *British Poultry Science* 34, 443–458.

Hocking, P.M., Maxwell, M.H. and Mitchell, M.A. (1996) Relationships between the degree of food restriction and welfare indices in broiler breeder females. *British Poultry Science* 37, 263–278.

Hocking, P.M., Maxwell, M.H., Robertson, G.W. and Mitchell, M.A. (2001) Welfare assessment of modified rearing programmes for broiler breeders. *British Poultry Science* 42, 424–432.

Hocking, P.M., Bernard, R. and Robertson, G.W. (2002a) Effects of low dietary protein and different allocations of food during rearing and restricted feeding after peak rate of lay on egg production, fertility and hatchability in female broiler breeders. *British Poultry Science* 43, 94–103.

Hocking, P.M., Maxwell, M.H., Robertson, G.W. and Mitchell, M.A. (2002b) Welfare assessment of broiler breeders that are food restricted after peak rate of lay. *British Poultry Science* 43, 5–15.

Katanbaf, M.N., Siegel, P.B. and Gross, W.B. (1987) Research note: prior experience and response of chickens to a streptococcal infection. *Poultry Science* 66, 2053–2055.

Katanbaf, M.N., Dunnington, E.A. and Siegel, P.B. (1989) Restricted feeding in early and late feathering chickens. 2. Reproductive responses. *Poultry Science* 68, 352–358.

Kostal, L., Savory, C.J. and Hughes, B.O. (1992) Diurnal and individual variation in behaviour of restricted-fed broiler breeders. *Applied Animal Behaviour Science* 32, 361–374.

Kubikova, E., Vyboh, P. and Kostal, L. (2001) Behavioural, endocrine and metabolic effects of feed restriction in broiler breeder hens. *Acta Veterinaria Brno* 70, 247–257.

Maxwell, M.H., Robertson, G.W., Spence, S. and McCorquodale, C.C. (1990) Comparison of haematological values in restricted- and ad libitum-fed domestic fowls: white blood cells and thrombocytes. *British Poultry Science* 31, 399–405.

Maxwell, M.H., Hocking, P.M. and Robertson, G.W. (1992) Differential leucocyte responses to various degrees of food restriction in broilers, turkeys and ducks. *British Poultry Science* 33, 177–187.

Mench, J.A. (1988) The development of aggressive-behavior in male broiler chicks – a comparison with laying-type males and the effects of feed restriction. *Applied Animal Behaviour Science* 21, 233–242.

Millman, S.T. and Duncan, I.J.H. (2000a) Effect of male-to-male aggressiveness and feed-restriction during rearing on sexual behaviour and aggressiveness towards females by male domestic fowl. *Applied Animal Behaviour Science* 70, 63–82.

Millman, S.T. and Duncan, I.J.H. (2000b) Strain differences in aggressiveness of male domestic fowl in response to a male model. *Applied Animal Behaviour Science* 66, 217–233.

Millman, S.T., Duncan, I.J.H. and Widowski, T.M. (2000) Male broiler breeder fowl display high levels of aggression toward females. *Poultry Science* 79, 1233–1241.

Muir, W.M. (1996) Group selection for adaptation to multi-hen cages: selection program and direct responses. *Poultry Science* 75, 447–458.

O'Sullivan, N.P., Dunnington, E.A., Smith, E.J., Gross, W.B. and Siegel, P.B. (1991) Performance of early and late feathering broiler breeder females with different feeding regimens. *British Poultry Science* 32, 981–995.

Pearson, R.A. (1983) Prevention of foot lesions in broiler breeder hens kept in individual cages. *British Poultry Science* 24, 183–190.

Robinson, F.E. and Robinson, N.A. (1991) Reproductive performance, growth rate and body composition of broiler breeder hens differing in body weight at 21 weeks of age. *Canadian Journal of Animal Science* 71, 1233–1239.

Robinson, F.E., Robinson, N.A., Hardin, R.T. and Wilson, J.L. (1995) The effects of 20-week body weight and feed allocation during early lay on female broiler breeders. *Journal of Applied Poultry Research* 4, 203–210.

Savory, C.J. and Lariviere, J.M. (2000) Effects of qualitative and quantitative food restriction treatments on feeding motivational state and general activity level of growing broiler breeders. *Applied Animal Behaviour Science* 69, 135–147.

Savory, C.J. and Mann, J.S. (1999) Stereotyped pecking after feeding by restricted-fed fowls is influenced by meal size. *Applied Animal Behaviour Science* 62, 209–217.

Savory, C.J. and Maros, K. (1993) Influence of degree of food restriction, age and time on behaviour of broiler breeder chickens. *Behavioural Processes* 29, 179–190.

Savory, C.J., Seawright, E. and Watson, A. (1992) Stereotyped behaviour in broiler breeders in relation to husbandry and opioid receptor blockade. *Applied Animal Behaviour Science* 32, 349–360.

Savory, C.J., Carlisle, A., Maxwell, M.H., Mitchell, M.A. and Robertson, G.W. (1993a) Stress arousal and opioid peptide-like immunoreactivity in restricted- and ad lib.-fed broiler breeder fowls. *Comparative Biochemistry and Physiology* 106A, 587–594.

Savory, C.J., Maros, K. and Rutter, S.M. (1993b) Assessment of hunger in growing broiler breeders in relation to a commercial restricted feeding programme. *Animal Welfare* 2, 131–152.

Savory, C.J., Hocking, P.M., Mann, J.S. and Maxwell, M.H. (1996) Is broiler breeder welfare improved by using qualitative rather than quantitative food restriction to limit growth rate? *Animal Welfare* 5, 105–127.

Wambeke, F. van and Okerman, F. (1976) The effect of rearing and qualitative feed restriction during the rearing period on the development and reproductive performances of broiler breeder males. *Archive für Geflügelkunde* 40, 64–71.

Whitehead, C.C., Herron, K.M. and Waddington, D. (1987) Reproductive performance of dwarf broiler breeders given different food allowances during the rearing and breeding periods on two lighting patterns. *British Poultry Science* 28, 415–427.

Wilson, H.R., Rowland, L.O. and Harms, R.H. (1971) Use of low protein grower diets to delay sexual maturity of broiler breeder males. *British Poultry Science* 12, 157–163.

3 Pododermatitis and Hock Burn in Broiler Chickens

C. BERG

Department of Animal Environment and Health, Faculty of Veterinary Medicine, Swedish University of Agricultural Sciences, Skara, Sweden

Introduction

The vast majority of the broilers grown in the world today are raised in large flocks under loose-housing conditions. In some countries the houses are climate-controlled and have heating systems: in other parts of the world cooling equipment is used. Some houses have advanced computer-controlled ventilation systems and an instant auxiliary power supply, while others rely on manually adjusted sidewall curtains. The birds can be grown to a market weight just over 1 kg, while in other instances they are kept until the live weight reaches 3 kg or more. The commercial hybrids used vary, and so do the feeding regimes, the vaccination strategies and the choice of lighting programmes. Regardless of this, all these birds have something in common: they will all spend their life in close contact with some type of litter material, and also in contact with droppings, which form part of the litter surface. If the litter conditions are suboptimal, there is a considerable risk that the birds will develop contact dermatitis on their feet, hocks and/or breast. This can have consequences both for bird welfare and for production.

Contact Dermatitis in Broilers

Footpad dermatitis, also known as plantar pododermatitis, is a condition characterized by lesions on the ventral footpads of poultry (Nairn and Watson, 1972; Wise, 1978; Martland, 1985; Schulze Kersting, 1996). Footpad dermatitis is a type of contact dermatitis (Nairn and Watson, 1972; Greene *et al.*, 1985) affecting the plantar regions of the feet (Fig. 3.1). The lesions can develop in less than a week (Ekstrand and Algers, 1997). In an early stage, discoloration of the skin is seen. Hyperkeratosis and necrosis of the epidermis can be seen histologically. In severe cases, the erosions are developed into ulcerations with inflammatory reactions of the subcutaneous tissue (Greene *et al.*, 1985). Crusts formed by exudates, litter and faecal material often cover the ulcerations (Martland, 1985). The lesions may heal if the litter quality is substantially improved (Martland, 1985), but under commercial conditions this is rarely the case, as the environmental conditions are not usually improved before the birds are sent off for slaughter. After healing, the footpad does not show the normal skin fissure pattern and has a slightly paler colour (Greene *et al.*, 1985).

Hock burn is another manifestation of the same disease (Greene *et al.*, 1985; Martland, 1985; Bruce *et al.*, 1990). The skin of the hocks becomes dark brown, and in severe cases scabs can be seen (Fig. 3.2). The same type of contact dermatitis can also be seen on the skin covering the ventral aspect of the bird's keel bone, and is then called breast blisters (Greene *et al.*, 1985; Martland, 1985; Bruce *et al.*, 1990). All these lesions, and especially the footpad lesions, are sometimes referred to as ammonia burns, which is slightly inaccurate as they are not caused by ammonia only (Martland, 1985). The histopathological changes observed in the skin are similar to those described in many other types of dermatitis, and no lesions specific for the disease have been observed (Greene *et al.*, 1985; Martland, 1985).

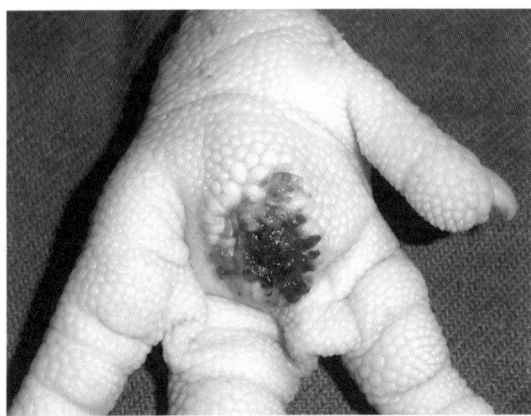

Fig. 3.1. Broiler foot with severe contact dermatitis.

Fig. 3.2. Broiler with severe footpad dermatitis and hock burn. (Photograph by G.S. Sanotra.)

Similar lesions can also be seen in fattening turkeys (Abbott *et al.*, 1969; Schmidt and Lüders, 1976; Harms and Simpson, 1977; Martland, 1984). The lesions in turkeys basically have the same aetiology and pathology as those seen in broilers (Martland, 1984; Gonder and Barnes, 1987; Berg, 1998). Whereas laying hens can also display foot and keel skin problems, these lesions are often of a different type. Layers can show hyperkeratosis of the footpads but more commonly suffer from bumble-foot (Gunnarsson *et al.*, 1995; Wang *et al.*, 1998), where an abscess is formed in the central footpad, causing a swelling that is often visible from a dorsal view (Tauson and Abrahamsson, 1994; Gunnarsson *et al.*, 1995). Bumble-foot has been shown to be correlated to the use of wet or non-optimally designed perches in cages or aviaries (Tauson and Abrahamsson, 1994; Wang *et al.*, 1998), and is very rarely seen in broilers. Laying hens housed with access to perches can also display keel bone bursitis, which is an inflammatory process in the bursa under the skin (Gunnarsson *et al.*, 1995).

The contact dermatitis lesions of broilers described above are thought to be caused by a combination of wet litter and unspecified chemical factors in the litter (Nairn and Watson, 1972; Harms *et al.*, 1977; Greene *et al.*, 1985; McIlroy *et al.*, 1987; Berg, 1998; Ekstrand *et al.*, 1998) (Fig. 3.3).

Reasons for Concern

In severe cases, dermatitis lesions may cause pain which, together with a deteriorated state of health, constitutes a welfare issue. It has been reported that birds with footpad dermatitis have an unsteady walk (Harms and Simpson, 1975), and it has also been reported that footpad dermatitis causes birds to walk with a hobbling gait (Hester, 1994). Nevertheless, it is very difficult to identify lameness caused by footpad

Fig. 3.3. If the litter is wet and sticky, there is an increased risk of contact dermatitis in the flock. The birds will also have dirty plumage. In this case, the underlying cause of wet litter was wet droppings related to an intestinal infection.

dermatitis in a commercial flock. As broilers with footpad dermatitis usually get the same kind of lesions on both feet, severely affected birds are rarely seen limping, but instead are less likely to move. Hock burn usually affects both legs simultaneously.

Apart from animal welfare aspects, contact dermatitis is relevant to the poultry meat industry for several reasons. It has been shown that broilers with severe footpad dermatitis have slower weight gain (Martland, 1985; Ekstrand and Algers, 1997), which has been suggested to be a result of pain-induced inappetance (Martland, 1985). If the problem is widespread in a flock, this can lead to substantially reduced profit for the producer. As flocks with a high incidence of footpad dermatitis often also show a high prevalence of other types of contact dermatitis, such as breast blisters and hock burns (Greene *et al.*, 1985; Martland, 1985), in addition to lower body weights, downgrading may adversely affect the profitability of these flocks (Wise, 1978; Cravener *et al.*, 1992). In many parts of the world, the feet are not processed for human consumption and thus little attention is paid to their condition. On the contrary, the hocks are commonly left on the carcass and displayed to the consumers, as is the breast, which means that lesions on the hocks and breast are more likely to cause concern at the processing plant.

Finally, the lesions can be a gateway for bacteria, which may spread haematogenously and cause joint lesions and impaired product quality in other ways (Schulze Kersting, 1996). In poultry, staphylococci are common inhabitants of the skin. Staphylococci can be found as secondary infections in footpad ulcers (Hester, 1994) and are involved in a number of different disease complexes (McCullagh *et al.*, 1997).

There is limited information available on the actual incidence of different types of contact dermatitis in broilers, and the figures generally only reflect the European situation. The few internationally reported surveys have concentrated on estimating the prevalence at the time of slaughter. For hock burn, figures from 7 to around 20% have been reported, and the prevalence of breast blisters has been estimated at between 0.0% and 0.2–0.3% (McIlroy *et al.*, 1987; Bruce *et al.*, 1990; Menzies *et al.*, 1998). Swedish surveys of broiler footpad dermatitis have reported an average flock prevalence of severe footpad lesions of 5–10% (Elwinger, 1995; Berg, 1998), with a range from 0–100% in different flocks (Ekstrand *et al.*, 1998).

In the EC report on the welfare of chickens kept for meat production (European Commission, 2000), it was concluded that contact dermatitis is a widespread problem in European broiler production, and that the problem could not easily be handled by breeding efforts or by changes in age or weight at slaughter within commercial ranges. It was also concluded that management practices would be of greatest importance in preventing broiler contact dermatitis.

The Epidemiology of Broiler Contact Dermatitis

A number of risk factors for contact dermatitis in broilers have been identified. In some cases the studies have focused on only one of type of skin lesion, whereas others have included footpad dermatitis, hock burn and breast blisters in the same investigation. It can generally be assumed that when a factor is identified as a risk factor for one of the manifestations, it is very likely that it will also influence the prevalence of the others. As the feet, in contrast to the hocks and breast, are in constant contact with the litter material 24 hours a day for the entire growing period, foot lesions can generally be expected to be more prevalent than hock and breast lesions. As a result of this, footpad status is a more sensitive indicator of litter problems than hock or breast status. Nevertheless, there are occasions where a broiler flock can display a very high prevalence of hock burn at the time of slaughter and still show no signs of footpad dermatitis (Fig. 3.4).

It has been shown in numerous experimental, on-farm and questionnaire-based studies that the prevalence of contact dermatitis in broilers is related to litter quality (Nairn and Watson, 1972; Greene *et al.*, 1985; McIlroy *et al.*, 1987; Ekstrand and Algers, 1997; Berg, 1998; Ekstrand *et al.*, 1998; Nielsen *et al.*, 2002). When the litter is wet, sticky and compact, dermatitis lesions are more commonly seen. This means that any factor causing wet or sticky litter will be a risk factor for contact dermatitis. Risk factors are discussed below, with comments on how they can be expected to influence the prevalence of contact dermatitis. One should keep in mind, however, that most of the studies on broiler contact dermatitis have been carried out in temperate climates, and that some risk factors of importance in tropical or subtropical regions may not yet have been identified.

Litter material

A type of litter that can absorb and hold moisture efficiently will not appear as wet as a litter material with a poorer absorption potential, and result in a lower incidence of contact dermatitis (Shanawany, 1992; Tucker and Walker, 1992). The type of litter used varies between different regions as a result of variation in availability and cost, and it is difficult to give advice that is generally applicable. Some litter materials, such as peat and sawdust, may have a high water-holding capacity but have other disadvantages, such as dust problems or bacterial contamination. However, it has been shown that using wood shavings as the litter material tends to result in a lower prevalence of footpad dermatitis than using straw (Ekstrand *et al.*, 1997; Sørensen *et al.*, 2002). It is also recommended that when straw is used it should be chopped very short, to increase the water-holding capacity. However, some studies have failed

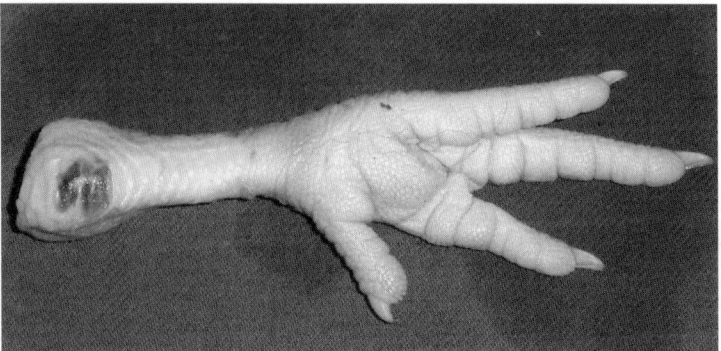

Fig. 3.4. Broiler leg with hock burn. In this case no footpad dermatitis lesion is present, which is unusual.

to find a difference between different lengths of straw with respect to footpad status (Sørensen *et al.*, 2002).

Permanent litter, in houses where litter material is added between batches but perhaps discarded only once a year, can be considered a risk factor if the climate or ventilation system does not dry out the litter completely between batches of birds.

Litter depth

Footpad dermatitis has been found to be more common in flocks reared on thick layers of litter material than in flocks on thinner layers (Ekstrand *et al.*, 1997). A possible explanation is that the chickens are less prone to pecking, scratching and turning the litter particles over if the litter is thick and compact, and are thereby less effective in ventilating the litter and keeping it dry. Other studies have reported somewhat contradictory results, with no effect of litter depth on the incidence of breast blisters in broilers (Stephenson *et al.*, 1960), and results obtained with different litter materials have varied (Tucker and Walker, 1992). This inconsistency may be explained by differences in the structure, particle size and other quality aspects of the litter materials tested, as well as differences in floor type.

Floor type

To prevent water coming up from underneath through a solid floor, the ground surrounding the house should be thoroughly drained and the floor must be properly cast and should not allow any passage of capillary water. Improved performance and improved footpad health in turkey poults reared in houses with floor heating has been reported (Berg, 2000). As a result, floor heating is more common in recently constructed broiler houses, and producers have reported improved foot health in these birds.

Stocking density

Stocking density has been reported to influence litter quality. Reduced litter quality at increased stocking densities (McIlroy *et al.*, 1987; Blokhuis and van der Haar, 1990; Gordon, 1992; Tucker and Walker, 1992) leads to an increased incidence of contact dermatitis (Proudfoot *et al.*, 1979; Cravener *et al.*, 1992; Gaardbo Thomsen, 1992; Martrenchar *et al.*, 1997; Hall, 2001; Sørensen *et al.*, 2002).

Drinker system

The type of drinker system, which is related to both water spillage and water consumption, is significantly associated with the prevalence of footpad dermatitis (Elson, 1989; Tucker and Walker, 1992; Cholocinska *et al.*, 1997; Ekstrand *et al.*, 1997), hock burn (Lynn and Elson, 1990) and breast blisters (Elson, 1989; Meijerhof,

1989). A water system with large cups instead of drinking nipples is likely to result in more spillage and, consequently, wet litter. The same applies if the design encourages the birds to play with the system, and any water system that leads to over-drinking may cause wet droppings and wet litter. A system consisting of water nipples with small drip cups is probably the most efficient way of reducing water spillage (Fig. 3.5).

Feed composition

The basic feed composition does influence the risk of contact dermatitis in broilers, and it was shown in the 1940s that nutritional deficiencies (for example, in biotin) could be of importance (Patrick *et al.*, 1942). It has been suggested that soybean-based rations may be involved in producing an irritant litter (Nairn and Watson, 1972) or that soybean meal included in the diet may affect the consistency of the faeces, resulting in sticky droppings and sticky litter and leading to an increase in the incidence of footpad dermatitis (Jensen *et al.*, 1970). There is an effect of dietary fat quality on litter surface friability (Bray and Lynn, 1986; Tucker and Walker, 1992), and over-drinking caused by too high salt content in the feed can also result in wet litter, which in turn may increase the prevalence of contact dermatitis. Furthermore, the type of wheat used when feeding whole wheat as a part of the diet can influence the incidence of footpad dermatitis (Sørensen *et al.*, 2002). It is also generally accepted that excess crude protein levels in the feed will increase the production of uric acid in the kidneys, leading to wet droppings that are rich in nitrogen, which will result in a high prevalence of contact dermatitis (Gordon *et al.*, 2003). Epidemiological studies have shown a significant association between feed manufacturer and footpad dermatitis in broilers (Ekstrand and Carpenter, 1998b; Ekstrand *et al.*, 1998; Sørensen *et al.*, 2002) and also a significant effect of feed manufacturer on the incidence of hock and breast lesions (McIlroy *et al.*, 1987; Bruce *et al.*, 1990).

Fig. 3.5. Nipple drinkers with drip cups have been shown to reduce water spillage compared with bell drinkers or nipples only, thereby decreasing the risk of wet litter and contact dermatitis.

Light

Recent studies indicate that the use of light programmes with long dark periods, which may have beneficial effects on other aspects of bird health, can lead to an increased prevalence of footpad dermatitis, hock burn and breast blisters (Renden *et al.*, 1992; Sørensen *et al.*, 1999) and an increased frequency of rejection due to skin lesions (not specified) at slaughter (Søholm Petersen, 2001). A possible explanation is that, even if the total activity is increased, the birds will spend longer sitting down continuously during the long dark periods (Renden *et al.*, 1992) and work the litter less, resulting in poorer litter conditions and consequently a higher prevalence of contact dermatitis.

It has also been shown that intermittent light will increase the prevalence of breast blisters compared with continuous light (Deaton *et al.*, 1978), but other studies have not found such an effect (Quarles and Kling, 1974).

The positioning of the light sources and the distribution of light within the house may also be of relevance, as this can influence the behaviour of the birds. If the light is very bright in certain areas or spots the birds may huddle there, and the litter can become locally wet and compact, thereby increasing the risk of contact dermatitis in the birds choosing these spots for resting (Søholm Petersen *et al.*, 2002).

Gender

Results of the effect of gender on the prevalence of contact dermatitis are somewhat contradictory. While some studies (Berg, 1998) have found no association between the sex of the birds and the prevalence of footpad dermatitis, other studies have shown such an association for other types of contact dermatitis, with a higher prevalence in male birds (Stephenson *et al.*, 1960; Harms and Simpson, 1975; Bruce *et al.*, 1990; Cravener *et al.*, 1992).

Age

Scientific results are also contradictory with respect to the effect of age. Several studies have found an increasing prevalence of contact dermatitis with age (Stephenson *et al.*, 1960; Bruce *et al.*, 1990; Sørensen *et al.*, 2002) while others have not found any effect of age (Martland, 1985; Ekstrand *et al.*, 1997). A possible explanation is that the age spans compared have been different, as slaughter age may vary considerably between different countries.

Hybrids

In some epidemiological studies a clear difference has been found in the prevalence of footpad dermatitis between different commercial hybrids at the same age (Ekstrand *et al.*, 1998; Sanotra and Berg, 2003), whereas other studies have found no such difference (Ekstrand *et al.*, 1997). Experimental studies have indicated a strain effect also for breast blisters (Renden *et al.*, 1992).

Health

Although rarely mentioned in the scientific literature, it is evident that the health of the broiler flock can influence the prevalence of contact dermatitis. The most obvious example is that if a flock suffers from diarrhoea, whether caused by poor feed quality, intestinal parasites such as coccidia, or by viral or bacterial infections, this will probably lead to wet litter (Neill *et al.*, 1984) and thus to problems with contact dermatitis. When enteric infections are likely to become a problem, vaccination should, if applicable, be used to prevent such outbreaks. However, vaccination against infectious bursal (Gumboro) disease may actually increase the incidence of footpad dermatitis (Sørensen *et al.*, 2002).

Similarly, it is likely that if a flock suffers from a high incidence of leg weakness, regardless of the underlying cause, affected birds will move around less and spend more time sitting down, and the risk of developing contact dermatitis will be increased.

Climate

The weather influences litter quality. High relative humidity, both outdoors (Payne, 1967; McIlroy *et al.*, 1987) and inside the house (Payne, 1967; Weaver and Meijerhof, 1991), is associated with poor litter quality. There is a significant seasonal effect on the prevalence of broiler contact dermatitis, with the highest prevalence found during the autumn and winter months in northern Europe (McIlroy *et al.*, 1987; Bruce *et al.*, 1990; Ekstrand and Carpenter, 1998a). These are the months with the highest relative humidity of the outdoor air in Sweden (Ekstrand and Carpenter, 1998a), and current ventilation systems are not always able to compensate for this. Producers also have to take increased heating costs into account when deciding to increase ventilation rates during the cold season. It has been shown that the type of heating system used can influence the incidence of contact dermatitis (Nielsen *et al.*, 2002) and it has also been found that if the temperature in the house is low when the litter is distributed before day-old chicks are placed, the prevalence of footpad dermatitis at slaughter will be higher (Sørensen *et al.*, 2002).

Intervention

In Sweden, a monitoring programme for footpad dermatitis was introduced in the mid-1990s. In this programme, trained inspectors classify the foot health of each broiler flock at the time of slaughter according to a scoring system developed at the Swedish University of Agricultural Sciences (Algers and Berg, 2001). The monitoring programme is linked to an advisory programme. The prevalence and severity of footpad dermatitis in broilers decreased over time when the surveillance programme was initiated and executed (Berg, 1998; Ekstrand *et al.*, 1998). A similar monitoring programme for footpad dermatitis has recently been started in Denmark (Sørensen *et al.*, 2002). Furthermore, an evaluation of the dermatitis control strategies implemented over a number of years in Northern Ireland has indicated a dramatic reduction in both hock burn and breast blisters (Menzies *et al.*, 1998).

Prevention of Contact Dermatitis

Improving ventilation, using floor-heating systems or manually turning the litter over to ventilate it can improve wet litter slightly. As mentioned above, lesions will generally heal with time, but in practice the birds are usually sent to slaughter before this happens. It can therefore be assumed that the point prevalence of dermatitis recorded at slaughter is a reasonable reflection of the true incidence during the rearing period. It can also be concluded that, to have an overall effect on the incidence of contact dermatitis, the measures taken must concentrate on preventing the occurrence of wet litter, not trying to improve litter that has already become poor.

As the conditions under which broilers are raised vary between different parts of the world, it is extremely difficult to give efficient general advice on how to prevent contact dermatitis. In each case the possible underlying causes of wet litter should be investigated and relevant action taken. This can be anything from changing the feed composition to changing the litter material used or the type of water equipment installed. Most often, a combination of different measures will prove to be successful.

It has been shown that a surveillance and advisory programme can be used successfully to decrease the incidence of footpad lesions in broiler populations and thus to improve the health and welfare of the birds (Algers and Berg, 2001). The prevalence of hock burn and breast blisters is often recorded at the slaughterhouse by veterinary inspectors or their assistants, and if these figures are used properly by the company's advisers they can be used to pinpoint the producers where the problem commonly occurs. Such control strategies have proved to be efficient in decreasing the prevalence of both breast blisters and hock burn (Menzies *et al.*, 1998).

Improved litter quality can have beneficial effects on broiler welfare in several ways: not only by decreasing the prevalence of contact dermatitis but also by leading to reduced ammonia emissions and thereby improved air quality, improved hygiene and thereby a decreased risk of disease outbreaks, and to general improvement in bird cleanliness.

From an animal welfare point of view and also for economic reasons it is vital to decrease the prevalence of contact dermatitis substantially in commercially grown broilers. To do this, investment in housing, water equipment and ventilation systems may be necessary. The responsibility of feed suppliers should not be forgotten and overstocking must be avoided, but the most important factor is to increase knowledge about the epidemiology of the disease among broiler producers. Given that the right measures are taken and that new investments are made with this disease complex in mind, contact dermatitis will be a disease rarely seen in the future (Fig. 3.6).

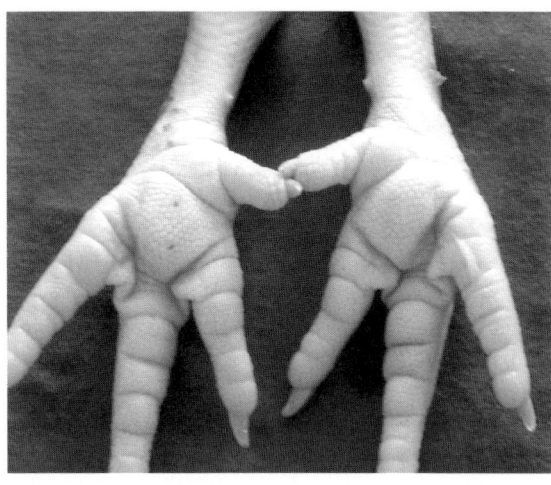

Fig. 3.6. Broiler feet without signs of contact dermatitis.

References

Abbott, W.W., Couch, J.R. and Atkinson, R.L. (1969) The incidence of foot-pad dermatitis in young turkeys fed high levels of soybean meal. *Poultry Science* 48(6), 2186–2188.

Algers, B. and Berg, C. (2001) Monitoring animal welfare on commercial broiler farms in Sweden. *Acta Agriculturae Scandinavica Section A, Animal Science Supplementum* 30, 88–92.

Berg, C. (1998) Footpad dermatitis in broilers and turkeys – prevalence, risk factors and prevention. PhD thesis, Swedish University of Agricultural Sciences, Uppsala, Sweden. *Acta Universitatis Agriculturae Sueciae, Veterinaria* 36.

Berg, C. (2000) Pågående forskning på kalkonsidan. *Näbben – Aktuell Information Från Svensk Fågel* 2, 2–3.

Blokhuis, H.J. and van der Haar, J.W. (1990) The effect of the stocking density on the behavior of broilers. *Archiv für Geflügelkunde* 54, 74–77.

Bray, T.S. and Lynn, N.J. (1986) Effects of nutrition and drinker design on litter condition and broiler performance. *British Poultry Science* 27, 151.

Bruce, D.W., McIlroy, S.G. and Goodall, E.A. (1990) Epidemiology of a contact dermatitis of broilers. *Avian Pathology* 19, 523–538.

Cholocinska, A., Wezyk, S., Herbut, E. and Cywa-Benko, K. (1997) Effect of a broiler watering system on the hygienic quality of litter. In: *Proceedings of the 9th International Congress in Animal Hygiene, Helsinki, Finland*, pp. 301–304.

Cravener, T.L., Roush, W.B. and Mashaly, M.M. (1992) Broiler production under varying population densities. *Poultry Science* 71, 427–433.

Deaton, J.W., Reece, F.N. and McNaughton, J.L. (1978) Effect of intermittent light on broilers reared under moderate temperature conditions. *Poultry Science* 57, 785–788.

Ekstrand, C. and Algers, B. (1997) The effect of litter moisture on the development of footpad dermatitis in broilers. In: *Proceedings of the 11th International Congress of the World Veterinary Poultry Association, Budapest, Hungary*, p. 370.

Ekstrand, C. and Carpenter, T.E. (1998a) Temporal aspects of footpad dermatitis in Swedish broilers. *Acta Veterinaria Scandinavica* 39, 229–236.

Ekstrand, C. and Carpenter, T.E. (1998b) Using a Tobit regression model to analyse risk factors for foot-pad dermatitis in commercially grown broilers. *Preventive Veterinary Medicine* 37, 219–228.

Ekstrand, C., Algers, B. and Svedberg, J. (1997) Rearing conditions and footpad dermatitis in Swedish broiler chickens. *Preventive Veterinary Medicine* 31, 167–174.

Ekstrand, C., Carpenter, T.E., Andersson, I. and Algers, B. (1998) Prevalence and prevention of footpad dermatitis in broilers in Sweden. *British Poultry Science* 39, 318–324.

Elson, H.H. (1989) Drinker design affects litter quality. *Poultry* 5 (1), 8–9.

Elwinger, K. (1995) Broiler production under varying population densities – a field study. *Archiv für Geflügelkunde* 59, 209–215.

European Commission (2000) The welfare of chickens kept for meat production (broilers). Report of the Scientific Committee on Animal Health and Animal Welfare. Brussels.

Gaardbo Thomsen, M. (1992) Influence of increasing stocking rates on performance and carcass quality of broilers. In: *Proceedings of the 4th European Symposium on Poultry Welfare.* UFAW, Edinburgh, pp. 285–287.

Gonder, E. and Barnes, H.J. (1987) Focal ulcerative dermatitis ('breast buttons') in marketed turkeys. *Avian Diseases* 31, 52–58.

Gordon, S.H. (1992) The effect of broiler stocking density on bird welfare and performance. *British Poultry Science* 5, 1120–1121.

Gordon, S.H., Walker, A.W. and Charles, D.R. (2003) Feeding and broiler welfare. In: *Proceedings of the Symposium Measuring and Auditing Broiler Welfare – A Practical Guide.* University of Bristol, UK, p. 19.

Greene, J.A., McCracken, R.M. and Evans, R.T. (1985) A contact dermatitis of broilers – clinical and pathological findings. *Avian Pathology* 14, 23–38.

Gunnarsson, S., Odén, K., Algers, B., Svedberg, J. and Keeling, L. (1995) Poultry health and behaviour in a tiered system for loose housed layers. Report 35. Department of Animal Hygiene, Swedish University of Agricultural Sciences, Skara, Sweden.

Hall, A.L. (2001) The effect of stocking density on the welfare and behaviour of broiler chickens reared commercially. *Animal Welfare* 10, 23–40.

Harms, R.H. and Simpson, C.F. (1975) Biotin deficiency as a possible cause of swelling and ulceration of foot pads. *Poultry Science* 54, 1711–1713.

Harms, R.H. and Simpson, C.F. (1977) Influence of wet litter and supplemental biotin on foot pad dermatitis in turkey poults. *Poultry Science* 56, 2009–2012.

Harms, R.H., Damron, B.L. and Simpson, C.F. (1977) Effect of wet litter and supplemental biotin and/or whey on the production of foot pad dermatitis in broilers. *Poultry Science* 56, 291–296.

Hester, P.Y. (1994) The role of environment and management on leg abnormalities in meat-type fowl. *Poultry Science* 73, 904–915.

Jensen, L.S., Martinson, R. and Schumaier, G. (1970) A foot-pad dermatitis in turkey poults associated with soybean meal. *Poultry Science* 49, 76–82.

Lynn, N.J. and Elson, H.A. (1990) Which drinkers reduce possible downgrades? *Poultry* 6 (1), 11–12.

Martland, M.F. (1984) Wet litter as a cause of plantar pododermatitis, leading to foot ulceration and lameness in fattening turkeys. *Avian Pathology* 13, 241–252.

Martland, M.F. (1985) Ulcerative dermatitis in broiler chickens: the effects of wet litter. *Avian Pathology* 14, 353–364.

Martrenchar, A., Morisse, J.P., Huinnic, D., Cotte, J.P. and Moinard, C. (1997) The effect of stocking density and group size on different behavioural and productivity traits of broilers. In: *Proceedings of the 5th European Symposium on Poultry Welfare.* Wageningen Agricultural University and ID-DLO, Wageningen, The Netherlands, pp. 153–154.

McCullagh, J.J., McNamee, P.T. and Ball, H.J. (1997) Application of pulsed field gel electro-phoresis (PFGE) to the typing of *Staphylococcus aureus* isolates collected from broiler houses and hatcheries in Northern Ireland. In: *Proceedings of the 11th International Congress of the World Veterinary Poultry Association, Budapest, Hungary,* p. 50.

McIlroy, S.G., Goodall, E.A. and McMurray, C.H. (1987) A contact dermatitis of broilers – epidemiological findings. *Avian Pathology* 16, 93–105.

Meijerhof, R. (1989) Are nipples better than cups? *Poultry* 5 (1), 11.

Menzies, F.D., Goodall, E.A., McConaghy, D.A. and Alcorn, M.J. (1998) An update on the epidemiology of contact dermatitis in commercial broilers. *Avian Pathology* 27, 174–180.

Nairn, M.E. and Watson, A.R.A. (1972) Leg weakness of poultry: a clinical and patho-logical characterisation. *Australian Veterinary Journal* 48, 645–656.

Neill, S.D., Campbell, J.N. and Greene, J.A. (1984) Campylobacter species in broiler chickens. *Avian Pathology* 13, 777–785.

Nielsen, B.L., Sørensen, P., Su, G., Søholm Petersen, J. and Eskildsen, B. (2002) Trædepudesvidninger hos slagtekyllinger. *Grøn Viden*. Husdyrbrug nr 28, Danmarks JordbrugsForskning, Foulum, Denmark.

Patrick, H., Boucher, R.V., Butcher, R.A. and Knandel, H.C. (1942) The nutritional significance of biotin in chick and poultry nutrition. *Poultry Science* 21, 476.

Payne, C.G. (1967) Factors influencing environmental temperature and humidity in intensive broiler houses during the post-brooding period. *British Poultry Science* 8, 101–118.

Proudfoot, F.G., Hulan, H.W. and Ramey, D.R. (1979) The effect of four stocking densities on broiler carcass grade, the incidence of breast blisters, and other performance traits. *Poultry Science* 58, 791–793.

Quarles, C.L. and Kling, H.F. (1974) The effect of three lighting regimes on broiler performance. *Poultry Science* 53, 1435–1438.

Renden, J.A., Bilgili, S.F. and Kincaid, S.A. (1992) Live performance and carcass yield of broiler strain crosses provided either sixteen or twenty-three hours of light per day. *Poultry Science* 71, 1427–1435.

Sanotra, G.S. and Berg, C. (2003) Investigation of lameness in the commercial production of broiler chickens in Sweden. Report, Department of Animal Environment and Health, Swedish University of Agricultural Sciences, Skara, Sweden.

Schmidt, V. and Lüders, H. (1976) Ulcerations of the sole and toe pads of fattened turkey cocks. *Berlin München Tierarztlicher Wochenschrift* 89, 47–50.

Schulze Kersting, I. (1996) Untersuchungen zur Einstreuqualität und Leistung in der Broilermast in Abhängigkeit von der Besatzdichte. PhD thesis, Institut für Tierernährung, Tierärtzliche Hochschule Hannover, Germany.

Shanawany, M.M. (1992) Influence of litter water-holding capacity on broiler weight and carcass quality. *Archiv für Geflügelkunde* 56, 177–179.

Stephenson, E.L., Bezanson, J.M. and Hall, C.F. (1960) Factors affecting the incidence and severity of a breast blister condition in broilers. *Poultry Science* 39, 1520–1524.

Søholm Petersen, J. (2001) Betydningen af mørkerperioder for slagtekyllingers velfærd og produktivitet. Farmtest Report 1–2001. Landbrugets Rådgivningscenter, Skejby, Århus, Denmark.

Søholm Petersen, J., David, B., Berg, C. and Fisker, C. (2002) Broiler welfare, gait and footpad scores. In: *Proceedings of the Nordic Advisory and Veterinary Seminar 2002, Bornholm, Denmark*.

Sørensen, P., Nielsen, B., Søholm Petersen, J., Eskildsen, B. and Su, G. (2002) Trædepudesvidninger hos slagtekyllinger. DJF Report Husdyrbrug nr 42, Danmarks JordbrugsForskning, Foulum, Denmark.

Sørensen, P., Su, G. and Kestin, S.C. (1999) The effect of photoperiod/scotoperiod on leg weakness in broiler chickens. *Poultry Science* 78, 336–342.

Tauson, R. and Abrahamsson, P. (1994) Foot and skeletal disorders in laying hens: effects of perch design, hybrid, housing system and stocking density. *Acta Agriculturae Scandinavica Section A, Animal Science* 44, 110–119.

Tucker, S.A. and Walker, A.W. (1992) Hock burn in broilers. In: Garnsworthy, P.C. *et al.* (eds) *Recent Advances in Animal Nutrition*. Butterworth Heinemann, Oxford, UK, pp. 33–50.

Wang, G., Ekstrand, C. and Svedberg, J. (1998) Wet litter and perches as risk factors for the development of foot-pad dermatitis in floor hens. *British Poultry Science* 39, 191–197.

Weaver, W.D. and Meijerhof, R. (1991) The effect of different levels of relative humidity and air movement on litter conditions, ammonia levels, growth, and carcass quality for broiler chickens. *Poultry Science* 70, 746–755.

Wise, D.R. (1978) Nutrition–disease interactions of leg weakness in poultry. In: Haresign, W. and Lewis, D. (eds) *Recent Advances in Animal Nutrition*. Butterworth Heinemann, Oxford, UK, pp. 41–57.

4 Evaluating the Impact of Metabolic Disorders on the Welfare of Broilers

R.J. JULIAN

Department of Pathobiology, Ontario Veterinary College, University of Guelph, Guelph, Ontario, Canada

Introduction

The ability to hide evidence of illness is a protective mechanism that chickens have developed over millions of years of evolution because any sign of weakness is quickly spotted by a predator. This instinctive behaviour adds to the difficulty of accurately assessing and measuring the welfare implications of metabolic disease. It is often difficult to tell whether a sick broiler is in physical or mental distress. Hypoglycaemic broilers are lethargic and uncoordinated. They may be in distress, but do not appear to be in pain. In humans, serious metabolic diseases such as diabetes and hypothyroidism may not be obvious to the affected person or the physician, even though they may be causing depression and malaise.

Metabolic disease in humans is usually classed as illness associated with a failure in one of the body hormonal or enzyme systems, storage disease related to lack of metabolism of secretory products, or the failure or reduced activity of some metabolic function. These conditions frequently have a genetic cause (Stanbury *et al.*, 1983). There are numerous genetic, metabolic disorders in poultry (Migaki, 1982) but, except for dwarfism, none has been reported in commercial broiler chickens. In poultry, we usually regard as metabolic diseases conditions that are associated with increased metabolism, rapid growth rate or high egg production, and that result in the failure of a body system because of the increased work-load on that organ or system. The difficulty of assessing the impact of metabolic disease on the welfare of broilers is complicated by the lack of agreement on which conditions should be included in the classification of metabolic disease in poultry. A broad interpretation of metabolic disorders within the following general areas will be used for this chapter: metabolic disorders that result from deficiency or failure in the production or transport of an enzyme, hormone or secretory mechanism; and metabolic disorders that result from rapid growth, high nutrient intake or high metabolic rate.

Metabolic Disorders that Result from the Failure or Deficiency in the Production of an Enzyme, Hormone or Secretory Mechanism

Fatty liver and kidney syndrome (FLKS) in broilers

FLKS is a biotin deficiency-related metabolic disease in broiler chicks aged 2–3 weeks, resulting in impaired hepatic gluconeogenesis and increased fat deposition. The problem is caused by low activity of the biotin-dependent enzyme pyruvate carboxylase. Birds die from hypoglycaemia and the clinical signs and deaths are related to hypoglycaemia (Whitehead *et al.*, 1978). The condition usually occurs suddenly, as an outbreak, associated with some management, feed (fat level) or environmental change that affects feeding. Affected broilers are usually well grown. Clinical signs include aphagia, lethargy and weakness with uncoordinated behaviour and head movement (sometimes classed as nervous signs). Affected chicks may lie on their breast with their neck and legs extended (Butler, 1976). Mortality can vary from 5 to 35%, and at necropsy the liver and kidneys are markedly enlarged, pale and fatty. Adequate dietary biotin will prevent FLKS.

Spiking mortality syndrome in broilers

Spiking mortality is the name given to a metabolic disorder of previously healthy, normal appearing, broiler chickens that experience a sudden increase in mortality between 12 and 18 days of age. Live chicks are found recumbent and uncoordinated, frequently lying on their breast with legs extended. Reported nervous signs are related to the inability to rise and to abnormal head movements. The clinical signs and death have been shown to be caused by hypoglycaemia (Davis and Vasilatos-Younken, 1995). Hypoglycaemia is a metabolic disorder and may be related to melatonin deficiency caused by lack of a long dark period. Melatonin does have an affect on metabolism (Apeldoorn *et al.*, 1999). Increasing the length of the dark period will usually prevent the problem.

Hypoglycaemia in people causes anxiety, sweating and increased heart rate. It is difficult to assess the welfare impact of the hypoglycaemia of FLKS and spiking mortality in broiler chickens.

Dwarfism

Using dwarf female breeders may affect growth rate in broilers but there is no evidence that it causes a welfare problem (Merat, 1984).

Metabolic Disease Resulting from Rapid Growth, High Nutrient Intake or High Metabolic Rate

Cardiovascular disease in broilers

Ascites syndrome (AS), pulmonary hypertension syndrome

Research on ascites in meat-type chickens reared at moderate and low altitude has shown that the pathogenesis is similar to the high-altitude disease. It is the result of pulmonary hypertension. Pulmonary hypertension is caused by the high oxygen requirement of rapid growth, which demands a high cardiac output. The problem occurs because the right ventricle has to work harder to pump more blood through the lungs to supply oxygen for metabolism in rapidly growing broiler chickens and because of insufficient vascular capacity in the lungs to carry that volume of blood (Julian, 1993; Wideman, 2001). In order to pump more blood, the muscular wall of the right ventricle becomes thicker and stronger (it hypertrophies), which increases the pressure in the vessels that carry blood through the lungs, thereby causing lung oedema. The valve that stops the back-flow of blood from the right ventricle when it contracts to pump blood to the lungs is made up of muscle from the right ventricle wall (Julian, 1996b) and is very different from the same valve in mammals. The valve also becomes thicker when the ventricular wall thickens and then ceases to work properly, and starts to leak (valvular insufficiency). This leaking valve allows blood under pressure to be forced backward from the heart through the portal vein to the liver, causing portal hypertension. The pressure in the liver causes increased leakage of fluid from the blood in the liver sinusoids into the lymph channels. The fluid then flows into the hepato-peritoneal sacs, where it is called ascitic fluid (Julian, 1993, 1994; Wideman, 2000).

The leaking valve also means that the right ventricle has to pump more blood (it has to pump the same blood twice), resulting in volume overload. The increased workload results in heart failure. Now the right ventricle can no longer pump as much blood through the lungs. The bird stops growing and appears uncomfortable. The veins become congested and filled with dark blood, the tissues become cyanotic, the abdomen becomes dilated with fluid and the bird has difficulty breathing. Some broilers die quickly from suffocation, but some will survive for several days or weeks (Julian, 1993, 1996a, 1998).

IMPACT ON WELFARE.   Affected broilers are in obvious physical distress and should be euthanized.

Sudden death syndrome, acute death syndrome, flip-over disease, heart attack

Sudden death syndrome (SDS) occurs in commercial broilers that are grown on free-choice feed that encourages high feed intake and rapid growth. Young, healthy, fast-growing broiler chickens die suddenly while standing, walking, sparring or feeding. They die with a short terminal wing-beating convulsion and are often found on their back. SDS can start as early as day 3; it continues for 8–12 weeks and the highest losses occur between days 9 and 21. The majority of affected birds are males. In good flocks, 2–4% of males may die from SDS. In a well-managed flock it is the major cause of mortality (Julian, 1986). SDS is a metabolic condition which appears to be related to high carbohydrate intake, rapid metabolic rate, defective cell membrane integrity and intracellular electrolyte imbalance. It can be prevented completely by feeding only 80% of the amount of feed the broilers would eat free-choice (Bowes *et al.*, 1988). Death probably results from ventricular fibrillation (Bowes and Julian, 1988; Julian, 1996a).

There are no diagnostic lesions at necropsy. SDS broilers are always well fleshed and appear normal. The abdomen is distended. There is usually feed in the digestive tract. The intestine is full of ingesta and mucus. The ventricles are contracted and the atria are dilated. The right ventricle is normal on cross-section. The condition was first reported to be caused by lung oedema (Hemsley, 1965) and the lungs are oedematous, except in freshly dead birds (Julian, 1996a). Slowing the growth rate by feeding mash, a long daily dark period (Classen and Riddell, 1989) and low-intensity light will reduce the incidence of SDS.

IMPACT ON WELFARE.   Because SDS causes rapid loss of consciousness in a previously healthy broiler (Newberry *et al.*, 1987), it should not be a welfare concern.

Musculoskeletal disease in broilers

'There is an unacceptably high incidence of bone and joint disorders in fast-growing strains of broiler chickens' (Webster, 1994). Most of the non-infectious causes of lameness and leg deformities that affect broiler chickens are related to rapid growth (Riddell, 1992; Sanotra *et al.*, 2001) and can therefore be classed as metabolic bone diseases. Lameness is most prominent in rapidly growing males. Slowing growth, particularly in the first 15–20 days of life (Classen and Riddell, 1989), will markedly reduce the incidence of angular bone deformity (valgus–varus), dyschondroplasia and spondylolisthesis (kinky back), which probably account for 65–80% of the leg deformity and lameness in broiler chickens on a high-density, nutritionally adequate ration.

Broiler chickens have short, thick bones. Rapid growth in these bones results in a wide, thick growth plate that models bones with inadequate strength under the growth plate at the diaphyses. Rapid growth results in tendon and bone that may not have sufficient strength to support the weight of heavy broilers. This may result in painful tearing of tissue in conditions such as spondylolisthesis, ruptured gastrocnemius tendon and separation of the proximal femoral epiphysis, backward bending of the proximal tibia in bones weakened by dyschondroplasia, epiphysolitis (osteochondrosis), and pressure-induced microfractures at the diaphyses of the proximal tibia, which cause pain when a heavy broiler stands and walks. For a recent review of lameness in broilers and its relation to welfare, see Bradshaw *et al.* (2002).

Spondylolisthesis (kinky-back)

This is the name for ventral dislocation of the anterior end of the articulating fourth thoracic vertebra with over-riding of the posterior end by the fifth to cause pinching of the spinal cord. Damage to the spinal cord causes leg weakness, and this is usually followed by partial posterior paralysis. Affected broilers are lame, sit on their tail with their feet extended or fall to one side. The lesion must be differentiated from scoliosis (which usually causes no clinical signs) and osteomyelitis or osteochondrosis of the vertebrae or growth plates (Riddell, 1996), which produce similar clinical signs. This can only be done at a post-mortem examination by cutting the vertebral column medially in the midline to show the lesion (Riddell, 1992; Julian, 1998; Crespo and Shivaprasad, 2003). Dislocation may also occur between other cervical and thoracic vertebrae. Spondylolisthesis is a frequent lesion in broiler chickens, being associated with rapid growth in genetically susceptible broilers (Riddell, 1993). It is more common in females. Osteochondrosis/epiphyseolitis in the T4 growth plates may cause cartilage to protrude into the vertebral canal, causing similar clinical signs; it is more frequent in males (Julian, 1994, 1998).

IMPACT ON WELFARE.   Spondylolisthesis is a major welfare concern in broilers because of the pain and distress associated with the lesion and because partial paralysis affects the ability to reach food or water.

Tibial dyschondroplasia (TD)

Failure of the change of proliferating avascular prehypertrophying growth plate cartilage to hypertrophying cartilage to allow it to be replaced by bone at the lower edge of the growth plate results in an abnormal mass of cartilage under the growth plate. This lesion is called dyschondroplasia (Farquharson and Jefferies, 2000). It is a specific form of growth plate abnormality found in commercial meat-type poultry. This mass of cartilage occurs most frequently in the proximal tibia, but may also be found at other growth plates, such as the proximal metatarsal and femoral head. If the lesion is small, bony change is minimal. If the lesion is large, the end of the affected bone enlarges and becomes weakened, allowing abnormal modelling. The tibia may be bent backward, as it grows, by the pull of the strong gastrocnemius muscle. If the backward bending is greater than a few degrees, the bird cannot stand and is down on its hocks (a creeper). Affected metatarsal bones may bow medially. The bone may fracture spontaneously or at processing, or occasionally necrosis develops around the cartilage plug and a sequestrum forms (Julian, 1998). This results in long-bone necrosis or fracture and severe lameness. As bone growth slows with approaching maturity, the lesion may be removed and the bone may be remodelled to appear normal. Thirty to fifty per cent of male, meat-type poultry may develop dyschondroplastic lesions, but lameness only occurs if there is deformity or enlargement with loss of bone strength, resulting in weakness, fracture or necrosis. A post-mortem examination or radiograph is required to diagnose dyschondroplasia. The lesion is seen as a white or discoloured mass of cartilage under the growth plate when the medial side of the proximal tibia is cut away (Julian, 1998).

The cause of dyschondroplasia is multifactorial, but rapid growth (particularly without a long daily rest period), a deficiency of growth factor, a high anion/cation ratio in the feed and genetic susceptibility are primary (Riddell, 1976, 1992, 1993; Thorp, 1992; Thorp *et al.*, 1993, 1995; Cook *et al.*, 1994; Leach and Twal, 1994; Rath *et al.*, 2000; Reddi, 2000). Dyschondroplasia causes 5–25% of the lameness in broiler chickens. It caused up to 50% of the lameness in some flocks of broiler chickens in countries where animal protein (which has a high anion/cation ratio) is used in the ration. Alkalizing the ration by removing 1 kg of NaCl and replacing it with 2 kg of $NaHCO_3$ per tonne will reduce the incidence of dyschondroplasia. The addition of 5–10 µg of 1,25-dehydroxycholecalciferol/kg of feed may prevent TD (Rennie *et al.*, 1993), but since vitamin D deficiency results in acidosis (Booth *et al.*, 1977), the pH of the growth plate may be more significant than has been assumed previously.

IMPACT ON WELFARE. TD can produce mild or severe chronic pain, or acute severe debility. Lameness caused by TD is a significant welfare concern.

Valgus–varus deformity, angular bone deformity, twisted legs

These are the terms used to describe lateral or medial deviation of the distal tibiotarsus, frequently with a corresponding deviation of the metatarsus and secondary displacement and sometimes complete slippage of the gastrocnemius tendon (Julian, 1984; Riddell, 1992; Thorp, 1992). Affected birds are bow-legged (varus) or knock-kneed (valgus) until they go down (twisted legs). This deformity usually starts before day 10 as a modelling defect but may be caused by a lack of remodelling as the bone grows. It may not result in significant deformity until after day 21. Growth plates are normal in the distal tibiotarsus but the proximal metatarsus may be enlarged. Intertarsal ligaments become stretched and the joint is slack. Bone strength is normal but spontaneous fracture may occur through the growth plate between the tibia and attached tarsal bones when there is severe deformity (Julian, 1984).

The aetiology of valgus–varus deformity is not clear, although the defect is related to overnutrition, rapid growth and management factors (Julian, 1984; Classen, 1992). It may be caused by continuous bone growth, without a long daily rest period to allow the bones to correct the misalignment. It may have to do with uneven growth of the two attached tarsal bones or of the growth plate at the end of the distal tibia, or asymmetrical tendon tension on fast-growing bones. It is the most frequent cause of lameness in broiler chickens, accounting for up to 60% of skeletal disorders. A similar chondrodystrophic lesion may be caused by some B vitamin or mineral deficiencies (Riddell, 1992, 1996). Prevention involves slowing the growth rate in weeks 1 and 2, or a long daily dark period (Classen and Riddell, 1989).

IMPACT ON WELFARE. This deformity can be assessed in the live broiler or post mortem, but the degree of deformity may not be related to the severity or duration of pain. Joint trauma with fracture, or swelling, may be a better indication of the welfare significance. The defect is painful if it puts tension on the joint, or if the broiler is walking on its hocks. Affected broilers gain less than their pen-mates. If the broiler goes down and is unable to rise, this is an immediate serious welfare concern and the bird should be culled.

Rupture of the gastrocnemius tendon

This is a common problem in heavy broilers. The rupture occurs above the hock and is usually primary, caused by excessive weight on tendons that have inadequate tensile strength. Rupture of the tendon of one leg puts stress on the other tendon and bilateral rupture is frequent. Affected birds are lame or may be down on their hocks, using their wings to assist movement (creepers) (Julian, 1994, 1998). Haemorrhage from the injury is visible as red, blue or green discoloration in the tissue above the hock on the back of the leg, and results in condemnation of the affected part at processing (red-leg, green-leg). The ruptured tendon and the fibrosis of attempted repair can be palpated as a hard mass on the back of the leg above the hock.

IMPACT ON WELFARE. Pain may be most severe as the tendon separates, but ruptured tendon is a chronic, debilitating condition and is considered an important welfare problem. Affected birds should be humanely culled.

Deep pectoral myopathy

Spontaneous deep pectoral myopathy (DPM) has been recognized in turkeys for many years. It is now being seen with increasing frequency in broilers, particularly heavy broilers grown for 'cut-up'. DPM is an exertional myopathy involving the supracoracoideus muscle, which is the muscle that raises the wing. The lesion can be easily induced by holding the broiler by its legs and allowing it to flap its wings, which is how it is done experimentally (Martindale *et al.*, 1979; Wight *et al.*, 1981; Crespo and Shivaprasad, 2003). Muscle activity increases the production of lactic acid, which if not carried away, results in damage to the muscle cells with swelling and oedema. The supracoracoideus muscle is held against the breastbone by a strong fascial sheath which prevents expansion of the muscle. When swelling causes pressure to rise within the muscle, the pressure restricts the blood flow into the muscle. All or part of the muscle may die.

IMPACT ON WELFARE. Although no one has described pain in affected broilers, this condition is likely to be a very painful condition during the acute phase because of the swelling and pressure. A similar condition (march gangrene, anterior tibial syndrome or compartment syndrome) in the peroneus muscle in people is reported to cause pain.

References

Apeldoorn, E.J., Schrama, J.W., Mashaly, M.M. and Parmentier, H.K. (1999) Effect of melatonin and lighting schedule on energy metabolism in broiler chickens. *Poultry Science* 78, 223–229.

Booth, B.E., Tsai, H.C. and Morris, R.C. Jr (1977) Metabolic acidosis in the vitamin D-deficient chick. *Metabolism* 26, 1099–1105.

Bowes, V.A and Julian, R.J. (1988) Organ weights of normal broiler chickens and those dying of sudden death syndrome. *Canadian Veterinary Journal* 29, 153–156.

Bowes, V.A., Julian, R.J., Leeson, S. and Stirtzinger, T. (1988) Research note: effect of feed restriction on feed efficiency and incidence of sudden death syndrome in broiler chickens. *Poultry Science* 67, 1102–1104.

Bradshaw, R.H., Kirkden, R.D. and Broom, D.M. (2002) A review of the aetiology and pathology of leg weakness in broilers in relation to welfare. *Avian and Poultry Biology Reviews* 13, 45–103.

Butler, E.J. (1976) Fatty liver diseases in the domestic fowl. A review. *Avian Pathology* 5, 1–14.

Chowdhury, S.D. (1988) Lathyrism in poultry – review. *World's Poultry Science Journal* 44, 7–16.

Classen, H.L. (1992) Management factors in leg disorders. In: Whitehead, C.C. (ed.) *Bone Biology and Skeletal Disorders in Poultry*. Carfax Publishing, Abingdon, UK, pp.195–211.

Classen, H.L. and Riddell, C. (1989) Photoperiodic effects on performance and leg abnormalities in broiler chickens. *Poultry Science* 68, 873–879.

Cook, M.E., Bai, Y. and Orth, M.W. (1994) Factors influencing growth plate cartilage turnover. *Poultry Science* 73, 889–896.

Crespo, R. and Shivaprasad, H.L. (2003) Developmental, metabolic and other noninfectious disorders. In: Saif, Y.M., Barnes, H.J., Glisson, J.R., Fadley, A.M., McDougald, L.R. and Swayne, D.E. (eds) *Diseases of Poultry*, 11th edn. Iowa State University Press, Ames, Iowa, pp.1055–1102.

Davis, J.F. and Vasilatos-Younken, R. (1995) Markedly reduced pancreatic glucagon levels in broiler chickens with spiking mortality syndrome. *Avian Diseases* 39, 417–419.

Farquharson, C. and Jefferies, D. (2000) Chondrocytes and longitudinal bone growth: the development of tibial dyschondroplasia. *Poultry Science* 79, 994–1004.

Hemsley, L.A. (1965) Causes of mortality in fourteen flocks of broiler chickens. *Veterinary Record* 77, 467–472.

Julian, R.J. (1984) Valgus–varus deformity of the intertarsal joint in broiler chickens. *Canadian Veterinary Journal* 25, 254–258.

Julian, R.J. (1986) The effect of increased mineral level in the feed on leg weakness and sudden death syndrome in broiler chickens. *Canadian Veterinary Journal* 27, 157–160.

Julian, R.J. (1993) Ascites in poultry. Review. *Avian Pathology* 22, 419–454.

Julian, R.J. (1994) Metabolic disorders and welfare of broilers. In: *Proceedings, 9th European Poultry Conference, Glasgow*. Vol. I. UK Branch, WPSA, Walker and Connell, Darvel, UK, pp. 64–69.

Julian, R.J. (1996a) Cardiovascular disease. In: Jordan, F.T.W. and Pattison, M. (eds) *Poultry Diseases*, 4th edn. W.B. Saunders, London, pp. 343–374.

Julian, R.J. (1996b) Cardiovascular system. In: Riddell, C. (ed.) *Avian Histopathology*, 2nd edn. American Association of Avian Pathologists, Kennett Square, Pennsylvania, pp. 69–88.

Julian, R.J. (1998) Rapid growth problems: ascites and skeletal deformities in broilers. *Poultry Science* 77, 1773–1780.

Leach, R.M. and Twal, W.O. (1994) Autocrine, paracrine and hormonal signals involved in growth plate chondrocyte differentiation. *Poultry Science* 73, 883–888.

Martindale, L., Siller, W.G. and Wight, P.A.L. (1979) Effects of subfascial pressure in experimental deep pectoral myopathy of the fowl: an angiographic study. *Avian Pathology* 8, 425–436.

Migaki, G. (1982) Compendium of inherited metabolic diseases in animals. In: Desnick, R.J., Patterson, D.F. and Scarpelli, D.G. (eds) *Animal Models of Inherited Metabolic Diseases: Proceedings of the International Symposium on Animal Models of Inherited Metabolic Disease, October 1981, Bethesda, Maryland*. Alan R. Liss, New York, pp. 473–501.

Newberry, R.C., Gardiner, E.E. and Hunt, J.R. (1987) Behaviour of chickens prior to death from sudden death syndrome. *Poultry Science* 66, 1446–1450.

Rath, N.C., Huff, G.R., Huff, W.E. and Balog, J.M. (2000) Factors regulating bone maturity and strength in poultry. *Poultry Science* 79, 1024–1032.

Reddi, A.H. (2000) Initiation and promotion of endochondral bone formation by bone morphogenetic proteins: potential implication for avian tibial dyschondroplasia. *Poultry Science* 79, 977–981.

Rennie, J.S., Whitehead, C.C. and Thorp, B.H. (1993) The effect of dietary 1,25-dehydroxycholecalciferol in preventing tibial dyschondroplasia in broilers fed on diets imbalanced in calcium and phosphorus. *British Journal of Nutrition* 69, 809–816.

Riddell, C. (1976) Selection of broiler chickens for high and low incidence of tibial dyschondroplasia with observations on spondylolisthesis and twisted legs (perosis). *Poultry Science* 55, 145–151.

Riddell, C. (1992) Non-infectious skeletal disorders of poultry: an overview. In: Whitehead, C.C. (ed.) *Bone Biology and Skeletal Disorders in Poultry*. Carfax Publishing, Abingdon, UK, pp. 119–145.

Riddell, C. (1993) Developmental and metabolic diseases of meat-type poultry. In: *Proceedings, Xth World Veterinary Poultry Association Congress, Sydney*, pp. 79–89.

Riddell, C. (1996) Skeletal system. In: Riddell, C. (ed.) *Avian Histopathology*, 2nd edn. American Association of Avian Pathologists, Kennett Square, Pennsylvania, pp. 45–60.

Sanotra, G.S., Lund, J.D., Ersboll, A.K., Petersen, J.S. and Vestergaard, K.S. (2001) Monitoring leg problems in broilers: a survey of commercial broiler production in Denmark. *World's Poultry Science Journal* 57, 55–70.

Stanbury, J.B., Wyngaarden, J.B., Fredrickson, D.S., Goldstein, J.L. and Brown, M.S. (1983) *The Metabolic Basis of Inherited Disease*, 5th edn. McGraw-Hill, New York.

Thorp, B.H. (1992) Abnormalities in the growth of leg bones. In: Whitehead, C.C. (ed.) *Bone Biology and Skeletal Disorders in Poultry*. Carfax Publishing, Abingdon, UK, pp. 147–166.

Thorp, B.H., Ducro, B., Whitehead, C.C., Farquharson, C. and Sørensen, P. (1993) Avian tibial dyschondroplasia: the interaction of genetic selection and dietary 1.25-dihydroxycholecalciferol. *Avian Pathology* 22, 311–324.

Thorp, B.H., Jakowlew, S.B. and Goddard, C. (1995) Avian dyschondroplasia: local deficiencies in growth factors are integral to the aetiopathogenesis. *Avian Pathology* 24, 135–148.

Webster, A.F.J. (1994) Consumer views and the realities of animal welfare. In: *Proceedings, 9th European Poultry Conference, Glasgow*. Vol. I. UK Branch, WPSA, Walker and Connell, Darvel, pp. 7–10.

Wideman, R.F. (2000) Cardio-pulmonary hemodynamics and ascites in broiler chickens. *Avian and Poultry Biology Reviews* 11, 21–43.

Wideman, R.F. (2001) Pathophysiology of heart/lung disorders: pulmonary hypertension syndrome in broiler chickens. *World's Poultry Science Journal* 57, 289–307.

Wight, P.A.L., Siller, W.G. and Martindale, L. (1981) The sequence of pathological events in deep pectoral myopathy of broilers. *Avian Pathology* 10, 57–76.

Whitehead, C.C., Bannister, D.W. and Cleland, M.E. (1978) Metabolic changes associated with the occurrence of fatty liver and kidney syndrome in chicks. *British Journal of Nutrition* 40, 221–234.

5 Infectious Disease: Morbidity and Mortality

A. BUTTERWORTH

School of Veterinary Science, University of Bristol, Langford, Bristol, UK

Introduction

Meat chickens are the most numerous farmed terrestrial animal, with an estimated annual production of 40 billion birds. Compare this with cattle, for example, of which the world population was believed to be about 1.5 billion, and sheep and goats 1.75 billion in 2001. From Siberia to South Africa, broiler chickens are produced under quite similar intensively housed conditions, achieving growth gains of up to 50 g per day and slaughter weight by 32–45 days of age. The global genetic pool is dominated by Ross-Aviagen, Cobb-Vantress, Arbor Acres and Hy-Line.

Many would argue that a farmed animal takes its chances in the gamble to be free from disease during its life, and that it is not the individual that should be considered, but the risk that it runs of succumbing to a given condition as a probability within the whole population. The poultry industry can justifiably crow about its success in controlling a number of infectious diseases, metabolic disorders and housing challenges, and it can show that, compared with other species, the individual animal's risk of dying from a given condition or of becoming chilled, wet or predated is very low compared with other species. However, assessment of the percentage risk fails to accommodate the 'magnifying effect of *n*'.

Here is an example of this effect. If the population wishes to eat chicken, it will, of course, require many more individual animals to produce 20 tons of chicken than 20 tons of beef. If it takes 10,000 chickens or 70 cattle to produce 20 tons of chicken/beef, then a condition affecting 0.5% of the population will affect 50 chickens (0.5% of 10,000) but only 0.35 cattle (0.5% of 70) – one every 3 years on a farm.

Stated simply, to produce meat, more chickens will suffer a given condition than will cows. If individuals count (there are ethical arguments about this), then poultry production invites higher levels of care than beef production because of the increased potential for individual suffering. This is the magnification of *n*.

The magnification of n creates the potential for conditions that may only be present at an irritant, economically insignificant level to cause suffering to huge numbers of birds.

It should not be forgotten that comparisons between different systems (or species) with respect to percentage morbidity or mortality ignore the effects of age: broiler chicken mortality is mortality within 6 weeks, layer mortality is reckoned over a period up to 76 weeks, and dairy cow mortality would be over 8 years or so (or is expressed as percentage of the herd per year rather than per crop or flock).

Farming is a practical business, and farms cannot produce livestock without producing some deadstock. Mortality is not a welfare issue *per se* since a dead animal cannot suffer. However, for intensively farmed species, it is the process by which the animal becomes a 'mort' that raises potential welfare concerns. If the sick animal is not seen by the stockperson and the animal dies from disease, dehydration, injury or starvation, and there is no active intervention from the stockperson through culling or euthanasia, then the number of deadstock may be an indicator of poor welfare.

However, if farms take active steps to cull, remove, rehabilitate or euthanize animals, then high mortality figures may reflect good welfare through reduction in the potential for unnecessarily prolonged suffering through humane culling. In these circumstances, nearly all animals will have been culled rather than being found dead. As mortality figures are invariably recorded, and are therefore easy to audit, how these figures are derived on each farm is an important consideration, as indicated in Table 5.1.

For example, broiler mortality on good broiler units can be 2.5%, and lamb mortality on good farms can be 2.5%. Assuming (unfortunately wrongly) that all the farms in the country are capable of the lowest rates of mortality that have been reported, 2.5% of 800,000,000 (UK broiler production) = 20 million broilers per year, and 2.5% of 19 million (UK lamb production) = 0.48 million lambs per year.

If individuals count – and existing definitions of welfare would suggest that welfare can best be assessed by examining the impact on individuals – then keeping large numbers of small animals like poultry will inevitably lead to greater individual suffering than keeping smaller numbers of larger animals. Is this a call for a move toward 'sizeism', in which small, numerous food animals are viewed as having a higher welfare risk, or simply a reminder that systems which use only percentage

Table 5.1. The relationship between the control of mortality, mortality records, and the likely impact on welfare.

Farm approach to diseased or injured animals	Mortality recorded (%)	Welfare is likely to be:
Active policy of humane culling, with almost no animals being found dead	High	High
	Low	High
Some animals found dead, some culled	High	Poor or acceptable
	Low	Acceptable
Almost all animals found dead, little attention to culling	High	Poor
	Low	Acceptable

control of morbidity and mortality as a measure of their effectiveness in animal welfare may fail to account for the large numbers of individuals involved.

The practical stockperson is likely to sigh and say 'Well, what can I do better? I control disease, I provide a warm, safe, growing environment, and I cull birds (lambs, fish, calves) if I don't think they will make the grade, but I won't kill animals unnecessarily as this costs me money. What more can I do?'

An auditor visiting different farms might record mortality and morbidity figures or have acceptable levels defined in a standard. Because of the significance of very small percentage changes for the number of animals affected in broiler grower units, auditors should be aware of, or even calculate, the number of animals that a difference of 0.5 or 1% in mortality figures reflects.

The numerical arguments outlined above lead only to further questions.

- Could we practically improve the conditions for numerous small food animals without such significant cost that the systems become untenable?
- Should we make *percentage* comparisons between species, or are the *numbers* affected the right indicator of the welfare challenge?
- Do we see these large numeric losses (but small percentage losses) and the potential for individual suffering that they carry with them as an issue?

Auditing of the welfare impact of different systems may help us answer some of these questions, since only by comparison of systems through uniform audits can we recognize critical control points ('hazard analysis and critical control points') and work out whether what we do makes a difference.

Auditing Disease

The word 'disease' implies a state lacking 'ease', or well-being. In comparison with other animal industries, the poultry industry is good at controlling major disease challenges through vaccination, integration, the selective use of antibiotics and anticoccidials, and management practices. Overall mortality, to 6 weeks of age, may be as low as 2.5%. However, given the same housing, air, feed and water, some birds may become diseased and some remain healthy. Intrinsic biological variation results in differences in nutrient reserves and immune status between individuals at hatching. Interactions among differences in genetic make-up, social status and behaviour, and environmental exposure provide different outcomes for the individual. In the pragmatic world of farming, flock health status may be frequently chosen as the index of welfare, but welfare is usually defined in terms either of an individual animal's ability to cope (Broom, 1986) or of health and behavioural needs.

For the individual animal, it matters not whether any disease it suffers from is common or rare, or whether it is created by the system it lives in or by its natural susceptibility to the disease. In terms of their impact on the welfare of the global population of poultry, common conditions such as lameness (Kestin *et al.*, 1992), enteritis, footpad lesions, cellulitis, distended crop and respiratory disease have the greatest significance because they are often considered as routine, and if they do not cause significant economic loss may be accepted as established hazards about which little can be done. By virtue of the numbers of animals affected – the magnification of *n* – these common irritant, non-life-threatening conditions may become very significant.

Auditable Indicators of Disease

The following signs are all indicative of a bird that is in a state of crisis. From a welfare point of view they are symptomatic of substantial stress and of a reduced ability to cope with the cumulative and combined effects of current and previous stressors. They reflect strategies that have evolved to conserve and redirect energy in the body towards combating disease and regaining health. Their presence indicates the need for human intervention to support the recovery of sick birds and, equally importantly, to try to prevent the spread of disease and to recognize possible causes, so that corrective and preventive action may be taken. Birds are seldom treated as individuals (except in the case of culling) but rather as populations, and the decision to treat whole groups will be based on the cost-effectiveness of treatment.

Malaise

Typical signs and symptoms of an unhealthy bird include the following.

- *Withdrawal.* The bird isolates itself as far as possible from other members of the flock and from interactions with them, so that the sick bird is to be found under feeder or drinker lines, or at the edges or corners of houses or cages. The bird also becomes less responsive to most external stimuli;
- *Hunched posture.* The neck is retracted down towards the body, the tail may droop, the general appearance is more rounded and contracted and the eyes are often closed;
- *Dull feathers.* The feathers no longer refract the light and (particularly in brown-feathered birds) look darker. This may be due to the bird's loss of interest in or lack of energy for preening and feather maintenance, but can also be a sign of an inadequate diet or severe parasite infestation. Figure 5.1 shows a moribund bird showing some of these signs.

Pain

In evolutionary terms, a prey animal with manifest signs of pain is more likely to be selected by a predator, and domestic fowl show few visible signs of pain on visual inspection. However, a bird that is in pain will indicate behavioural distress (including escape behaviour) or may become quiet, withdrawn and depressed, and may vocalize if the affected area is gently palpated. In poultry there is evidence of pain or severe discomfort in several musculoskeletal disorders, with vocalization on manipulation of the hock, stifle joint or hip (A. Butterworth, personal observation). It is apparent that broiler chickens will avoid aversive and damaging stimuli, such as

Fig. 5.1. A moribund bird showing hunched, collapsed posture with eyes closed. The bird shows no interest in the presence of the stockperson.

having their toes trodden on by the stockman as he moves around the house, or being crushed in a corner by the weight of a number of birds. In these circumstances, the birds will vocalize and make efforts to escape. These sounds and behaviours will be very familiar to humans and recognizable as avoidance of discomfort, damage or pain, and can be audited as such without recourse to the complex arguments about how birds feel and experience pain.

Dehydration and emaciation

These are always symptomatic of poor welfare and a state of disease. As modern systems of poultry husbandry usually provide adequate feed and water *ad libitum*, emaciation and dehydration reflect inability of the individual bird to access these resources. Occasionally this is due to social stress, but more often to lameness (Butterworth, 1999; Kestin, 2001; Butterworth *et al.*, 2002) or to morbidity from serious diseases, such as septicaemia, respiratory disease, ascites or injury. Sometimes, access to water becomes difficult for small birds if the drinkers are raised during the growth cycle. The height and number of drinkers are easy to audit; what is more difficult is to be sure that all the birds, even small birds, have easy access to water, and that the height that suits the general population does not leave runted birds surviving only on water splashed from nipple cups (Fig. 5.2) (drinker height may have been used in the past to eliminate runt birds by preventing them from reaching water and thus causing their eventual demise). Figure 5.3 indicates the relationship between blood osmolality (a measure of hydration), the time period over which water was withdrawn and lameness on farms where severe lameness resulted in larger birds having difficulty accessing water.

Fig. 5.2. Small birds may get left behind as drinkers are raised during the production cycle, and thus may remain small as they fail to compete for resources.

Decreased productivity

A reduction in eggshell quality, egg output in broiler breeders or growth rate in meat birds may indicate that the birds are redirecting nutrients to the repair of damaged tissues. Whereas high productivity does not in itself equate with good health and welfare, reduced productivity (weight gain or maintenance of weight), in the same thermal conditions and with equal access to food water and space, is often a sign of disease or distress. Figure 5.2 shows the difference in size between two birds in a flock, and it is common for a (usually small) part of the flock to be runted. Runted birds may not eventually be able to reach the drinkers and so remain small without easy access to a resource, and hence their growth is retarded. No producer wants to see a spread of bird sizes, and electrical stunning at the slaughter plant may be of reduced effectiveness for small (or very large) birds.

Immunosuppression and reduced liveability

Good flock liveability is both a welfare and an economic goal, and there has been much research into nutrients that enhance immune defences. Trace elements such as zinc, iron, copper, selenium and manganese are essential for resistance to disease and normal immune function (Fletcher *et al.*, 1988). The particular importance of zinc

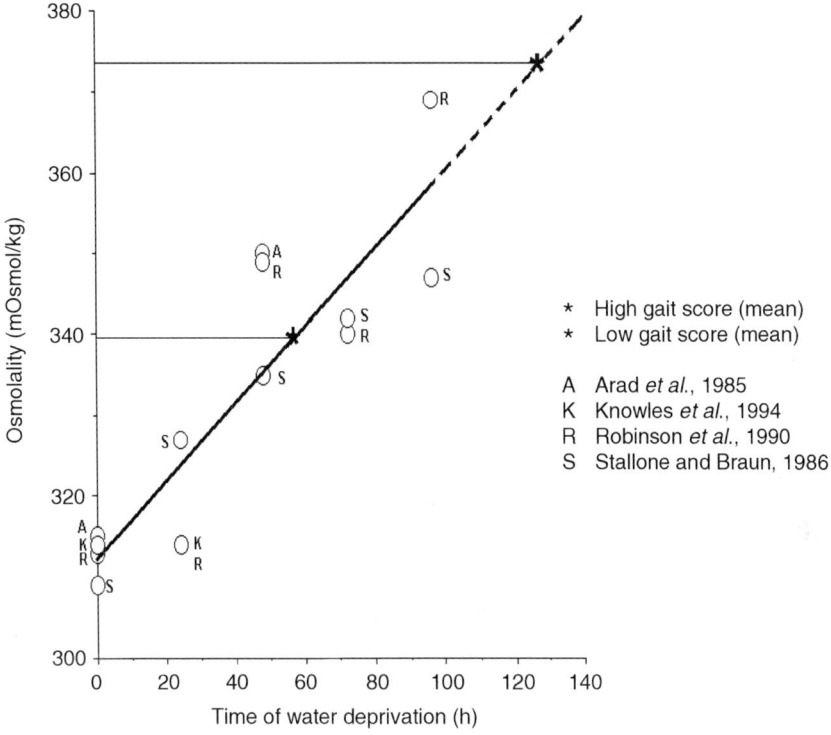

Fig. 5.3. The effect of time of dehydration on plasma osmolality in domestic fowl. A best fit line, *x–y*, has been drawn for the data available from previous work (Arad *et al.*, 1985; Stallone and Braun, 1986; Robinson *et al.*, 1990; Knowles *et al.*, 1994). The mean plasma osmolality values for the birds with high gait score and low gait score in this case study are also indicated.

in poultry was reviewed by Kidd *et al.* (1996). Birds are susceptible to a number of diseases which result in immunosuppresssion, including Gumboro disease, chicken infectious anaemia, oncogenic viruses and fungal toxins.

The use of intermittent lighting schedules that also reduce daily illumination may improve liveability and reduce mortality in laying hens (Lewis *et al.*, 1996). Immune function was improved in a line of White Leghorn hens that was selected for longevity (Cheng *et al.*, 2001). As well as living longer, the birds also showed reduced cannibalism and flightiness, and improved feather score.

Auditing Disease Incidence and the Impact of Disease

There are numerous diseases of broilers, but the most common are those that are most likely to affect welfare by their persistent low grade effects (although some uncommon diseases can affect huge numbers of animals acutely and can have significant but sporadic effects on bird welfare). A list of the more common generalized conditions of broilers would include septicaemia, viral, nutritional and bacterial enteritis, respiratory disease, pododermatitis, cellulitis, crop distension, cardiovascular disease (including ascites), hock burn, skeletal and infectious lameness and skeletal injuries. Any disease will affect the bird, but it is a combination of severity and duration that affect the outcome for the individual. To illustrate an auditable approach to the effects of the severity and duration of disease on the individual bird, consider the following examples.

Engorged crop

The formation of an over-large, fluid-filled crop leads to poor growth and chronic ill-thrift. Crop distension might sporadically affect 0.1% of birds in a broiler flock and sometimes bigger numbers in turkey flocks (Peckham, 1984). This is not usually a life-threatening condition, but will compromise the bird and will be reflected in reduced growth. A small bird may find it harder to access resources in the shed, and if the bird is significantly smaller it may be culled on-farm (there are some welfare concerns about methods for on-farm slaughter) or may be too small to be effectively stunned in the electrical stunning bath at the slaughter plant. This condition is a low-level, sporadic condition of little economic impact but may, for the individual bird, seriously reduce the quality of its life.

- Severity is moderate (affects the bird's ability to feed and metabolize food, may impair walking and may (?) be a cause of discomfort through stretching of tissues).
- Duration is moderate (days, weeks).
- Overall (combination of severity and duration) it has a moderate impact on the bird.

Kinky back

Deformity and osteomyelitis of the third, fourth or fifth thoracic vertebra lead to spinal cord compression and eventual immobility in affected birds (Fig. 5.4). Kinky back is believed to affect about 0.3% of meat birds.

- Severity is high (malaise, dehydration, starvation, possibly pain).
- Duration is moderate (days/weeks).
- Overall (combination of severity and duration) it has a high/moderate effect.

Ascites

Ascites is an accumulation of fluid in the peritoneal cavities, most commonly caused by increased pressure in the blood vessels, which forces out excess fluid and inhibits reabsorption of tissue fluid. Another common cause is right ventricular failure. Most aspects of the disease are considered by Julian (1993). The condition may be reduced or prevented by slowing metabolism via genetic, dietary or husbandry measures, such as increased periods of darkness (Gordon, 1997). In their review of causal

Fig. 5.4.　Immobile bird showing the 'back on the hocks' stance associated with kinky back.

factors, Scheele *et al.* (2003) suggest that a fundamental cause is the imbalance between oxygen supply and the amount of oxygen required to sustain rapid growth and high food efficiency. They also provide evidence for the close association of the quantity, form and quality of the diet and the incidence of ascites. Ascites is believed to affect about 0.15% of meat birds, but there are significant regional and genotypic differences.

- Severity is very high (severe malaise, dehydration, starvation and possibly pain).
- Duration is moderate (days/weeks).
- Overall (combination of severity and duration) it has high/moderate effect.

Summary

Disease, morbidity and mortality are auditable measures on broiler farms, yet simple percentage incidence rates for disease and percentage mortality figures fail to provide the whole picture for a number of reasons. First, mortality figures alone tell you little about the degree of control of on-farm mortality – did the animals merely die or were they culled?

Secondly, for some of the common low-grade diseases, such as footpad dermatitis, hock burn and cellulitis, their chronic nature may have a greater cumulative impact on large numbers of animals than that of acute conditions such as enteritis or ascites. A severity/duration audit, whilst at present only a tool to understand the impact on individual animals, provides the framework for a specific type of audit: a disease impact audit. Other authors in this book expand upon the possibility of welfare indices, and disease impact auditing is a potential part of this process (see Chapters 7, 23 and 14).

The magnification of n effect suggests that broilers need to be audited with extra care when different broiler flocks and farms are being compared, as small percentage differences can make dramatic changes to the numbers of animals with potentially compromised welfare.

References

Arad, Z., Arnason, S.S., Chadwick, A. and Skadhauge, E. (1985) Osmotic responses to heat and dehydration in the fowl. *Journal of Comparative Physiology* 155, 227–234.

Broom, D.M. (1986) Indicators of poor welfare. *British Veterinary Journal* 142, 524–525.

Butterworth, A. (1999) Infectious components of broiler lameness: a review. *World's Poultry Science Journal* 55, 327–352.

Butterworth, A., Weeks, C.A., Crea, P.R. and Kestin, S.C. (2002) Dehydration and lameness in a broiler flock. *Animal Welfare* 11, 89–94.

Cheng, H.W., Eicher, S.D., Singleton, P. and Muir, W.M. (2001) Effect of genetic selection for group productivity and longevity on immunological and haematological parameters of chickens. *Poultry Science* 80, 1079–1086.

Fletcher, M.P., Gershwin, M.E., Keen, C.L. and Hurley, L. (1988) Trace element deficiencies and immune responsiveness in humans and animal models. In: Chandra, R.K. (ed.) *Nutrition and Immunology*. Alan R. Liss, New York, pp. 215–239.

Gordon, S.H. (1997) Effect of light programmes on broiler mortality with reference to ascites. *World's Poultry Science Journal* 53, 68–70.

Julian, R.J. (1993) Ascites in poultry. *Avian Pathology* 22, 419–454.

Kestin, S.C., Knowles, T.G., Tinch, A.E and Gregory, N.G. (1992) Prevalence of leg weakness in broiler chickens and its relationship with genotype. *Veterinary Record* 131, 191–194.

Kestin, S.C., Gordon, S., Su, G. and Sørensen, P. (2001) Relationships in broiler chickens between lameness, liveweight, growth rate and age. *Veterinary Record* 148, 195–197.

Kidd, M.T., Ferket, P.R. and Qureshi, M.A. (1996) Zinc metabolism with special reference to its role in immunity. *World's Poultry Science Journal* 52, 309–324.

Knowles, T.G., Warriss, P.D., Brown, S.N., Edwards, J.E. and Mitchell, M.A. (1994) Responses of broilers to deprivation of food and water for 24 hours. *British Veterinary Journal* 151, 197–202.

Lewis, P.D., Morris, T.R. and Perry, G.C. (1996) Lighting and mortality rates in domestic fowl. *British Poultry Science* 37, 295–300.

Peckham, M.C. (1984) Vices and miscellaneous diseases and conditions. In: Calnek, B.W. (ed.) *Diseases of Poultry*, 8th edn. Iowa State University Press, Ames, Iowa, pp. 741–782.

Robinson, B., Koike, T.I., Kinzler, S.L. and Nekdon, H.L. (1990) Arginine vasotocin and mesotocin in the anterior hypothalamus, neurohypophysis, proventriculus and plasma of leghorn cockerels during dehydration. *British Poultry Science* 31, 651–659.

Scheele, C.W., Van Der Klis, J.D., Kwakernaak, C., Buys, N. and Decuypere, E. (2003) Haematological characteristics predicting susceptibility for ascites. 1. High carbon dioxide tensions in juvenile chickens. *British Poultry Science* 44, 476–483.

Stallone, J.N. and Braun, E.J. (1986) Regulation of plasma arginine vasotocin in conscious water-deprived domestic fowl. *American Journal of Physiology* 250, 644–685.

6 Abnormal Behaviour and Fear

G.S. SANOTRA[1] AND C.A. WEEKS[2]

[1]Department of Animal Science and Animal Health, Division of Ethology and Health, The Royal Veterinary and Agricultural University, Copenhagen, Denmark; [2]School of Veterinary Science, University of Bristol, Langford, Bristol, UK

Behaviour

Young broiler chickens show relatively normal behaviour during the first 2–3 weeks of their life, similar to that of their ancestral red jungle fowl (Kruijt, 1964). The behavioural repertoire includes eating, drinking, sleeping, idling, preening, running, jumping, food-running, scratching, ground-pecking, wing-flapping, wing or leg stretching, dustbathing, agonistic encounters and vocalizing. Several of these activities can be performed while either lying or standing, but the normal posture for ground-pecking, eating and preening is standing. During the second half of their life, many broilers perform most behaviours whilst lying and there is a significant decline in many activities, particularly in lame birds (Weeks *et al.*, 2000). These observations of Weeks and colleagues revealed that sound and lame broilers averaged 76 and 86% of their time lying down, respectively. Sound broilers spent significantly twice as long standing idle: 7.2% against 3.5% for moderately lame birds. The sound birds stood to preen for an average of 3.5% of the day, which declined significantly to only 1.33% for lame birds. With access to feed at a lower height than the commercial norm, lame birds lay to eat for 2.6% and stood to eat for 3.1% of the day, differing significantly from the sound broilers, which predominantly ate from the normal standing posture (4.7% of the day and only 0.6% eating whilst lying). Lying increased but walking declined with age in both groups. Walking activity due to lameness was significantly reduced to 1.5%.

Diet and genotype also affect behaviour, and Weeks (2002) noted that birds on a reduced energy and protein diet (Label Rouge) were more active and spent significantly longer feeding, ground-pecking and walking. Her observational study compared 14 genotypes at 6 weeks of age, all managed as broilers, and found substantial differences in behaviour between genotypes. Figure 6.1 illustrates that lighter, slower-growing genotypes, such as the Light Sussex and Ixworth, were more active than meat-type birds, such as the Redbro and Ross broilers. Less active birds

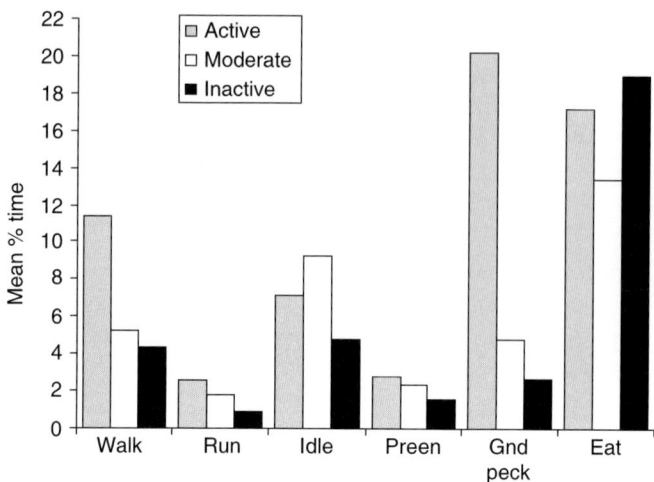

Fig. 6.1. Comparison of some standing activities for active, slow-growing genotypes, moderately active, medium-growing genotypes and inactive, fast-growing genotypes at 6 weeks of age on a relatively low-density broiler diet. (From Weeks, 2002.)

will inevitably be selected by breeding companies aiming for good feed conversion efficiency.

There are reasonable grounds for believing that reduction in load-bearing on the legs of young chicks enhances locomotor activity (Rutten *et al.*, 2002). As well as lameness and increasing body weight, other factors contribute to the decline with age of active behaviours in broilers. These include the increase in stocking density rate, which creates a 'barrier effect' that makes it harder for birds to move past each other. The reduced mobility with crowding leads to fewer aggressive interactions (Pettit-Riley *et al.*, 2002). At high stocking density the environmental temperature tends to be high (Blokhuis and van der Haar, 1990; Bessei, 1992; Reiter and Kutritz, 2001), which reduces the inclination to be active. Light is a dominant factor in the regulation and control of the behaviour and health of most animals. Diurnal rhythms, sleep and synchronized behaviour patterns are examples of behaviours controlled at least in part by the photoperiod. Continuous light without dark periods may disrupt the natural synchronization of behaviour in broiler chickens, disrupting the normal sleep patterns and reducing the general level of activity. Lighting regimes that include a dark period increase general activity in broilers (Renden *et al.*, 1996; Sanotra *et al.*, 2002), which in turn can affect leg health and bone growth (Classen, 1991). Sanotra and colleagues (2002) compared several behaviour activities in broiler chickens reared with and without light programmes and under high and low rates of stocking density. The results are illustrated in Fig. 6.2.

Dustbathing is an important maintenance or 'comfort' behaviour that occurs in virtually all birds, including the domestic hen and the red jungle fowl (Kruijt, 1964). The principal functions of dustbathing behaviour are the removal of surplus lipid from the plumage (Borchelt and Duncan, 1974), improvement of feather structure (Healy and Thomas, 1973) and the removal of ectoparasites (Simmons, 1964). Dustbathing behaviour comprises a number of components, and the bird's whole body (including the legs, the wings, the head and the neck) is involved.

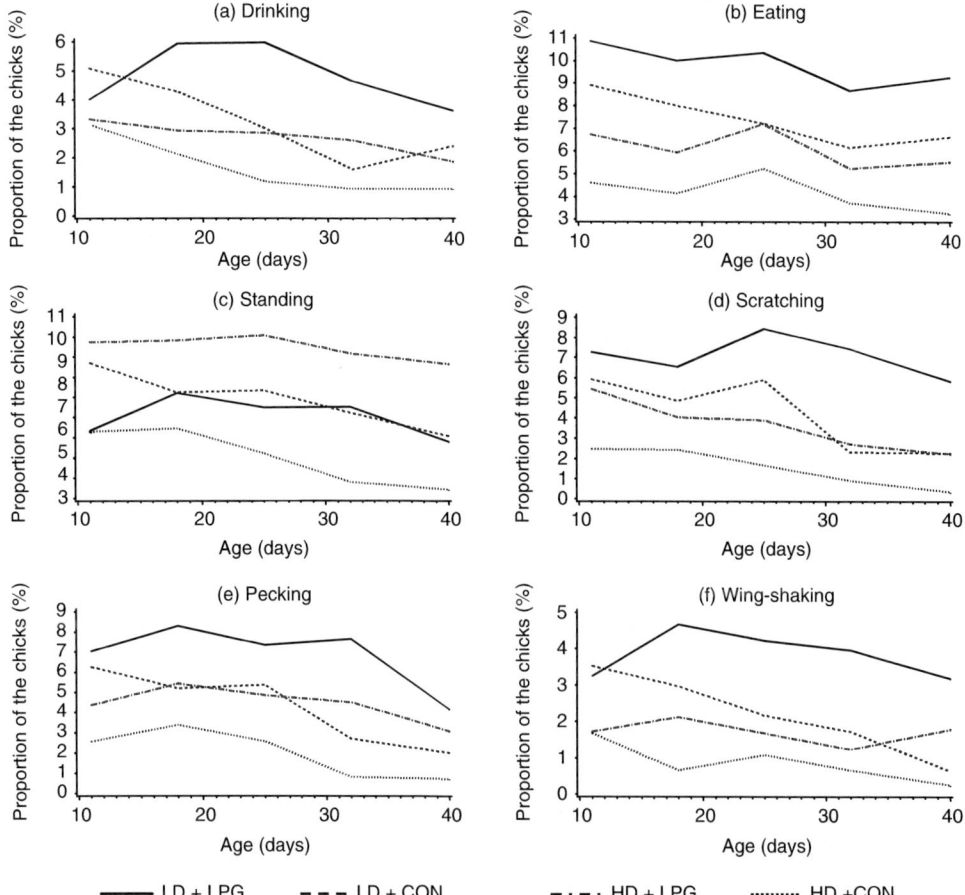

Fig. 6.2. Changes in the amount of behavioural activities with age in broiler chickens reared at low (LD) and high (HD) stocking density and with (LPG) and without (CON) 8 hours of darkness during the night. (From Sanotra *et al.*, 2002.)

Dustbathing in broiler chickens is a behavioural need that is often neglected, especially when they are kept at high stocking densities and without a suitable dark period during the night, which is needed for the maintenance of the diurnal rhythm (Gordon, 1994; Sanotra *et al.*, 2002). Most commercial broilers are reared under barren environmental conditions with no attractive stimuli. The litter is often wet and dirty, which does not promote the motivation to dustbathe. Under optimal conditions broiler chickens perform similar levels of dustbathing behaviour (Fig. 6.3) as layers.

Broiler chickens with leg problems may not dustbathe (Sanotra *et al.*, 2001). A possible explanation for this is that the vigorous leg movements involved in dustbathing could cause discomfort or pain in some lame broilers. Dustbathing activity also decreases with age and with the increase in stocking density levels (Blokhuis and van der Haar, 1990).

Fig. 6.3.   Broiler chickens dustbathing in trays with sand. (Photograph by G.S. Sanotra.)

Fear

In simple terms, chickens express fear by freezing, escape or flight behaviour, which may be seen in chicks by 3–5 days of age (Kruijt, 1964; Salzen, 1979). Fear is regarded as an undesirable emotional state, although ideally fear behaviour functions to protect the animal from injury. Jones (1987a) defines fear as 'the adaptive psychological response to perceived danger'. Murphy and Wood-Gush (1978) and Duncan (1985) mention fear as a state of suffering in domestic fowl that has been investigated widely. Various factors, such as environment, novelty, genetic factors (including strain and gender), age, human–bird interaction, handling and transport, maturation and experience, are obviously very important in the ontogeny of fear (Gallup, 1977; Suarez and Gallup, 1981; Cashman *et al.*, 1989; Hemsworth and Coleman, 1998; Jones and Hocking, 1999). Under natural conditions, the mother hen is always with her flock of chicks to protect, teach and help them in dangerous situations. Intensive production systems leave no room for chicks to learn such things because they are reared in huge flocks in a controlled environment. However, these husbandry systems protect the birds from both predators and climatic conditions. But the birds under such protected environments show varying degrees of fearfulness in terms of prolonged duration of tonic immobility (TI). This is a relatively prolonged state of freezing and is also described as a catatonic state (Fig. 6.4).

Leg health problems such as tibial dyschondroplasia can prolong the duration of tonic immobility (Vestergaard and Sanotra, 1999), as shown in Fig. 6.5.

Additionally, both the bird's age and high stocking density influence and prolong tonic immobility (Sanotra *et al.*, 2001). Farm animals handled aversively may become highly fearful of humans (Hemsworth and Gonyou, 2000). In a review of fear in poultry, Jones (1996) indicated that positive human–animal relationships enabled birds to conserve their mental and physiological resources to compensate for environmental fluctuations. Furthermore, regular visual contact is as effective as

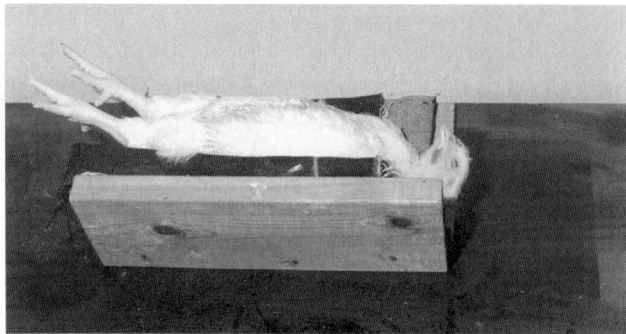

Fig. 6.4. A broiler chick in a state of tonic immobility with its eyes closed. (Photograph by G.S. Sanotra.)

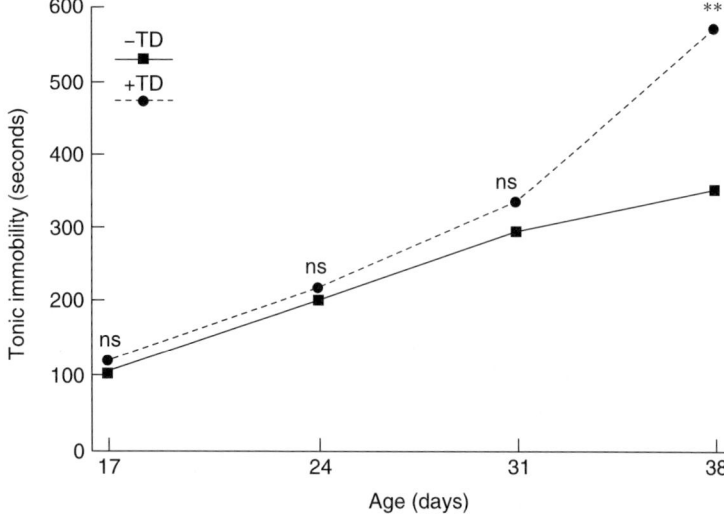

Fig. 6.5. The effect of tibial dyschondroplasia and age on the duration of tonic immobility. Broilers with tibial dyschondroplasia (+TD) showed a significantly longer duration of tonic immobility at 38 days of age than those without tibial dyschondroplasia (−TD) ($P < 0.001$).

actual physical handling in reducing fear of humans. Birds reared with outdoor access and at low stocking densities are generally less fearful than intensively reared chickens (Grigor *et al.*, 1995; Sanotra *et al.*, 1995; Scott *et al.*, 1998). Environmental enrichment, such as novel objects or even video stimuli during the rearing period, can reduce the underlying fearfulness of domestic fowls, as revealed by their responses in a range of laboratory tests (Jones and Waddington, 1992; Jones, 1996).

The multiple novel stressors associated with catching and transporting broilers to slaughter evoke a strong fear response that is relatively independent of their rearing environment and prior experience (see review by Weeks and Nicol, 2000). One study (Cashman *et al.*, 1989) found that broilers arriving at the slaughter plant had an average duration of tonic immobility of 12.6 minutes, which was comparable to that of broilers exposed to high-intensity electric shock (Gallup, 1973). Weeks and

Nicol (2000) suggested several ways of reducing the stress and fear of transport; automated catching and minimal time in transit are key points to note when auditing welfare. Shackling of live birds appears to be a highly aversive procedure and may also be painful; the birds respond by trying to escape by struggling and flapping their wings (Sparrey and Kettlewell, 1994). From a welfare perspective it is best avoided. This may easily be achieved using gas stunning/killing.

References

Bessei, W. (1992) Das Verhalten von Broilern unter intensiven Haltungsbedingungen. *Archiv für Geflügelkunde* 56, 1–7.

Blokhuis, H.J. and van der Haar, J.W. (1990) The effect of stocking density on the behaviour of broilers. *Archiv für Geflügelkunde* 54, 74–77.

Borchelt, P.L. and Duncan, L. (1974) Dustbathing and feather lipid in bobwhite quail (*Colonius virginianus*). *Condor* 76, 471–472.

Cashman, P.J., Nicol, C.J. and Jones, R.B. (1989) Effects of transportation on the tonic immobility fear reactions of broilers. *British Poultry Science* 30, 211–221.

Classen, H.K. (1991) Increasing photoperiod length provides better broiler health. *Poultry Digest* 50, 14–28.

Duncan, I.J.H. (1985) How do fearful birds respond? In: Wegner, R.M. (ed.) *Second European Symposium on Poultry Welfare*. WPSA, Celle, pp. 96–106.

Gallup, G.G. (1973) Tonic immobility in chickens: is a stimulus that signals shock more aversive than the receipt of shock? *Animal Learning and Behavior* 1, 228–232.

Gallup, G.G. Jr (1977) Tonic immobility: the role of fear and predation. *Psychological Record* 27, 41–61.

Gordon, S.H. (1994) Effects of day-length and increasing day-length programs on broiler welfare and performance. *World's Poultry Science Journal* 50, 269–282.

Grigor, P.N., Hughes, B.O. and Appleby, M.C. (1995) Effects of regular handling and exposure to an outside area on subsequent fearfulness and dispersal in domestic hens. *Applied Animal Behaviour Science* 44, 47–55.

Healy, W.M. and Thomas, J.W. (1973) Effects of dustbathing on plumage of Japanese quail. *Wilson Bulletin* 85, 442–448.

Hemsworth, P.H. and Coleman, G.J. (1998) *Human–Livestock Interactions. The Stockperson and the Productivity and Welfare of Intensively Farmed Animals*. CAB International, Wallingford, UK.

Hemsworth, P.H. and Gonyou, H.W. (2000) Human contact. In: Appleby, M.C. and Hughes, B.O. (eds) *Animal Welfare*. CAB International, Wallingford, UK, pp. 205–217.

Jones, R.B. (1987) The assessment of fear in the domestic fowl. In: Zayan, R. and Duncan, I.J.H. (eds) *Cognitive Aspects of Social Behaviour in the Domestic Fowl*. Elsevier, Amsterdam, pp. 82–149.

Jones, R.B. (1996) Fear and adaptability in poultry: insights, implications and imperatives. *World's Poultry Science Journal* 52, 131–174.

Jones, R.B and Hocking, P.M. (1999) Genetic selection for poultry behaviour: big bad wolf or friends in need? *Animal Welfare* 8, 343–359.

Jones, R.B. and Waddington, D. (1992) Modification of fear in domestic chicks *Gallus gallus domesticus* via regular handling and early environmental enrichment. *Animal Behaviour* 43, 1021–1034.

Kruijt, J.P. (1964) *Ontogeny of Social Behaviour in Burmese Red Junglefowl (Gallus gallus spadiceus)*. E.J. Brill, Bonnaterre.

Murphy, L.B. and Wood-Gush, D.G.M. (1978) The interpretation of the behaviour of domestic fowl in strange environments. *Biology of Behaviour* 3, 39–61.

Pettit-Riley, R., Estevez, I. and Russek-Cohen, E. (2002) Effects of crowding and access to perches on aggressive behaviour in broilers. *Applied Animal Behaviour Science* 79, 11–25.

Reiter, K. and Kutritz, B. (2001) Behaviour and leg weakness in different broiler breeds. *Archiv für Geflügelkunde* 63, 137–141.

Renden, J.A., Moran, E.T. Jr and Kincaid, S.A. (1996) Lighting programs for broilers that reduce leg problems without loss of performance or yield. *Poultry Science* 75, 1345–1350.

Rutten, M., Leterrier, C., Constantin, P., Reiter, K. and Bessei, W. (2002) Bone development and activity in chickens in response to reduced weight-load on legs. *Animal Research* 51, 327–336.

Salzen, E.A. (1979) The ontogeny of fear in animals. In: Sluckin, W. (ed.) *Fear in Animals and Man.* Van Nostrand Reinhold, New York, pp. 125–163.

Sanotra, G.S., Vestergaard, K.S. and Thomsen, M.G. (1995) The effect of stocking density on walking ability, tonic immobility, and the development of tibial dyschondroplasia in broiler chicks. In: *Proceedings of the 29th International Congress of the International Society for Applied Ethology, Exeter, UK, August 3–5,* pp. 221.

Sanotra, G.S., Lund, J.D., Vestergaard, K.S. and Flagstad, A. (2001) Association between leg pain and dust-bathing behaviour in broiler chickens. In: *Proceedings of the 35th International Congress of the ISAE, University of California, Davis, USA, August 4–9,* pp. 215.

Sanotra, G.S., Lund, J.D. and Vestergaard, K.S. (2002) Influence of light-dark schedules and stocking density on behaviour, risk of leg problems and occurrence of chronic fear in broilers. *British Poultry Science* 43, 344–354.

Scott, G.B., Connell, B.J. and Lambe, N.R. (1998) The fear levels after transport of hens from cages and a free-range system. *Poultry Science* 77, 62–66.

Simmons, K.E.L. (1964) Feather maintenance. In: Thompson, A.L. (ed.) *A New Dictionary of Birds.* New York, McGraw-Hill, pp. 278–286.

Sparrey, J.M. and Kettlewell, P.J. (1994) Shackling of poultry – is it a welfare problem? *World's Poultry Science Journal* 50, 167–176.

Suarez, S.D. and Gallup, G.G. Jr (1981) Predatory overtones of open-field testing in chickens. *Animal Learning and Behaviour* 9, 153–163.

Vestergaard, K.S. and Sanotra, G.S. (1999) Relationship between leg disorders and changes in the behaviour of broiler chickens. *Veterinary Record* 144, 205–209.

Weeks, C.A. (2002) Some behavioural differences between fast and slow-growing strains of poultry. *British Poultry Science* 43, S24–S26.

Weeks, C. and Nicol, C. (2000) Poultry handling and transport. In: Grandin, T. (ed.) *Livestock Handling and Transport,* 2nd edn. CAB International, Wallingford, UK, pp. 363–384.

Weeks, C.A., Danbury, T.C., Davies, H.C., Hunt, P. and Kestin, S.C. (2000) The behaviour of broiler chickens and its modification by lameness. *Applied Animal Behaviour Science* 67, 111–125.

2 Inputs (Resource-based)

7 The Significance of Biosecurity to Broiler Welfare

M. PATTISON

Rossmoyne, Barbon, Carnforth, UK

The recent avian influenza outbreak in Europe has brought disease control – and the impact of disease on bird welfare – strongly into perspective, as never before. Disease outbreaks reinforce the need to view biosecurity in total: from country and company to individual farm. A notifiable disease such as avian influenza or Newcastle disease has requirements for compulsory slaughter, which may involve the humane killing of thousands of birds whether they have the disease or are simply 'in contact'. This must be done by people who have been properly trained and who use methods that will minimize the suffering of the birds. This task should not be underestimated, and putting in place a biosecurity system, which could help to prevent such a situation, should be included in all company budgets.

Biosecurity is integral to meeting the expectations of the industry and the consumer. Within the poultry industry, the requirement is good technical results according to the standards for a given product. Predictable results form the base for effective planning of production, bird numbers, profit and future investment. Only when effective planning has been done can the best financial results be achieved. For birds to achieve their genetic potential they need to be kept free from diseases, and this, historically, was the primary goal of biosecurity. There is now a requirement to provide reassurance about freedom from zoonoses (infections which can be transmitted from animals to man) in food animals. Food products, including chicken, must also be free of all residues, including antibiotics.

A good biosecurity system does have a cost, but this is easily justified against the cost of an outbreak of disease. If the disease causes mortality there is clearly an adverse effect on bird welfare, which can be measured. However, the disease may be subclinical (not readily apparent), causing poor performance and affecting welfare. This may not be easily measured and requires greater skill from the stockperson to recognize.

Biosecurity requirements vary depending on the stock and the requirements for the progeny. For example, it is very important to keep grandparent or elite stock free of diseases that are transmitted vertically. Although egg-layers and broilers may be affected by a slightly different set of diseases, the basic principles of disease control apply to all stock.

Objectives

The objective of disease control in a flock is to prevent the entry of pathogens that may cause clinical disease or subclinical disease, or lead to a risk of zoonoses. This should provide a health status that is suitable for attaining optimal production and welfare of the flock and its progeny.

Disease Control in Practice

Before deciding what protocols are required, it is necessary to do some form of auditing or 'hazard analysis and critical control point' (HACCP) analysis to identify the vulnerable points. Key control points would include:

- Movements of people and equipment.
- Location of other livestock.
- Direct contacts with other farms.
- Health status of stock coming on to the farm.

Biosecurity can only be as good as the weakest point in the chain.

There are essentially three means of controlling disease. These are *physical barriers*, such as farm geography, building design, the use of showers, and protective clothing; *biological barriers* in the form of vaccines and competitive exclusion products; and *pharmacological barriers*, such as antibiotics and coccidiostats.

Physical barriers

These include farm location and its isolation or proximity to other poultry farms. The design of the farm, its internal organization, the control of access, changing of clothes and the management of the staff are all critical factors.

It is important for breeder farms to be as isolated as possible and to be well separated from concentrations of commercial or free-range poultry. Investment in a shower and changing facility is a good way to control the access of people to the buildings (Fig. 7.1). The farm should also be designed to ensure that the poultry are kept free from exposure to rodents, wild birds and possibly insects. It is vital that the movement of people, both within the farm and between farms, is carefully controlled.

Correctly set up and managed, physical barriers are by far the best way of controlling disease effectively over time.

Fig. 7.1.   Changing and shower facilities for all personnel to use before entering a broiler house reduce the risk of introducing infections. (Reproduced with permission of Aviagen Ltd, Newbridge, UK.)

Biological barriers

Vaccines cannot be expected to give complete protection and their efficacy depends very much on good storage and administration. Vaccines have to work in an environment where viruses may be mutating; an example is the emergence of variants of infectious bronchitis. There are also new or emerging diseases for which vaccines are unavailable. Competitive exclusion products, which are a mixture of 'friendly' bacteria (also known as probiotics), are a modern and elegant method of disease control and are used to help prevent intestinal infections. Competitive exclusion can be viewed as a local physical barrier.

Pharmacological barriers

Antibiotics rarely offer a biosecurity barrier. Their use should be limited to short periods to treat disease, as antibiotic resistance may develop and reduce their effectiveness. The primary breeder has a responsibility not to transfer antibiotic resistance to day-old chicks. Coccidiostats are essential to the broiler industry and provide effective protection against coccidiosis. Vaccines are used to prevent coccidiosis in breeders.

Feed and water

It is important to select good raw materials for feed. It is important that the feed is of known microbiological status. During manufacture it should go through an effective decontamination process, usually heat treatment. Some pathogens, such as clostridial spores, are difficult to kill. After treatment, the correct handling of feed is essential to prevent recontamination and infection of the birds.

Drinking water is a common source of pathogens and can be the means of transmission of viral and bacterial diseases such as Newcastle disease and avian influenza. Water must be clear and clean. In-line ultraviolet treatment can be used to kill most viruses and bacteria, though staphylococci are partly protected by their cell wall. Chlorine is effective and will also control contamination of the biofilm in pipes. To ensure that chlorine is at the required concentration, it should be measured at the end of the drinker lines; 2–5 p.p.m. is desirable, but where chlorine is present in drinking water skimmed milk powder must be added to ensure live vaccines are effective. Drinking systems must also be kept clean, and there are many effective water sanitation products available for this.

Monitoring

The purposes of monitoring are to measure and record health status by identifying the diseases both present and not present; to ensure vaccines work; to establish the effectiveness of biosecurity; and to trigger an effective investigation when the monitoring system detects a problem. There are many well-designed monitoring systems in poultry operations. Many companies operate an internal auditing system. These are designed to check the following:

- Routine data, such as mortality, production performance, and processing results. For example, the measurement of hock scabs on carcasses is an excellent indicator of litter management on the farm and hence is an indirect measure of welfare.
- Inputs to the system, such as feed, water and staff.
- Health status checking using bacteriology, serology (blood tests) and virology.

Blood tests are used to check for the presence of virus diseases and mycoplasma and they also indicate the efficacy of the vaccination programme by the measurement of antibody titres. Culture tests are carried out to look for bacteria such as *Salmonella*.

The successful control of *Salmonella* within broiler integrations in the UK is a good example of an effective biosecurity system which has had many benefits in controlling other disease conditions.

Each operation must define its own programme based on its challenges, resources and objectives, and must use the most cost-effective procedures and data management tools to interpret results and set goals. This task requires a team effort between veterinarians, laboratory technicians and company managers.

Auditing

The purpose of auditing is to set standards, to identify areas of weakness and to identify areas to improve. By carrying out the process of auditing you should also create an understanding of how an operation works and its strong and weak points.

Audits may be used to assess the following areas:

- Farm biosecurity and infrastructure.
- Biosecurity of feed from raw material to farm.
- Hatchery biosecurity.
- Transport quality and practice.

To be of greatest value, an individual company should develop its own audit system. An example of an auditing check list is shown in Box 7.1. Scores 1–5 are given according to how effectively each category is being organized.

People

People are the key to effective biosecurity. They must understand why biosecurity is required and how to achieve it. It is sensible to include training in welfare at the same time, because this is so closely linked to biosecurity. The people working on the farm and those associated closely with farms, such as catchers, are the people who have the biggest impact on welfare. However, training in welfare and biosecurity must include all company staff. Nobody can be missed out. There are examples of electricians introducing disease by not changing clothing before entering a house to deal with emergencies.

Following best biosecurity practice must be automatic and therefore part of the daily routine of all employees. The control of movement of all people, equipment and livestock must be considered and correct.

Box 7.1. Farm biosecurity and infrastructure quality audit.

Farm location
Visit procedure
Protective clothing
Farm hygiene
Egg disinfection/storage
Stocking density
Health monitoring
Farm monitoring
Staff sampling
Cleanout procedure
Litter disposal
Predelivery samples
Pest control
House surrounds
Sampling records
Score each 1–5 (1 is very poor, 5 is excellent)

8 Feed

S.H. GORDON[1], A.W. WALKER[1] AND D.R. CHARLES[2]

[1]ADAS Gleadthorpe, Meden Vale, Mansfield, UK; [2]DC R&D, Loughborough, UK

Introduction: What is the Link Between Feeding and Welfare?

Broiler chickens are grown in a closed system and so there is a need to provide all of the bird's energy and nutrient requirements for maintenance, growth and health. The link between health and welfare is apparent, for example, if a bird is lame because of mineral and/or vitamin deficiencies and there is a consequent inability to feed and drink. This exacerbates the nutrient deficiency and growth is retarded, often so much that the bird is less able to feed and drink as the feed and water points are too high for it to reach. The welfare of such a bird is poor and there is a need either to cull it or to segregate it from the flock and to provide an environment and nutrient supply that will aid its recovery.

However, there are ways in which the feed may affect welfare, other than through direct effects on health. For example, if a diet provides excess protein there will be an increase in uric acid excretion, and this must be voided by the bird in the droppings. Not only will the droppings be nitrogen-rich but they will also be wetter. This increases the risk of wet capped litter, and when this happens new droppings are deposited on the litter surface. It is then not possible for the bird to avoid contact with the wet droppings, either when standing or when sitting. The feathers will become soiled and the risk of birds developing hock burn or pododermatitis will be increased. Good management skills affecting the litter and house environment will be needed in these circumstances. If these are not forthcoming, the welfare of the birds may be further compromised due to high atmospheric ammonia concentrations within the house.

Some of the means by which welfare may be influenced by the feed, either through effects on health or otherwise, are discussed in more detail below. There is, however, evidence that some nutrients, and the method of feed presentation, affect behaviour, and in this way there may also be effects on the birds' welfare.

Feed Effects: Direct and Indirect

Direct effects on welfare

The effects of nutrient supply on the healthy growth of bone tissue and on the avoidance of leg disorders has long been recognized. Edwards (1992), in a comprehensive review, quoted Hart *et al.* (1922) and Mitchell *et al.* (1923) on the subject, and there is even a chapter in a 19th century textbook by Lewis (1871). Edwards (1992) went on to cite no fewer than six review articles published during the period from 1982 to 1989. The following points are summarized from Edwards's review.

A deficiency of vitamin A causes poor bone calcification and a staggering gait, but excess vitamin A can also cause lameness.

Fish liver oils were used for several decades to supply vitamin D_3 in poultry diets, but it is now recognized that the active metabolite is 1,25 dihydroxycholecalciferol $(1,25(OH)_2D_3)$. This stimulates calcium absorption and the differentiation of cells in bone development. Due to biopotency problems, it has been suggested that feed formulators should use more than one source of vitamin D.

Bone calcification is depressed in the presence of excess vitamin E if the diet is also deficient in calcium or vitamin D. Nicotinic acid deficiency can cause perosis lameness. The dietary supply of tryptophan affects the requirement for nicotinic acid. Perosis is also caused by folic acid deficiency. Lameness caused by biotin deficiency was recognized during the 1960s and 1970s, when it was realized that biotin availability was low in some cereal feed ingredients. There may be interactions with dietary fatty acid composition. Choline deficiency has also been implicated in the leg abnormality tibial dyschondroplasia.

Clearly, bone development requires an adequate supply of calcium, and tibial dyschondroplasia has been associated with calcium shortage, particularly if phosphorus levels are high. In addition to the six reviews quoted by Edwards (1992), Simons (1986) quantitatively reviewed the calcium requirements of growing broilers and developed partition equations. He also reviewed phosphorus requirements and pointed out the need to be cautious about phytin phosphorus due to poor availability at high calcium levels unless phytase is present. He also noted that the supply of magnesium is likely to be adequate in normal diets and the real danger is excess. Edwards (1992) added that iron and aluminium could interfere with phosphorus utilization.

Interactions between chloride, sodium and potassium have been noted in the context of tibial dyschondroplasia. Other minerals reviewed by Edwards (1992) were manganese, zinc and copper, as well as the trace elements boron, fluoride, silicon and vanadium. He reported his own work on reducing tibial dyschondroplasia in broilers with a supplement containing the elements B, Ni, Al, Sr, Br, V, Si, Sn, Cr, F, Mo, Li, Mn, Zn, Fe, Cu and I.

Consequential effects on welfare

Individual birds with serious leg abnormalities may have difficulty with locomotion, and therefore with walking towards the feeders and drinkers. They may also be less able to reach feeders and drinkers, leading to the risk of dehydration and starvation.

Indirect effects on welfare

It may be tempting to assume that, provided the supply of protein is adequate for the demands of growth and maintenance, further precision in its supply is unlikely to improve welfare. However, this is not so: excess protein must be broken down and eliminated, thus increasing uric acid excretion in the droppings. In a programme of work in the UK on litter condition and hock burn in the 1980s at ADAS Gleadthorpe, some effects of protein supply on litter nitrogen content were quantified (Lynn *et al.*, 1991; Tucker and Walker, 1992). Not only did excess feed protein or poor feed protein quality cause higher litter nitrogen levels, but they also caused higher litter moisture levels. Salt levels were also shown to affect litter condition. Wet droppings and greasy or wet litter can cause soiling of the feathers.

Who are the Influencers?

The dietary factors described above suggest that there are potential direct effects of feeding on welfare. However, indirect effects may exacerbate the problem if management is suboptimal. Management factors that could be relevant include some obvious examples, such as the supply, design and distribution of feeders and drinkers. Tucker and Walker (1992) described the effects of drinker design principles on water spillage, and hence on wet litter. They also reported the effects of stocking density on litter friability, and the effects of feed fat quality on litter wetness and litter greasiness.

Tucker and Walker (1992) emphasized the importance of ventilation rate and house psychrometrics, and of wall and roof insulation. This is because it is important to avoid condensation on building surfaces and in the litter, leading to damp litter. The physics of house heat and moisture balance was discussed in detail by Clark and McArthur (1994), and factors affecting practical ventilation rate requirements for poultry were reviewed by Charles and Walker (2002).

Classen (1992) reviewed management factors affecting leg disorders. Short daylengths were mentioned as a means of slowing growth. Intermittent lighting was also reviewed, but the effects were confusing due to the wide variety of light/dark patterns used. He considered that slowing whole body growth at young ages, to encourage skeletal growth, offered promise.

In *ad libitum*-fed birds, photoperiods shorter than 23 h have been found to reduce the incidence of total mortality and in some studies fewer birds were culled due to gait abnormalities (reviewed by Gordon, 1994, 1997). A slower growth rate during early life was thought to benefit liveability by reducing metabolic demands on the bird. The functional loading of the locomotor system during activity is less in lighter birds, and this may benefit joint health and bone development. A photoperiod shorter than 23 h also allows birds to have a common daily rest period.

Among those who affect the relevant variables are nutritionists and feed suppliers; house designers and builders; equipment designers, installers and suppliers; ventilation designers, installers and suppliers; and researchers, developers and providers of the supporting technologies. Those whose on-farm skills are relevant include owners, company and site managers, and stockpeople, who have control over the provision and use of the feed and equipment, and can monitor the behaviour and condition of the birds.

Aspects of Feed

It is clear from this chapter so far that feed formulation is important, but feed quality control and monitoring also have parts to play. Ingredient monitoring for nutrient content is supported by the avoidance of deterioration of nutrients during storage. During feed handling and transport it is necessary to avoid contamination and rancidity and to maintain pellet integrity.

Measuring and Auditing the Effects of Feed on Welfare

The measuring and auditing of welfare includes procedures imposed from outside the farm or the company, such as welfare inspections for compliance with both statutory and retail rules and codes of practice. Less formal procedures initiated and imposed within the farm or the company might include the following.

Sample weighings can be used to enable comparisons of farm data with the expectations of the breeder or the broiler integrator, and for assessing the evenness of flock growth. However, it is important to remember that broilers are living things and cannot be expected to be identical. Nevertheless, variation should be within normal expectations. Post-mortem analyses may be required if there is mortality beyond normal expectations. Methods of monitoring and analysing litter were mentioned by Tucker and Walker (1992), who provided subjective litter wetness and greasiness scores. The objective properties of litter that they analysed were moisture, nitrogen and ether extract content. They suggested litter sampling at both the surface and the core. They also published body condition score guidelines. The incidence of condemnations and the reasons for them should be recorded.

Feed Presentation: Links with Behaviour, Performance and Welfare

Feeding equipment

Robinson (1948) described feed troughs for chicks, complete with devices based on revolving bars over the troughs to prevent the chicks getting into the troughs and fouling the feed. Wire grilles were also used to reduce wastage and fouling. Hand-filled hopper feeders were available at that time.

Elson (1996) described two types of feeders for modern broiler houses. Flat-chain feeders have a chain running within a continuous trough of rectangular cross-section, open at the top. These feeders pick up feed as the chain passes through a hopper. The birds feed by pecking directly into the trough. He described pan feeders as the successors of the hand-filled hopper feeders of the 1950s. The pans are supplied by delivery tubes containing auger devices to distribute feed around the house from holding bins. The birds peck the feed from round pans, and the final delivery from the auger tubes to the pans is often through some kind of level control device. The aim is to maintain full but not overflowing pans. In some designs, wasteful flicking of the feed may be minimized by a grille through which the birds may peck.

When chicks are first placed it is common to provide extra feed at floor level and under bright light, in order to encourage them to find the feed. Elson described devices that flood the pan feeders at this stage, making the feed readily accessible.

Feeding equipment design has been shown to affect apparent feed intake, and work at ADAS Gleadthorpe on this subject was reported by Elson (1996). Some of the effect is probably due to spillage and wastage, but there could conceivably also be effects on genuine feed consumption. Whatever the partition between these two, there are consequences for practical feed conversion efficiency and for behaviour at the feeder. Pan feeders gave results similar to those of chain feeders in which the feed was at a low level in the trough. In the case of chain feeders, important features were the depth of feed in the trough and the design of the feed depth control device (Elson, 1996). It is important to ensure sufficient trough allowance (centimetres per bird) and feed depth in the trough to permit all birds to have effective *ad libitum* access, but without unacceptable levels of feed wastage and spoilage.

Feeder location

In intensive systems the UK *Code of Recommendations for the Welfare of Livestock* (DEFRA, 2002) requires that no bird should have to travel more than 4 metres to reach feed and water.

Methods of presentation of the feed mix

For many years the standard method of presenting the feed to the birds was to provide a complete feed *ad libitum*, supplied as mash, pellets or crumb. It was usual to provide a stepwise sequence of formulations of declining protein content in order to accommodate the changing rate of protein growth as a proportion of total growth.

It has been argued that there are three potential sources of imprecision with this approach. Firstly, it has frequently been pointed out (e.g. Emmans, 1976; Emmans and Fisher, 1986) that the needs of individuals within a population of poultry vary. Secondly, a stepwise change in the provision of protein cannot correctly meet the smoothly changing needs of the birds (Emmans, 1976). Just before a change there is likely to be some undersupply of protein, thus limiting growth, and just after a change there will often be some wasteful oversupply. Thirdly, ingredient analyses are not necessarily always constant. Apart from natural variations between samples, there may be systematic variations between crop varieties, such as those described for barley by Jeroch and Danicke (1995).

Undersupply and oversupply are not just commercial inefficiencies: they may also have welfare implications. Undersupply may be stressful to the birds if it leads to excessive searching behaviour and competition at the trough. An oversupply of protein necessarily raises exogenous nitrogen excretion rates, placing extra load on the excretory systems. The welfare effects of this are not clear.

Choice feeding

The recognition of these problems led to several attempts to introduce free-choice feeding, though the technique is perhaps better documented for layers (e.g. Emmans, 1976). Perhaps free-choice feeding can be regarded as being closer to the natural habits of an omnivore than the feeding of ready-mixed whole diets.

Forbes and Covasa (1995) reviewed the broad topic of diet selection and the biological factors influencing it. Hruby *et al.* (1996) found poorer body weights at 6 weeks of age on free-choice feeding than on a whole-feed treatment, and Olver and Jonker (1997) found that small groups of broilers were able to make appropriate choices, even using whole grain options. MacLeod and Dabutha (1997) found that Japanese quail maintained their growth rate over a wide range of temperatures (from 20 to 35°C) by selecting a diet mix which maintained protein intake despite changing energy requirement. They made the important point, quoting the work of Cowan and Michie (1978), Mastika and Cumming (1987) and Sinurat and Balnave (1986), that previous studies indicate that, while broilers sometimes maintain their growth rate over a range of temperatures, their response is inconsistent. This may be because broilers exhibit heat stress at higher temperatures.

Owing to practical difficulties, the provision of two feeds in separate feeders has never been commercially popular, though it may still merit research attention. However, the feeding of grain separately (see below) could perhaps be regarded as a version of choice feeding. Practical difficulties also include the dynamics of appetite control. Forbes and Shariatmadari (1996) fed high- or low-protein feeds for 10 minutes after a fast and found that the birds selected more of the opposite feed during the next hour.

Blending

The problem of matching the smooth change in the requirement of a flock while avoiding stepwise reformulation has been addressed by the blending of two diets dynamically, in order to provide an appropriate mix for the feed intake expected as the birds grow (Filmer, 1993).

Whole-cereal feeding

The practice of feeding a whole cereal, usually wheat, in addition to a balanced broiler feed has been much discussed in the UK industry and worldwide, and benefits over a balanced ration alone are sometimes claimed, beyond the obvious benefit of reduced milling cost. There could conceivably be behavioural benefits in providing the opportunity for the expression of searching and choosing behaviour. Cumming (1996) suggested some possible anatomical benefits. He reported that birds fed whole grains developed large, active gizzards, and was even of the opinion that the gizzard size currently deemed to be normal is probably atrophied. He suggested that large, active gizzards appear to be helpful in coccidiosis control. Forbes and Covasa (1995) have provided a comprehensive review, including regimes in which conventional feeds were diluted progressively with up to 35% cereal.

They pointed out several potential practical difficulties. There are a number of items of legislation relevant to the practice. Untreated cereals cannot be mixed with the compound by the manufacturer because of the risk of disease transmission. Also, birds eating less of the compound will receive less of any medication being applied. Forbes and Covasa (1995) also pointed out that for feeding systems involving a choice, the birds require a learning period. Jensen (1994) noted that if feeding systems involving choice resulted in a lower protein intake, the abdominal fat content of the carcass was significantly increased. More recently, Walton *et al.* (1996) found a slightly lower yield of breast meat in birds fed whole wheat. There was also a non-significantly higher fat content in the breast.

At first sight it is difficult to see how an effective dilution of the compound feed with a cereal could give comparable results, as is sometimes claimed, since the intake of the non-energy nutrients must be reduced at a given total appetite. There are several possibilities which could offer an explanation.

- The presence of whole grain could slow the rate of passage through the gut and thereby improve digestibility. Forbes and Covasa (1995) reviewed some evidence in support of such a proposition. They quoted some evidence that fibre improved gizzard development, prevented hypertrophy of the gizzard and reduced the incidence of coccidiosis. However, it should be noted that excessive crop development might cause processing problems at slaughter. Svihus *et al.* (1997b) found that gizzards were larger in broilers fed whole barley, and Olver and Jonker (1997) found larger gizzards in broilers fed whole maize and sorghum. They offered some evidence suggesting that the energy of whole grain was utilized more efficiently than that in mash, though similarly to that in pellets. They postulated that gizzard grinding may stimulate the flow of digestive juices, create larger surface areas for exposure to enzymes and also slow the rate of passage.
- Researchers and practical growers claiming greater productivity may sometimes have been feeding a compound feed of higher specification than merited by the particular batch of birds.
- The selection of the whole grains differently by individual birds of different growth potential constitutes a version of choice feeding. This would be expected to give greater benefits in as-hatched batches than with single sex growing.
- Increasing the cereal intake may merely increase the effective metabolizable energy (ME) level of the total diet, thereby increasing ME intake slightly, though this seems unlikely.
- Growers claiming unqualified benefits of cereal dilution may be ignoring possible effects on carcass quality.

Manipulating the growth curve

The practice of *ad libitum* feeding of a single complete diet throughout life was questioned above on the grounds of imprecision, but it has also been questioned for other reasons. Nir *et al.* (1996) made the interesting point that the *ad libitum* feeding of a single diet is an artificial condition, the natural condition being the sporadic avail-ability of feed. Yet in a large number of experiments on alternating periods of feed availability and deprivation which they reviewed, without exception the increased intake during the feeding period failed to compensate for the lack of intake during deprivation, and growth rate was depressed. However, recently there has been interest in the possible merits of some deliberate limitation of growth during early life.

Walker (1996) described techniques for controlling the rate of growth in early life by mealtime feeding, by withdrawing the feed for up to 8 hours per day between days 5 and 37 and permitting compensatory growth later. The effects included better feed conversion efficiency and lower mortality. He suggested that achieving the early restriction through energy allowance, rather than protein, reduces the risk of increasing fatness. The welfare implications are uncertain. Superficially it is tempting to suppose that continuous feed availability would provide better welfare, but it is probably an unnatural condition. The UK *Code of Recommendations for the Welfare Of Livestock* (DEFRA, 2002) states that feed for meat chickens should not be withheld for more than 12 hours before slaughter.

The dual phenomena of compensatory growth and the effect of early growth on later body composition have been recognized for some time. McCance (1960) pointed out that restriction of nutrition in many species had by then been under investigation by scientific methods for over 100 years. He reviewed the classic work of the 1940s and 1950s on reversing the planes of nutrition in farm animals in order to investigate the effects on growth and carcass composition. In his own work he held the weight of cockerels constant for 6 months and then allowed them full feeding. Although the undernourished birds grew rapidly on rehabilitation, after such a severe restriction they failed to reach the same final weight as the controls. During restriction they were restless and particularly susceptible to cold.

Wilson and Osborn (1960) reviewed 135 references on compensatory growth in farm animals. They concluded that homeostasis, or the preservation of constancy, applies to body weight. After a period of undernutrition, growth was found to occur at a rate appropriate to the physiological age rather than the chronological age. Waves of growth were thought to occur, starting with nervous tissue, then skeletal, then muscular and finally adipose tissue. Least recovery was observed to occur in the body regions that would have been growing fastest when restriction occurred. Retardation could permanently alter the final body proportions. There was concern about recovery from unavoidable undernourishment in the early life of animals and children, due to periods of food scarcity, rather than the fine-tuning of the growth of well fed animals.

More recently Plavnik and Hurwitz (1988), after quoting Osborne and Mendel (1915) on accelerated growth after restriction, restricted chicks to a maintenance diet during early growth (starting from 3 days to 11 days of age and applying the restriction for 3–7 days). At 54–59 days of age the males had less abdominal fat and had grown with a more efficient feed conversion. The effects were less clear with females, though the effects were achieved if restriction was confined to 3–5 days. The classic work on planes of nutrition in farm animals had highlighted effects on carcass fatness. Hammond (1960), in a book originally published in 1940, suggested that in the early stages of growth, increase in size is due mainly to cell multiplication, cell enlargement taking place later. If this is so, it supports the notion that the early restriction of broilers might reduce the number of fat cells, though the reduced abdominal fat observed by Plavnik and Hurwitz (1988) is difficult to reconcile with the concept of adipose tissue developing last, as proposed by the early authors.

The early work on such issues was aimed at the prevention of excessive carcass fat from the point of view of production performance and product quality. There may be welfare implications of the proportions of fat in the body since it is possible that organisms strive to achieve some ideal fat content and body fat distribution. In mammalian farm animals, breed differences in the typical fat partition index have been documented (Lawrence and Fowler, 1997); the fat partition index was defined as the quotient obtained by dividing the weight of dissectible subcutaneous fat by the sum of the weights of intermuscular, perinepheric and inguinal fat. Emmans and Fisher (1986) reviewed suggestions that in poultry there is a systematic increase in body lipid content during development.

There have been suggestions that feed restriction in early life may reduce the risk of ascites (e.g. Albers *et al.*, 1990). However it seems that care should be taken that the restriction is not too severe, because of the risk of increased susceptibility to ascites on reintroduction of full feeding (Jones, 1995).

Zubair and Leeson (1996) reviewed the principles and practice of early restriction followed by compensatory growth as researched to that date. They reported that feed limitation should be for not more than 7 days for males and 5 days for females. They reviewed the modes of action, including the suggestion that there may be changes in the enzymes associated with hepatic lipogenesis. Lower metabolic rate was thought unlikely to account for improvements in feed efficiency. They warned that the results of studies on the effects of early restriction on compensatory growth, feed conversion efficiency and fatness were inconsistent, probably because of factors such as differences in the methods used and in the duration of restriction.

Moran (1996) found that the simple expedient of omitting the fat from the diet was an effective method of limiting early growth and minimizing leg problems in male broilers without affecting processing yields. The low-fat diets were low in nutrient density but at the same energy/protein ratio as the controls. However, final weights were not equal until 50 days of age.

The early work on this topic was mainly aimed at performance improvement, but in recent times interest in the manipulation of the patterns of early growth and in the theoretical aspects of growth patterns, as outlined above, has been targeted at improving leg health in order to improve bird welfare.

Soaking and wet feeding

It has been standard practice during the entire history of intensive production to offer the birds dry feed. By contrast, earlier traditional systems often included wet mash (e.g. Howes, 1939; Robinson, 1948), though the practice was sometimes questioned. There has lately been some suggestion that wet feeding may be worth a second look.

Yalda and Forbes (1995), using small numbers of birds, reported improved feed efficiency with wet feeding, with no effect on carcass quality. The proportion of water made little difference to the results in the range 640–723 g water per kg of feed/water mixture. Earlier work had shown that smaller proportions of water (470 g/kg) gave no effect (Yalda and Forbes, 1991), and at even lower proportions of water Abasiekong (1989) found a depression in performance due to wet feeding at 20°C, though feed intake and weight gain benefited at 37°C. Hill (1977) had shown some years earlier that feed intake and water intake were strongly correlated in individual laying hens, though in larger numbers of birds the effect was often masked by the effects on feed intake of other variables, such as temperature, production level and time of day. The wet feeding of poultry on a large scale is the subject of patent application (PCT/GB92/01134, University of Leeds).

The effect is probably more subtle than merely increasing weight gain by virtue of increased feed intake. Thus, Yasar and Forbes (1996) found not only that both feed intake and weight gain were higher with wet feeding, but also that there was a lower viscosity of the intestinal contents. Yalda and Forbes (1995) found that digestibility increased from 0.65 to 0.73, yet Yasar and Forbes (1998) found no significant increase in apparent digestibility of dry matter, organic matter or crude protein. Yasar *et al.* (1996) found that wet feeding led to a reduction in intestinal mucosal cell proliferation rates. Yasar and Forbes (1997a) found that wetting the feed reduced the viscosity of the digesta. Yasar and Forbes (1997b) found that gut length and weight were increased by wet feeding, and there were larger villi. Work on wet feeding in pigs has also shown improved digestibility, and it has been suggested that the effects may include the activation of endogenous enzymes (Brookes and Carpenter, 1990). Yalda and Forbes (1996) found that the prolonged soaking of broiler feed was not necessary, though the nominal zero soaking time treatment effectively averaged 12 hours in the troughs. They suggested that enzymic effects did not account for the improved weight gain observed.

Svihus *et al.* (1997a) found that soaking or germinating whole barley decreased its β-glucan content and increased the dry matter of the gut contents.

A Note on Terminology

The phenomenon of fast growth after a period of restriction was generally referred to as 'compensatory growth' in the early literature (e.g. Wilson and Osborn, 1960). It later became conventional to call it 'catch-up growth' on the grounds that growth rate can never exceed the genetic potential and can only lead to a catching up. Some publications on human nutrition (for example, World Health Organization, 1985) use the term 'catch-up growth'. Recently there has been a tendency in poultry science to revert to the use of the original term, 'compensatory growth' (e.g. Zubair and Leeson, 1996).

References

Abasiekong, S.F. (1989) Seasonal effect of wet rations on performance of broiler poultry in the tropics. *Archives of Nutrition* 39, 507–514.

Albers, G., Barranon, Z. and Ortiz, C. (1990) Correct feed restriction prevents ascites. *Poultry (Misset)* April/May, 22–23.

Brookes, P.H. and Carpenter, J.L. (1990) The water requirements of growing–finishing pigs: theoretical and practical considerations.

In: Haresign, W. and Cole, D.J.A. (eds) *Recent Advances in Animal Nutrition*. Butterworths, London, pp. 115–136.

Charles, D.R. and Walker, A.W. (2002) *Poultry Environment Problems: A Guide to Solutions*. Nottingham University Press, Nottingham.

Clark, J.A. and McArthur, A.J. (1994) Thermal exchanges. In: Wathes, C.M. and

Charles, D.R. (eds) *Livestock Housing*. CAB International, Wallingford, UK, pp. 97–122.

Classen, H.L. (1992) Management factors in leg disorders. In: Whitehead, C.C. (ed.) *Bone Biology and Skeletal Disorders in Poultry*. Carfax, Abingdon, UK, pp. 195–211.

Cowan, P.J. and Michie, W. (1978) Environmental temperature and choice feeding of the broiler. *British Journal of Nutrition* 40, 311–315.

Cumming, R.B. (1996) Opportunities for sustainable management practices in the poultry industry. In: *Proceedings of Poultry Industry Conference, New Plymouth*. World's Poultry Science Association (New Zealand Branch), pp. 5–12.

DEFRA (2002) *Code of Recommendations for the Welfare of Livestock. Meat Chickens and Breeding Chickens*. Department for Environment, Food and Rural Affairs, London.

Edwards, H.M. (1992) Nutritional factors and leg disorders. In: Whitehead, C.C. (ed.) *Bone Biology and Skeletal Disorders in Poultry*. Carfax, Abingdon, UK, pp. 167–193.

Elson, H.A. (1996) Broiler feeding systems. *ADAS Poultry Progress*, May.

Emmans, G.C. (1976) Problems in feeding laying hens: can a feeding system based on choice solve them? [abstract]. *World's Poultry Science Journal* 31, 311.

Emmans, G.C. and Fisher, C. (1986) Problems in nutritional theory. In: Fisher, C. and Boorman, K.N. (ed.) *Nutrient Requirements of Poultry and Nutritional Research*. Butterworths, London, pp. 9–40.

Filmer, D.G. (1993) Applying nutrient allowances in practice – a new approach to growing poultry. In: *Eighth International Poultry Breeders Conference, Glasgow*.

Forbes, J.M. and Covasa, M. (1995) Application of diet selection by poultry with particular reference to whole cereals. *World's Poultry Science Journal* 51, 149–166.

Forbes, J.M. and Shariatmadari, F. (1996) Short-term effects of food protein content on subsequent diet selection by chickens and the consequences of alternate feeding of high- and low-protein foods. *British Poultry Science* 37, 597–607.

Gordon, S.H. (1994) Effects of daylength and increasing daylength programmes on broiler welfare and performance. *World's Poultry Science Journal* 50, 269–282.

Gordon, S.H. (1997) Effect of light programmes on broiler mortality with reference to ascites. *World's Poultry Science Journal* 53, 68–70.

Hammond, J. (1960) *Farm Animals, Their Breeding, Growth and Inheritance*, 3rd edn. Edward Arnold, London.

Hart, E.B., Halpin, J.G. and Steenbock, H. (1922) The nutritional requirements of baby chicks. II. Further study of leg weakness in chickens. *Journal of Biological Chemistry* 52, 379–386.

Hill, J.A. (1977) The relationship between food and water intake in the laying hen. PhD thesis, Huddersfield Polytechnic, UK.

Howes, H. (1939) *Modern Poultry Management*. Macmillan, London

Hruby, M., Hamre, M.L., Zollitsch, W. and Coon, C.N. (1996) The feeding of complete broiler diets calculated from the results of a free choice experiment. *Journal of Applied Poultry Research* 5, 297–303.

Jensen, J.F. (1994) Choice feeding in practice. In: *Proceedings of the 9th European Poultry Conference, Glasgow*. World's Poultry Science Association, pp. 223–226.

Jeroch, H. and Danicke, S. (1995) Barley in poultry feeding: a review. *World's Poultry Science Journal* 51, 271–292.

Jones, G.P.D. (1995) Manipulation of organ growth by early life food restriction: its influence on the development of ascites in broiler chickens. *British Poultry Science* 36, 135–142.

Lawrence, T.L.J. and Fowler, V.R. (1997) *Growth of Farm Animals*. CAB International, Wallingford, UK.

Lewis, W.M. (1871) *The People's Practical Poultry Book: Breeds, Breeding, Rearing and General Management of Poultry*. D.D.T. Moore, New York.

Lynn, N.J., Tucker, S.A. and Bray, T.S. (1991) Litter condition and contact dermatitis in broiler chickens. In: Vijtenboogaart, T.G. and Veerkamp, C.H. (eds) *Quality of Poultry Products. Poultry Meat. Proceedings of Spelderholt Jubilee Symposium, Doorwerth*.

Macleod, M.G. and Dabutha, L.A. (1997) Selection between a 'high protein' and a 'high energy' diet by Japanese quail in relation to ambient temperature and metabolic weight. In: *Proceedings of Spring Meeting, Scarborough*. World's Poultry Science Association (UK Branch), pp. 43–44.

Mastika, M. and Cumming, R.B. (1987) Effect of previous experience and environmental variations on the performance and pattern of feed intake of choice-fed and complete-fed broilers. In: Farrell, D.J. (ed.) *Recent Advances in Animal Nutrition in Australia.* University of New England Publishing Unit, Armidale, NSW, Australia, pp. 260–282.

McCance, R.A. (1960) Severe undernutrition in growing and adult animals. 1. Production and general effects. *British Journal of Nutrition* 14, 59–72.

Mitchell, H.H., Kendall, F.E. and Card, L.E. (1923) The vitamin requirements of growing chickens. *Poultry Science* 2, 117–124.

Moran, E.T. (1996) Body composition. In: Hunton, P. (ed.) *Poultry Production.* World Animal Science Subseries C9. Elsevier, Oxford, pp. 139–156.

Nir, I., Nitsan, Z., Dunnington, E.A. and Siegel, P.B. (1996) Aspects of food intake restriction in young domestic fowl: metabolic and genetic considerations. *World's Poultry Science Journal* 52, 251–266.

Olver, M.D. and Jonker, A. (1997) Effect of choice feeding on performance of broilers. *British Poultry Science* 38, 571–576.

Osborne, T.B. and Mendel, L.B. (1915) The resumption of growth after long continued failure to grow. *Journal of Biological Chemistry* 23, 439–454.

Plavnik, I. and Hurwitz, S. (1988) Early feed restriction in chicks: effects of age, duration and sex. *Poultry Science* 67, 384–390.

Robinson, L. (1948) *Modern Poultry Husbandry.* Crosby Lockwood, London.

Simons, P.C.M. (1986) Major minerals in the nutrition of poultry. In: Fisher, C. and Boorman, K.N. (eds) *Nutrient Requirements of Poultry and Nutritional Research.* Butterworths, London, pp. 141–154.

Sinurat, A.P. and Balnave, D. (1986) Free-choice feeding of broilers at high temperatures. *British Poultry Science* 27, 577–584.

Svihus, B., Newman, R.K. and Newman, C.W. (1997a) Effect of soaking, germination, and enzyme treatment of whole barley on nutritional value and digestive tract parameters of broiler chickens. *British Poultry Science* 38, 390–396.

Svihus, B., Herstad, O., Newman, C.W. and Newman, R.K. (1997b) Comparison of performance and intestinal characteristics of broiler chickens fed on diets containing whole, rolled or ground barley. *British Poultry Science* 38, 524–529.

Tucker, S.A. and Walker, A.W. (1992) Hock burn in broilers. In: Garnsworthy, P.C. and Wiseman, J. (eds) *Recent Advances in Animal Nutrition.* Nottingham University Press, Nottingham, pp. 33–49.

Walker, A.W. (1996) Effect of periodic fasting on broiler performance. *British Poultry Science* 37 (Supplement), 79–80.

Walton, A.M., Perry, G.C. and Richardson, R.I. (1996) Whole wheat grain addition to broiler rations. Effects on carcass yield, composition and sensory attributes of broiler meat. In: *World's Poultry Science Association (UK Branch), Proceedings of Spring Meeting, Scarborough,* pp. 77–78.

Wilson, P.N. and Osborn, D.F. (1960) Compensatory growth after undernutrition in mammals and birds. *Biological Reviews* 35, 324–363.

World Health Organization (1985) Energy and protein requirements. Report of a joint FAO/WHO/UNU expert consultation. World Health Organization, Geneva.

Yalda, A.A. and Forbes, J.M. (1991) The effect of adding water to the diet on body fatness and food intake in broiler chickens. *Proceedings of the Nutrition Society* 50, 98.

Yalda, A.A. and Forbes, J.M. (1995) Food intake and growth in chickens given food in the wet form with and without access to drinking water. *British Poultry Science* 36, 357–369.

Yalda, A.A. and Forbes, J.M. (1996) Effects of food intake, soaking time, enzyme and cornflower addition on the digestibility of the diet and performance of broilers given wet food. *British Poultry Science* 37, 797–808.

Yasar, S. and Forbes, J.M. (1996) Nutritional value of wet and dry grain based diets for broiler chickens. In: *World's Poultry Science Association (UK Branch), Proceedings of Spring Meeting, Scarborough,* pp. 60–61.

Yasar, S. and Forbes, J.M. (1997a) Viscosity of digesta in stomachs and intestines of broilers with water and guar gum addition to the diet. In: *World's Poultry Science Association*

(UK Branch), Proceedings of Spring Meeting, Scarborough, pp. 27–28.

Yasar, S. and Forbes, J.M. (1997b) Effects of wetting and enzyme supplementation of wheat-based foods on performance and gut responses of broiler chickens. In: *World's Poultry Science Association (UK Branch), Proceedings of Spring Meeting, Scarborough*, pp. 56–57.

Yasar, S. and Forbes, J.M. (1998) Apparent nutrient digestibility of cereal grain-based foods soaked in water for broiler chickens.

In: *World's Poultry Science Association (UK Branch), Proceedings of Spring Meeting, Scarborough*, pp. 58–59.

Yasar, S., Forbes, J.M. and McClean, D. (1996) Gut histo-morphology of broiler chickens with wet feeding. In: *World's Poultry Science Association (UK Branch), Proceedings of Spring Meeting, Scarborough*, pp. 62–63.

Zubair, A.K. and Leeson, S. (1996) Compensatory growth in the broiler chicken: a review. *World's Poultry Science Journal* 52, 189–201.

9 Light

N.B. Prescott[1], H.H. Kristensen[1,2] and C.M. Wathes[1]

[1]Silsoe Research Institute, Bedford, UK; [2]Royal Veterinary and Agricultural University, Groennegaardsvej, Frederiksberg C, Denmark

Introduction

The majority of broilers produced in the UK are reared in environmentally controlled buildings where artificial light is provided. However, these environments differ radically from natural daylight and some authors have suggested that they may represent poor welfare.

In this chapter we explore the effect of artificial light environments on the welfare of broilers as mediated through vision. While the pineal gland and the skin are also photoreceptors, we presume that the eyes are the main sense organs and vision the main sense for broilers. This seems reasonable, given their size and prominence and the proportion of the avian brain devoted to visual processing. Light environments that restrict the efficacy of visual processing may also reduce welfare if important visual information is lost or corrupted by the environment. For example, birds may be unable to recognize important features of other birds, navigate their way around the featureless landscape of a poultry house, recognize and respond appropriately to humans, or see their food and water clearly.

The Physical Light Environment in Broiler Houses

There are a number of features of the physical light environment of a broiler house that may affect the birds' welfare; the most important of these are described below.

Light levels

Light levels (also described as light intensity or illuminance) in environmentally controlled broiler houses are usually very low in comparison with natural daylight and the levels that humans would consider adequate in areas they inhabit. The range of light levels in different environments is given in Table 9.1.

Colour

Light sources in broiler houses may be either incandescent or of a range of fluorescent types. Cool white fluorescent types have a greater proportion of blue wavelengths than warm white fluorescent types, which contain proportionally more red wavelengths. None of the commonly used types of fluorescent light emits appreciable amounts of ultraviolet A light (UVA, λ 320–400 nm). The relative proportions of wavelengths in all these sources are very different from those in natural light, as shown in Fig. 9.1. Daylight has a relatively even distribution of wavelengths between 400 and 700 nm, although UVA becomes progressively attenuated as wavelength shortens. Incandescent lights contain an abundance of red wavelengths but are relatively depleted in blue wavelengths. Fluorescent lights emit a characteristic spiky discharge; the positions of the spikes and their relative sizes depend on the composition of the phosphor mix lining the inside of the tube, which is responsible for producing visible light. The perception of these light sources by poultry is not known, although clearly there is scope for colour signals to be lost or corrupted under artificial lighting. For example, red cues would be transmitted well under incandescent light but poorly under fluorescent lighting, relative to daylight.

Photoperiod

The ancestors of broiler chickens (red jungle fowl, *Gallus gallus*) evolved in the equatorial jungle, with around 12 hours of light and 12 hours of darkness in each 24-hour period. This is in sharp contrast to most current broiler production houses, where birds are kept with a much shorter dark period. The photoperiodic regime defines the number of hours of light (and dark) within each 24-hour period and the most commonly used regimes are described in Table 9.2. Various photoperiodic regimes have been applied and tested over the years and almost all of them have been shown to improve broiler welfare compared with conventional (near-)continuous

Table 9.1. Comparative illuminance levels.

Location	Illuminance (lx)
Direct sun	100,000
Overcast sky	1,000
Business office	250
Laying hen houses	≈ 20
Good street lighting	20
Twilight	10
Broiler houses	≈ 32–0
Turkey houses	≈ 21–0
Overcast night	0.0001

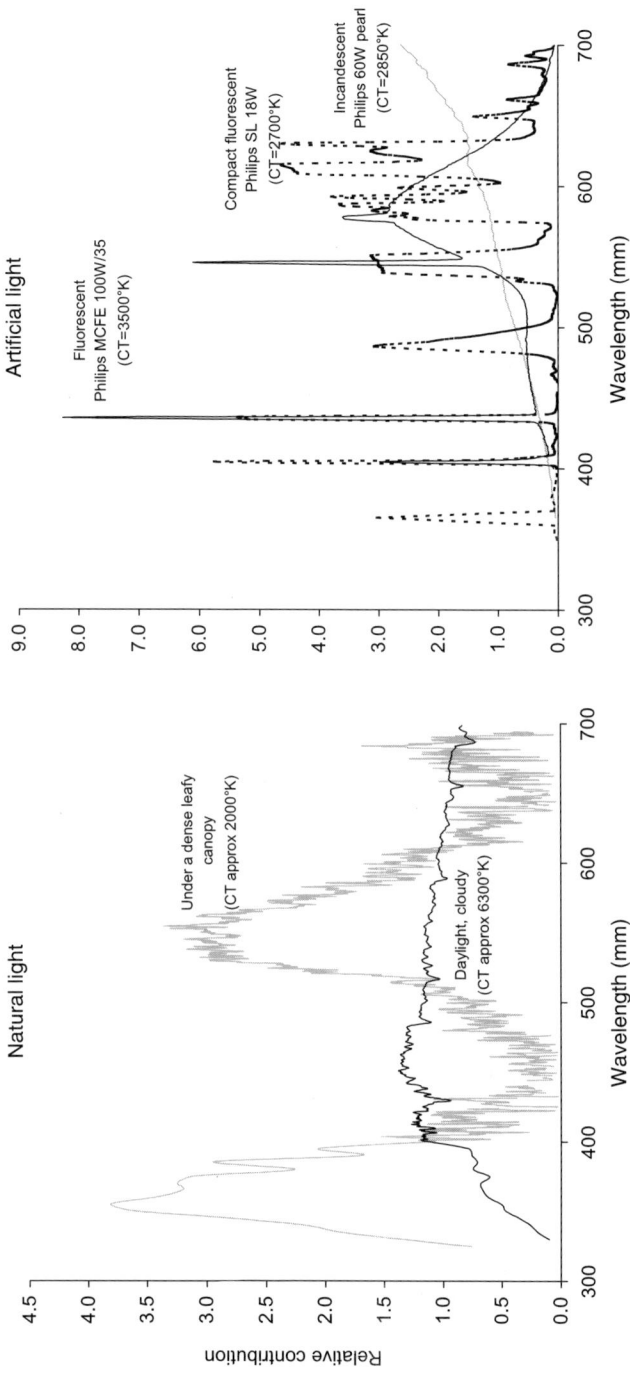

Fig. 9.1. Spectral power emissions for natural and artificial light sources.

Table 9.2.   The most commonly used photoperiodic regimes in broiler production.

Photoperiodic regime	Description	Example*
Continuous light	24 hours of continuous light. Some call 23L:1D continuous	24L:0D or 23L:1D
Fixed/restricted	One uninterrupted dark period within each 24-hour period	16L:8D
Intermittent	Several dark periods within each 24-hour period, these can be of the same or of varying lengths	3L:1D or 16L:2D:1L:2D:1L:2D
Biomittent	One light and one dark period every hour	0.25L:0.45D
Step-up/increasing	After 1–7 days of near-continuous lighting, the photoperiod is abruptly reduced. After this, the photoperiod is gradually increased over the rest of the growing period. The light regime during the step-up can be fixed, intermittent or a combination	1–7 days: 23L:1D 7–12 days: 16L:8D 12–36 days: +1L per 4 days 36–41 days: 22L:2D
Step-down/decreasing	Same principle as in the increasing regimes but the photoperiod is decreased gradually after 1–7 days so that the broilers have shorter photoperiods at the end than at the beginning of the growing period	1–7 days: 23L:1D 7–35 days: −1L per 4 days 35–42 days: 16L:8D

*23L:1D means 23 hours of light (L) and 1 hour of darkness (D).

lighting (Gordon, 1994; Kristensen, 1999). Increasingly, producers are now moving away from the misconception that more hours of light would enable the chickens to eat more and hence grow faster. It has been shown that broilers have a higher feed conversion ratio, fewer leg problems, lower mortality from ascites and more natural visual ability when they are given an uninterrupted dark period than when they are in continuous light (e.g. Classen, 1991; Charles *et al.*, 1992; Blair *et al.*, 1993; Gordon and Tucker, 1995, 1997). A dark period is also thought to help the chickens synchronize their behaviours, and hence requires high management standards to maintain good litter quality and enough feeder and drinker space for the chickens to simultaneously sit, eat and drink in larger numbers. Poorly managed litter may cause hock burn and pododermatitis, resulting in both welfare problems and rejection at the processing plant (Ekstrand *et al.*, 1997). Discussions and work continue in an attempt to define the optimal photoperiodic regime for broiler chickens. Results to date suggest an absolute minimum uninterrupted dark period of 4 hours should be given, although the requirements for sleep may be higher at some stages of the growing period than at others (Blokhuis, 1983). There have been reports that broilers reared with an uninterrupted dark period are more active than broilers reared in continuous light (e.g. Classen, 1991). Although this is probably a result (or a cause) of better leg health in the broilers, reflecting improved welfare, it can make broilers harder to catch prior to processing.

Dawn and dusk periods

In many broiler houses the transition between the light and dark phases of the photoperiod is abrupt. Savory (1976) suggests that dawn and dusk periods may help the birds to predict the onset of a dark and light period and encourage them to take a meal before the dark period, which may increase their feed conversion efficiency overnight. The provision of simulated dawn and dusk can further encourage more natural settling behaviour of the birds, in which they prepare for night. The provision of dawn and dusk periods (transition periods of gradual brightening or dimming) is a feature of some guidelines (Royal Society for the Prevention of Cruelty to Animals, 1999) and is increasingly common in the broiler industry.

Flicker

Conventional fluorescent lights flicker at twice the duty cycle of the electrical supply: flicker rates are 100 Hz in the UK and 120 Hz in the USA and Canada. The flicker modulation of these sources (change in the light level through a flicker cycle) is generally 20–40%. To overcome the discomfort that some human users find with these sources, high-frequency systems have recently become available. These flicker at many kHz and are usually described by humans as being more comfortable. Incandescent lights also flicker in response to the alternating electricity supply, but their flicker modulation is very low (5–10%) and imperceptible. Despite claims to the contrary, it now appears unlikely that poultry find the flicker from fluorescent lights aversive, or even that they can perceive it. Equally, there appears to be no benefit in providing high- rather than low-frequency lighting for poultry (see 'Vision', below).

Measuring the Light Environment

Measuring the various properties of light is relatively straightforward. The only parameter mentioned in the previous section that requires measurement is the light level. Colour and the flicker rate can be obtained either from the installer or the manufacturer of the lighting system. The photoperiod and the properties of any dawn or dusk period will be specified by the person responsible for the birds. Light level is measured using the lux unit (lx), which is a measure of illuminance (simply put, the amount of light incident on a surface per unit area). Illuminance can be measured using inexpensive light meters, but in our experience many broiler farms either do not routinely use these devices or do not even possess one. In a survey of duckling and turkey poult rearing houses, Barber *et al.* (2003) showed that farm managers' estimates of light levels in their houses were often inaccurate; given the specificity of the guidelines, this is a cause for concern. In the section below on vision we suggest that the lux unit is inappropriate for measuring light levels as they are perceived by poultry, and offer an alternative unit, the 'clux'.

Measuring the light level

Care must be taken to ensure that the sampling is frequent enough and in the correct places to accurately measure the light environment as experienced by the birds. In a floor-reared broiler shed the levels will need to be taken at the height of the bird's eye and in a pattern that measures all the variation in light level at that height. For buildings with regularly spaced light fittings this is relatively simple, but where light fittings are not regularly spaced or where different light types are used in the same building more frequent samples will need to be taken. Published methods for sampling light levels in buildings exist (e.g. IES, 1966).

The orientation of the sensor head is also important. Light meters collect light from a hemispheric field of view and three methods of collection are used. First, the sensor may be held horizontally. Secondly, the sensor may be inclined in the direction of maximum illuminance, which is usually in the direction of the nearest light source. Thirdly, and less frequently, six readings may be taken from each sampling point parallel to the faces of a notional cube. For human purposes, the orientation of the light meter should be normal to the plane of the surface that will be used (British Standard, 1985). By way of example, for an office desk this will mean that the sensor is held horizontally above the desk's surface, but for a drawing board in a design office the sensor will be inclined from the horizontal at the angle of the drawing board's surface. This means that measurement of illuminance, and presumably to a great extent the subjective assessment of the quality of illuminance, is specific to the orientation of the visual task being performed. What that orientation should be for a broiler in a broiler shed is difficult to determine.

It should be remembered that measures of illuminance merely give an indication of how well objects in an environment will be perceived. The actual quality of perception of objects, measured as luminance, is a function of the light that falls on them and the amount that is then reflected and perceived by the eye. Dark, unreflective objects in a bright environment may be perceived as well as bright, reflective objects in a dim environment. However, because of the difficulty in taking luminance measurements and the broad applicability of illuminance measures, luminance is rarely used to assess the suitability of light environments.

Legislation and guidelines

Various retailers, welfare groups and legislative bodies specify light environments to safeguard the welfare of broilers. In the UK, for example, the Department for Environment, Food and Rural Affairs states that under the Welfare of Farmed Animals (England) Regulations 2000 (S.I. 2000 No. 1870):

- Animals kept in buildings shall not be kept in permanent darkness.
- Where the natural light available in a building is insufficient to meet the physiological and ethological needs of any animals being kept in it then appropriate artificial lighting shall be provided.
- Animals kept in buildings shall not be kept without an appropriate period of rest from artificial lighting.

As guidelines they recommend that:

- Chickens should be housed at light levels which allow them to see clearly and which stimulate activity. This should be provided by lighting systems designed, maintained and operated to give a minimum light level of 10 lux at bird eye height. Illumination of the house to at least 20 lux will further encourage activity. Houses should have a uniform level of light. If a behavioural problem such as cannibalism occurs, it may be necessary to dim the lights for a few days.

- Meat chickens which do not have access to daylight should be given at least 8 hours of artificial lighting each day. It is important for bird welfare to provide them with a period of darkness (not less than 30 minutes) in each 24-hour cycle. This ensures the birds become used to total darkness and helps to prevent panic in the event of a power failure. Longer periods of darkness can reduce mortality and improve leg health (DEFRA, 2002).

These guidelines are broadly consistent with suggestions made by the Farm Animal Welfare Council (FAWC, 1992) the Royal Society for the Protection of Cruelty to Animals through the Freedom Foods scheme (RSPCA, 1999) and the Assured Chicken Production scheme. The Freedom Foods scheme also includes provision of dawn and dusk periods and requires more than 6 hours of continual darkness. The legal requirements have passed into UK law through the EU implementation of the Council Directive 98/58/EC of 20 July 1998: Concerning the Protection of Animals Kept for Farming Purposes.

The Role of Light in the Vision and Welfare of Broilers

Vision

Poultry possess simple, diurnal eyes, rather like a human's. The mechanisms subserving colour, temporal and spatial vision, whilst generally similar to those of humans, are subtly different. It is likely that the features incorporated in a chicken's eye are to some degree an adaptation to the ecological niche of their progenitor species, the red jungle fowl.

Domestic fowl have a number of adaptations to their colour perception apparatus that are not shared by humans. First, they possess three types of photoreceptor compared with just rods and cones in humans (King-Smith, 1971). The additional photoreceptor is a double cone, the function of which is unclear, though it does respond to incident light. Secondly, fowls have four photoreactive pigments associated with cone cells, which are responsible for photopic colour vision (Yoshizawa, 1992), compared with three in humans. These are maximally sensitive at wavelengths of 415, 455, 508 and 571 nm versus 419, 531 and 558 nm in humans (Dartnall *et al.*, 1983). Thirdly, fowls possess coloured oil droplets in their cone cells, which filter incident light before it reaches the photoreactive pigments. The droplets are associated variously with individual cone cell species (Bowmaker and Knowles, 1977). The spectral sensitivity curves derived for the fowl by Prescott and Wathes (1999), using a behavioural test, and by Wortel *et al.* (1987), using an electrophysiological test, differ from that of the human: the relative response is broader, and UVA radiation (λ 320–400 nm) can be perceived. The overall effect of these anatomical differences is a visual system that is well adapted to collecting spectral information. However, the penalty may be that a high level of illuminance is required for the system to work to its full potential.

The major implication of the fowl's spectral sensitivity is that the unit with which we have traditionally measured illuminance in poultry houses, the lux, may be inaccurate. For animals such as poultry, with spectral sensitivities different from that of humans, the lux unit will not correlate well with the perceived brightness of different light sources, since the lux uses the human's spectral sensitivity. For example, we calculate that, for typical fluorescent and incandescent luminaires illuminated to the same lux level (and consequently isoluminant for humans) fowls would perceive the incandescent bulb as about 20% brighter than the fluorescent tube. Alternative units for measuring fowl-perceived illuminance, the 'clux' or 'galluminance', have been derived by Prescott and Wathes (1999) and Nuboer *et al.* (1992a) respectively. Within species, the link between measured illuminance and perceived brightness is presumed to be approximately linear. It may be that the double cone has a more prominent role in luminance perception (Osorio *et al.*, 1999), but this novel theory requires further proof. Since additional factors relating to the perception of brightness may be involved in different species, it is not possible with these data to compare perceived brightness between fowls and humans for a particular light environment.

The role of temporal (flicker) sensitivity is important in broiler welfare for one important reason: conventional (low-frequency) fluorescent lights flicker at 100 Hz, which, if perceived by the bird, could be detrimental to its welfare in the same ways that perceived flicker is detrimental to humans.

Using a psychophysical method, Nuboer *et al.* (1992b) found that some fowls may have been able to perceive blue light ($\lambda \approx 476$ nm) flickering up to 105 Hz but were less sensitive to other colours (maximum sensitivity for ultraviolet, 370 nm, 70 Hz; indigo, 430 nm, 75 Hz; green, 556 nm, 95 Hz; red, 670 nm, 85 Hz). Also, flicker from a white compact fluorescent luminaire at 80 Hz was perceived, but higher flicker rates were not tested. The maximum frequencies perceived by humans are generally quoted to be between 50 and 60 Hz (Brundrett, 1974). However, flicker sensitivity in humans depends critically on two factors. First, flicker is perceived better at higher than at lower mean illuminance. Secondly, flicker with high modulation, a measure of the magnitude of illuminance change through a flicker cycle, is better perceived than flicker of low modulation (De Lange, 1958). The conventional Philips fluorescent and compact fluorescent luminaires mentioned previously and shown in Fig. 9.1 possessed flicker modulation depths of 23 and 39% respectively. The sensitivity to changing modulation was not characterized by Nuboer and colleagues, who used a single value of 95% (Nuboer *et al.*, 1992b). In a recent study, Jarvis *et al.* (2002) found that poultry cannot detect 100 Hz flicker at 100 lx, but may be able to at very much higher illuminances. In a tightly controlled experiment, Boshouwers and Nicaise (1992) found that, at an illuminance of 90 lx, broilers exposed to 100 Hz flicker exhibited less 'activity' than control birds exposed to flicker at 26 kHz. This finding is contrary to the work of Jarvis *et al.* (2002) but may reflect suprathreshold effects, significant deviations from sine-wave flicker or some other effect. At an illuminance of approximately 14 lx, however, Widowski and Duncan (1996) found that laying hens had no preference for fluorescent light flickering at a low frequency of 120 Hz or a high frequency between 20 and 60 kHz. In less controlled but nevertheless useful studies, Widowski *et al.* (1992) and Sherwin (1999) found that hens preferred fluorescent luminaires, flickering at 120 and 100 Hz and illuminated to approximately 12 and 10 lx, respectively, over incandescent luminaires, although both authors commented that colour, temperature and illuminance were confounded between their treatments. In addition, the illuminance and modulation depths for these fluorescent luminaires may have been so low as to preclude the perception of flicker by the birds. As a whole, these data suggest that the flicker from fluorescent lighting is probably not perceived, and if it is there are no marked behavioural indicators of aversion.

Acuity is a measure of spatial resolution or the level of detail detected in visual images. This is determined largely by the optical clarity and precision of the optical system and the density of rod and cone cells in the retina. Acuity falls rapidly once the near and far limits of accommodation are exceeded. Often gratings are employed to measure acuity, with the resolution limit defined as the minimum grating fineness that can just be distinguished from an isoluminant, uniform grey stimulus; the unit of measurement at this threshold is the number of bars or cycles per degree of visual angle. Spatial acuity is also dependent upon two other parameters: the luminance of the stimuli (L, cd/m^2) and the contrast (C, %) between light and dark bars. In fowls, acuity has been measured variously as 1.5 cycles/degree ($L = 39$ cd/m^2, $C = 80\%$; Over and Moore, 1981), 4–6 cycles/degree ($L = 2.7$ cd/m^2, $C = 93$–100%; DeMello *et al.*, 1992) or 7 cycles/degree ($L = 12.1$ cd/m^2, $C =$ unknown; calculated by DeMello *et al.* (1992) from Johnsen, 1914). However, it is less than for humans, with 30 cycles/degree (Spence, 1934). Variations between the results of these investigators may be due to the differences in the stimulus luminance and/or contrast. In crude terms, an acuity of 30 cycles/degree viewed at the human's near point (approximately 12.5 cm from the eye) would allow black dots 70 μm in diameter and separated by 70 μm to be resolved against a light background. An acuity of 5 cycles/degree seen at a chicken's near point (assumed to be approximately 5 cm) would allow a similarly presented line of dots 170 μm in diameter to be resolved. The two- or threefold better human acuity at the near point may reflect the action of the specialized fovea in humans. However, acuity falls rapidly as distance from the fovea increases; acuity has declined by 50% at 5° from the centre of the fovea, and declines by more than 90% at 30° (Coren *et al.*, 1979). Fowls also possess an area of high cone cell density (the area centralis) (Morris, 1982), which probably serves as a region specialized in discerning detail, though it is less specialized than a fovea. For a chick with its head erect and viewed from the side, the area centralis has two extensions, determined by mapping the subserving ganglion cells (Ehrlich, 1981). The central extension receives images from just above the central point of that eye's hemispheric field of view (the central field). The lateral extension extends from this, receiving images from a band running slightly downwards towards the beak (into the inferofrontal field). The central extension may be used for detailed imaging of objects in the upper visual field, such as potential predators, while the lateral extension may be used to image objects in the lower myopic field, such as food and small prey (Ehrlich, 1981). The limited reduction in ganglion cell density with increasing eccentricity from this region implies a less severe reduction in acuity than that encountered in humans. Also, due to the shape of the fowl's eyeball, which is flattened in comparison with the human's, all images are equally well focused upon the retina. This is dissimilar to humans, where images focused on the fovea cause the rest of the field of view to become defocused (King-Smith, 1971). This may mean that, although the maximum spatial acuity is very much higher for humans than for fowls, the mean spatial acuity around the whole field of view may be similar for the two species or even better for chickens.

Preference tests

Preference testing provides an important tool in animal welfare research and gives insight into what the animals want in a given situation and with given resources. However, there are several important features that may affect the preferences of the animals and hence may affect the results and applicability of a preference test. First, it is important to consider whether it is relevant to test animals singly or in a social group. Broilers are social animals and testing them singly may not only cause fear and distress due to separation from the group, but may also yield different results from group testing. It is important to consider which method would yield the most valid results for the purpose of the study. However, the social nature of broilers also causes birds in a group to affect each other's choices or preferences for resources, and hence the results of individual birds are not independent of those of other birds. In these circumstances the group should be the experimental/analysed unit rather than the individual. Thus, testing the preferences of broiler chickens in a group will necessitate the use of more birds than testing them individually. Secondly, the previous experience of an animal will probably affect its preferences. In poultry, the familiar resource is often preferred initially, although this preference may change with time and experience of other resources. It may therefore be important to ensure that all the resources are equally familiar to the broilers before enabling them to choose between them. In terms of lighting, this may be done by rearing the birds from hatch in a light environment that alters between the different options given later. For example, Davis *et al.* (1999) reared broiler chicks in a room where the light environment altered between the four light intensities tested at a later date, thereby ensuring equal experience of all the options. Consideration may also be given to the time or order of the exposure to the options, as this could potentially influence the choice made in a later preference test. The birds may need to be familiar with the test apparatus, since the preferences for light environments may otherwise be confounded with exploration of the test apparatus itself. Thirdly, one may consider assessing the initial choices made by the animal or the chronic choices, for which the options are available to the animal over a longer period of time. In a chronic preference experiment, the animals live in the test apparatus during the test and it is thus possible to study whether the preferences change with the diurnal rhythm or with the age of the birds. A combination of the above considerations was used by both Davies *et al.* (1999) and Kristensen and colleagues (H. Kristensen, N. Prescott, G. Perry, J. Ladewig and C. Wathes, in preparation), whereby groups of broiler chickens were given a choice of four different, equally familiar light environments (light intensities, Davies *et al.*; light colours, Kristensen *et al.*) over a period of 6–10 days at two ages (1–2 and 5–6 weeks). In both experiments, the broilers chose differently at the two ages, suggesting a shift in preference for both light intensity and light colour between the beginning and the end of the growing period in broilers. A summary of the results of a variety of preference tests is shown in the Table 9.3.

Some of the limitations of preference tests can be overcome by imposing a price on choosing a particular treatment. In this way, the cost that an animal is prepared to pay is related to the strength of motivation to gain some reward. Few motivation tests with broilers and lighting have been conducted though. In one, Savory and Duncan (1982/83) showed that broilers would work for a short period of darkness whilst

Table 9.3. Results of a variety of light preference tests in broiler chickens.

Choice offered	Preference	Author
Gradient of illuminance	Preferred bright light (20–25 lx) at 1 day old, declined to 10–15 lx by 14 days old	Alsam and Wathes, 1991
Red, green, blue or white light	Preferred green and blue in preference to red or white	Prayitno *et al.*, 1997
20 or 0.05 lx environments	Preferred 20 lx when young but increasing use of 0.05 lx after 4 weeks old	Berk, 1995
6, 20, 60 or 200 lx environments (with prior experience)	Preferred to rest in 200 lx at 2 weeks and 6 lx at 6 weeks. Active behaviours performed preferentially in 200 lx	Davis *et al.*, 1999
20 or < 1 lx environments (no prior experience)	Up to slaughter (7 weeks) spent 96% of their time in 20 lx	Novell *et al.* (unpublished data)
Choice of incandescent, spectral sensitivity match, warm white or full-spectrum fluorescent,	No preference at 1 week old but at 5 weeks old preferred daylight and warm white fluorescent	Kristensen *et al.*, 2002
Operant task to turn lights on and off	Preferred light to dark, not prepared to work hard for dark	Savory and Duncan, 1982/83

Novell and colleagues (E. Novell, N. Prescott and C. Wathes, unpublished data) showed that broilers were willing to go without food for more than 4 hours rather than venture into a dark area containing a food source.

Lighting and fear

Low light levels and blue lights are commonly used to reduce fear and distress during handling; for example, during the depopulation of broiler sheds or during 'hanging-on' at the processing plant. Perkins and colleagues (G. Perkins, N. Prescott, H. Kristensen and C. Wathes, unpublished data) found some evidence that both blue light and dim light reduced fearfulness in broilers in both an open-field and an approach test. The mechanism by which these manipulations reduce fear is not known, although it may well be through the visual channel. For example, these manipulations may obscure threatening features of the environment or have some causative effect on the bird's behaviour, shifting the repertoire away from alert and fear behaviours to resting behaviour. The photoperiod also affects stress and fear responses. Zulkifli *et al.* (1998) showed that broilers reared under continuous light had a higher heterophil and lymphocyte ratio and spent longer in tonic immobility, both indicative of stress and fear, than birds reared under a photoperiod of 12 hours light: 12 hours darkness.

Lighting and eye development

Some types of lighting can cause an abnormal development of the eyes in poultry. Several of these abnormalities may impose welfare problems for the birds, as they may cause pain or affect the visual abilities of the animals. In particular, inappropriate lighting may cause bupthalmos, glaucoma, myopia and/or retinal degeneration in poultry (e.g. Cummings *et al.*, 1986; Li *et al.*, 1995). Bupthalmos is a deviation from normal in the dimensions and weight of the eye and is usually measured post mortem. Glaucoma is described as an elevated intra-ocular pressure, which is manifested by the accumulation of fluid, resulting in enlargement of the eyeball and an increase in the weight of the eyeball (e.g. Shivaprasad, 1999). The increased intra-ocular pressure is easily measured in live animals, whereas the increase in eyeball weight can be measured during a post-mortem examination, in which flattening of the cornea may also be found in response to some light environments. Continuous lighting with a very short or no dark period, as well as very dim lighting, appears to cause buphthalmia and/or glaucoma in turkeys (Ashton *et al.*, 1973; Davis *et al.*, 1986), laying hens (e.g. Jensen and Matson, 1957) and broiler chickens (e.g. Cummings *et al.*, 1986; Li *et al.*, 1995). Since broilers are commonly reared in dim, near-continuous lighting, it is possible that a large number of birds in commercial production may suffer from light-induced changes in eye morphology. The light-induced changes in eye morphology may reduce the welfare of broiler chickens in several ways. First, it is known that glaucoma causes discomfort and, later, pain in humans and other mammals. It would be obvious to presume that broilers with severe glaucoma would be in pain, but reports of behavioural observations of birds with glaucoma are still sparse. In addition to potential discomfort and pain associated with light-induced glaucoma and buphthalmos, the changes in eye morphology may also affect the visual abilities of broilers (e.g. Lauber, 1987), which may have negative effects on welfare if the birds fail to identify the identity or intent of other birds or orientate themselves within their environment.

Lighting, lameness and mortality

Lighting is a strong exogenous factor in the control of many behavioural and physiological processes. Hence, the light environment can affect lameness and mortality through many potential routes. First, light intensity and colour and the photoperiodic regime can affect the physical activity of broiler chickens (see review by Lewis and Morris, 1998). Since increased physical activity can stimulate bone development, it may improve the leg health of broiler chickens. There is much evidence that photoperiods providing the broilers with an uninterrupted dark period in every 24-hour period can improve the leg health of the birds. This could be due to the secretion of certain growth hormones during specific phases of sleep, as well as the effects on activity described above. Aspects of lighting may also affect lameness indirectly via properties of the litter quality, since the particular type of lighting may cause birds to be more or less active (e.g. foraging; scratching in the litter can improve the litter quality, as suggested by Bizeray *et al.* (2002)). Patches of light in a poultry house may also cause crowding of chickens, thereby reducing the litter quality in those areas. Some photoperiodic regimes have been shown to affect mortality due to ascites (e.g. Buyse *et al.*, 1996); these are mainly intermittent light regimes in which several light and dark periods are applied within each 24-hour period. Mortality may also be caused by severely lame birds dying from starvation or dehydration if not found and culled promptly by the producer. Lighting has also been shown to affect stress responses in chickens, and since stress is related to immune function, suboptimal lighting may lead to higher susceptibility to common diseases in poultry houses.

Recommendations and Conclusions

A wholly artificial light environment should allow animals to see well enough to carry out critical visual tasks and prevent disturbance of the functional development of vision. Clearly, it must also allow efficient production. One reason for providing dim, artificial environments is to control pecking damage and cannibalism, but they also allow the application of very long light periods that have no natural equivalent. However, our impression of modern broiler farming is that feather pecking and cannibalism are rare and the use of unnatural photoperiods is becoming less common. Given these trends, the case for providing wholly artificial lighting in broiler houses is diminished, and windowed houses (perhaps with supplementary lighting to modify the natural photoperiod where desirable) become a reasonable alternative. These would still allow all the advantages of artificial environments in terms of the thermal and aerial environment but avoid the potential welfare consequences of providing dim, artificial lighting. We strongly recommend that this type of lighting regime be evaluated. The increasingly favoured free-range systems also avoid these welfare difficulties, although the artificial environment provided in the first few weeks of birds destined for these systems is still a cause for concern.

For artificial light environments, we can say with some certainty that bright light environments with relatively long dark periods would accomodate broilers' preferences and promote the normal development of the eye. We are therefore fully supportive of the recent trends in legislation and various guidelines that recommend higher light levels and longer dark periods. However, a consistent and robust method for measuring the light environment is necessary in order to underpin these positive trends. We can also say that low-frequency fluorescent lights are unlikely to affect a broiler's welfare. Promising recent work indicates that broilers have preferences for warm white and full-spectrum fluorescent lighting over incandescent lighting and that blue lighting may reduce stress during handling. Full-spectrum lighting may also have some benefits on mating behaviour in broiler breeders, although the economic and welfare consequences are unclear.

Acknowledgements

Much of this work was funded through the Silsoe Research Institute's Competitive Strategic Grant from the BBSRC. HHK is currently in receipt of a PhD studentship grant from the Royal Veterinary and Agricultural University, Copenhagen. We would like to thank our colleagues N. Davis, E. Jones, G. Perkins and E. Novell, whose work we have quoted liberally, and many others who have worked with us, provided resources and ideas.

References

Alsam, H. and Wathes, C.M. (1991) Conjoint preferences of chicks for heat and light intensity. *British Poultry Science* 32, 899–916.

Ashton, W.L.G., Pattison, M. and Barnett, K.C. (1973) Light-induced eye abnormalities in turkeys and the turkey blindness syndrome. *Research In Veterinary Science* 14, 424–426.

Barber, C.L., Prescott, N.B., LeSeuer, C. and Perry, G.C. (2003) Characteristics of the light environment in commercial duckling and turkey poult houses and the preferences of these species for different illuminances in relation to age and behaviour. *Animal Welfare* (in press).

Berk, J. (1995) Light-choice by broilers. In: Rutter, S.M., Rushen, J., Randle, H.D. and Eddison, J.C. (eds) *Proceedings of the 29th International Congress of the International Society for Applied Ethology.* Universities Federation for Animal Welfare, Potters Bar, UK.

Bizeray, D., Leterrier, C., Constantin, P., Le Pape, G. and Faure, J.M. (2002) Typology of activity bouts and effect of fearfulness on behaviour in meat-type chickens. *Behavioural Processes* 58, 45–55.

Blair, R., Newberry, R.C. and Gardiner, E.E. (1993) Effects of lighting patterns and dietary tryptophan supplementation on growth and mortality in broilers. *Poultry Science* 72, 4955–4962.

Blokhuis, H.J. (1983) The relevance of sleep in poultry. *World's Poultry Science Journal* 39, 333–337.

Boshouwers, F.M.G. and Nicaise, E. (1992) Responses of broiler chickens to high-frequency and low-frequency fluorescent light. *British Poultry Science* 33, 711–717.

Bowmaker, J.K. and Knowles, A. (1977) The visual pigments and oil droplets of the chicken retina. *Vision Research* 17, 7557–7564.

British Standard (1985) BS 8206. Lighting for buildings. Part 1. Code of practice for artificial lighting. British Standards Institution, London.

Brundrett, G.W. (1974) Human sensitivity to flicker. *Lighting Research and Technology* 6, 127–143.

Buyse, J., Simons, P.C.M., Boshouwers, F.M.G. and Decuypere, E. (1996) Effect of intermittent lighting, light intensity and source on the performance and welfare of broilers. *World's Poultry Science Journal* 52, 1211–1230.

Charles, R.G., Robinson, F.E., Hardin, R.T., Yu, M.W., Feddes, J. and Classen, H.L. (1992) Growth, body composition and plasma androgen concentration of male broiler chickens subjected to different regimes of photoperiod and light intensity. *Poultry Science* 71, 1595–1605.

Classen, H.L. (1991) Increasing photoperiod length provides better broiler health. *Poultry Digest* 50, 142–148.

Coren, S., Porac, C. and Ward, L.M. (1979) *Sensation and Perception.* Academic Press, London.

Cummings, T.S., French, J.F. and Fletcher, O.J. (1986) Ophthalmopathy in a broiler breeder flock reared in dark-out housing. *Avian Diseases* 30, 609–612.

Dartnall, H.J.A., Bowmaker, J.K. and Mollon, J.D. (1983) Human visual pigments: microspectrophotometric results from the eyes of seven persons. *Proceedings of the Royal Society of London Series B Biological Sciences* 220, 115–130.

Davis, G.S., Siopes, T.D., Peiffer, R.L. and Cook, C. (1986) Morphological changes induced by photoperiod in eyes of turkey poults. *American Journal of Veterinary Research* 47, 953–955.

Davis, N.J., Prescott, N.B., Savory, C.J. and Wathes, C.M. (1999) Preferences of growing fowls for different light intensities in relation to age strain and behaviour. *Animal Welfare* 8, 193–203.

DEFRA (2002) Meat chickens and breeding chickens: code of recommendations for the welfare of livestock. Department for Environment, Food and Rural Affairs, London.

De Lange, H. (1958) Research into the dynamic nature of the human fovea–cortex system with intermittent and modulated light. *Journal of the Ophthalmological Society of America* 48, 777–784.

DeMello, L.R., Foster, T.M. and Temple, W. (1992) Discriminative performance of the domestic hen in a visual acuity task. *Journal of the Experimental Analysis of Behaviour* 58, 147–157.

Ehrlich, D. (1981) Regional specialization of the chick retina as revealed by the size and density of neurones in the ganglion cell layer. *Journal of Comparative Neurology* 195, 643–657.

Ekstrand, C., Algers, B. and Svedberg, J. (1997) Rearing condition and foot-pad dermatitis in Swedish broiler chickens. *Preventive Veterinary Medicine* 31, 167–174.

FAWC (1992) Report on the welfare of broiler chickens. Farm Animal Welfare Council, Ministry of Agriculture, Fisheries and Food, London.

Gordon, S.H. (1994) Effects of daylength and increasing daylength programmes on broiler welfare and performance. *World's Poultry Science Journal* 50, 269–282.

Gordon, S.H. and Tucker, S.A. (1995) Effects of daylength on broiler welfare [abstract]. *British Poultry Science* 36, 844–845.

Gordon, S.H. and Tucker, S.A. (1997) Effect of light programme on broiler mortality, leg health and performance. *British Poultry Science* 38, s6–s7.

Jensen, L.S. and Matson, W.E. (1957) Enlargement of avian eye by subjecting chicks to continuous incandescent illumination. *Science* 125, 741.

Johnsen, H.M. (1914) Visual pattern-discrimination in the vertebrates. *Journal of Animal Behaviour* 6, 169–188.

King-Smith, P.E. (1971) Special senses. In: Bell, D.J. and Freeman, B.M. (eds) *Physiology and Biochemistry of the Domestic Fowl*, Volume 2. Academic Press, London.

Kristensen, H.H. (1999) Lysprogrammers paavirkning af slagtekyllingers adfaerd og velfaerd. Dyrenes Beskyttelse, Denmark.

Kristensen, H.H., Prescott, N.B., Ladewig, J., Perry, G.B., Johnsen, P.F. and Wathes, C.M. (2002) Light quality preferences of broiler chickens. In: *Proceedings of the Spring Meeting of the World's Poultry Science Association, UK Branch, 9–10 April 2002, York, UK.*

Lauber, J.K. (1987) Review: light-induced avian glaucoma as an animal model for human primary glaucoma. *Journal of Ocular Pharmacology* 3, 77–100.

Lewis, P.D. and Morris, T.R. (1998) Responses of domestic poultry to various light sources. *World's Poultry Science Journal* 54, 72–75.

Li, T., Troilo, D., Glasser, A. and Howland, H.C. (1995) Constant light produces severe corneal flattening and hyperopia in chickens. *Vision Research* 35, 1203–1209.

Morris, V.B. (1982) An afoveate area centralis in the chick retina. *Journal of Comparative Neurology* 210, 198–203.

Nuboer, J.F.W., Coemans, M.A.J.M. and Vos, J.J. (1992a) Artificial lighting in poultry houses: are photometric units appropriate for describing illumination intensities. *British Poultry Science* 33, 135–140.

Nuboer, J.F.W., Coemans, M.A.J.M. and Vos, J.J. (1992b) Artificial lighting in poultry houses: do hens perceive the modulation of fluorescent lamps as flicker. *British Poultry Science* 33, 1231–1233.

Osorio, D., Vorobyev, M. and Jones, C.D. (1999) Colour vision of domestic chicks. *Journal of Experimental Biology* 202, 2951–2959.

Over, R. and Moore, D. (1981) Spatial acuity of the chicken. *Brain Research* 221, 424–426.

Prayitno, D.S., Phillips, C.J.C. and Stokes, D.K. (1997) The effects of colour and intensity of light on behaviour and leg disorders in broiler chickens. *Poultry Science* 76, 1674–1681.

Prescott, N.B. and Wathes, C.M. (1999) Spectral sensitivity of the domestic fowl (*Gallus g. domesticus*). *British Poultry Science* 40, 332–339.

Prescott, N.B., Jarvis, J.R. and Wathes, C.M. (2003) Light, vision and the welfare of poultry. *Animal Welfare* 12, 269–288.

Rogers, L.J. (1982) Light experience and asymmetry of brain function in chickens. *Nature* 297, 223–225.

Rogers, L.J. (1995) *The Development of Brain and Behaviour in the Chicken.* CAB International, Wallingford, UK.

RSPCA (1999) Welfare standards for chickens. Royal Society for the Prevention of Cruelty to Animals, Horsham, UK.

Savory, C.J. (1976) Broiler growth and feeding behaviour in three different lighting regimes. *British Poultry Science* 17, 557–560.

Savory, C.J. and Duncan, I.J.H. (1982/83) Voluntary regulation of lighting by domestic fowl in Skinner boxes. *Applied Animal Ethology* 9, 73–81.

Sherwin, C.M. (1999) Domestic turkeys are not averse to compact fluorescent lighting. *Applied Animal Behaviour Science* 64, 47–55.

Shivaprasad, H.L. (1999) Poultry ophthalmology. In: Gelatt, K.N. (ed.) *Veterinary Ophthalmology,*

3rd edn. Williams and Wilkins, Philadelphia, Pennsylvania, pp. 1177–1185.

Simmons, P.C.M. (1982) Effects of lighting regimes on twisted legs, feed conversion and growth of broiler chickens [abstract]. *Poultry Science* 61, 1546.

Stewart, M.G., Rogers, L.J., Davies, H.A. and Bolden, S.W. (1992) Structural asymmetry in the thalamofugal visual projections in 2-day-old chick is correlated with a hemispheric difference in synaptic density in the hyperstriatum accessorium. *Brain Research* 585, 381–385.

Widowski, T.M. and Duncan, I.J.H. (1996) Laying hens do not have a preference for high-frequency versus low-frequency compact fluorescent light sources. *Canadian Journal of Animal Science* 76, 177–181.

Widowski, T.M., Keeling, L.J. and Duncan, I.J.H. (1992) The preferences of hens for compact fluorescent over incandescent lighting. *Canadian Journal of Animal Science* 72, 203–211.

Wortel, J.F., Rugenbrink, H. and Nuboer, J.F.W. (1987) The photopic spectral sensitivity of the dorsal and ventral retinae of the chicken. *Journal of Comparative Physiology* 160, 151–154.

Yoshizawa, T. (1992) The road to colour vision: structure, evolution and function of chicken and gecko visual pigments. *Photochemistry and Photobiology* 56, 859–867.

Zulkifli, I., Raseded, A., Syaadah, O.H. and Morma, M.T.C. (1998) Daylength effects on stress and fear responses in broiler chickens. *Asian-Australian Journal of Animal Science* 11, 751–754.

10 Air Hygiene

C.M. WATHES

Silsoe Research Institute, Bedford, UK

Introduction

We now know from numerous surveys that the air within a broiler house seethes with a dense miasma of gases and particulate matter. What is remarkable is that most of these aerial pollutants appear to be of little consequence to broiler welfare, but some do pose a hazard: certainly, the progenitor species of the domestic fowl was not exposed to such high concentrations of so many gases, microorganisms and dusts. There are additional reasons why this miasma should be cleared from the building's atmosphere, possibly with benefits for broiler chickens. When faced with the question of an acceptable standard of air hygiene for broiler chickens, the farmer and his advisers must consider the mechanisms by which poor air hygiene affects the health and welfare of his chickens because this information will produce guidelines or limits for exposure. It would be fortunate indeed if one pollutant could serve as an index of air quality because this would simplify the management of air quality.

Three topics are covered in this brief review of air hygiene in broiler houses. First, a picture is painted of the common aerial pollutants in English broiler houses. This demonstrates the complexity of the miasma, which explains, in part, why progress has been so slow in determining standards of air hygiene. Secondly, the mechanisms by which aerial pollutants affect the health and welfare of broiler chickens are considered. Finally, the minimum standards of air quality for broiler chickens are discussed. The importance of these topics is demonstrated in the Welfare of Farmed Animals (England) Regulations 2000 (SI 2000 No. 1870), which states that 'air circulation, dust levels, temperature, relative air humidity and gas concentrations shall be kept within limits which are not harmful to the animals'. The associated Welfare Code (DEFRA, 2002a) is more specific and advises that 'air quality, including dust level and concentrations of carbon dioxide, carbon monoxide and ammonia, should be controlled and kept within limits where the welfare of the birds is not negatively affected'. What these limits are is the crucial question.

Guaranteeing an acceptable standard of air hygiene in broiler houses is undoubtedly one of the Ten Commandments of animal husbandry.

The Natural History of Aerial Pollutants in Broiler Houses

Once a flock of broiler chickens is housed on litter at a high stocking density in a building with a slow air exchange rate, it is inevitable that aerial pollutants will build up to high, unnatural concentrations. The main sources are the feed, the litter and the chickens themselves. There are, of course, differences in the types and concentrations of aerial pollutants in a broiler house according to the building's design and management. Either indirectly or directly, season, disease, nutrition and the ambient environment affect the strength of the various sources of aerial pollutants and their clearance from the atmosphere of a broiler house.

Measurement of air quality in a broiler house normally requires sophisticated, expensive equipment if reliable data are to be gathered and is usually restricted to research organizations (Wathes, 1995). An estimate of the bird's exposure to gaseous pollutants such as ammonia or carbon dioxide can be gained using low-cost gas diffusion tubes (e.g. Draeger tubes). These can be deployed at a few locations and samples taken, typically over 8 h. Continuous monitoring of gases is not affordable on broiler farms with current sensors, though some technical development is under way. Low-cost means of monitoring airborne dust in poultry houses are not available, though inspection of settled dust on surfaces will indicate the dust concentration.

Table 10.1 shows the results of a large northern European study of aerial pollutants in livestock buildings in which common methods were employed. Measurements were made in four typical buildings in winter and summer in each country over 24 h at six locations for pollutant concentration and also at one location for pollutant emission, i.e. the product of concentration and ventilation rate at the building's exhaust (for an overview see Wathes *et al.*, 1998). The results for laying hens are shown for comparison; they demonstrate that air quality is poorest in broiler houses. The buildings included in the survey were representative of typical commercial practice in the mid 1990s and it is conceivable that changes in environmental management and husbandry have improved air quality since these measurements were made. However, while a survey of a similar scale is under way in the USA (H. Xin, Iowa State University, personal communication), none is planned in Europe.

The mean concentrations given in Table 10.1 are the averages over 24 hours. The short-term maximum concentration of ammonia reached 56 p.p.m. in the English broiler houses and 50, 40 and 43 p.p.m. in the Dutch, Danish and German houses (Groot Koerkamp *et al.*, 1998). The method used to measure dust concentration precluded an estimate of its hourly maxima. As indicated later in this chapter (see Air Quality Standards in Broiler Houses; and Table 10.5), occupational exposure limits for humans recognize the importance of acute exposure over 15 minutes, but this distinction has yet to be made for livestock, so the interpretation of hourly maxima is unclear. Breakdown of the outer cell wall of Gram-negative bacteria gives rise to endotoxins, and high concentrations of these were recorded.

Table 10.1. Mean concentrations and emissions of aerial pollutants in chicken housing in northern Europe. (From Groot Koerkamp et al., 1998; Seedorf et al., 1998; Takai et al., 1998.)

	Mean concentration					Mean emission rate		
	Inhalable dust (mg/m³)	Respirable dust (mg/m³)	Inhalable endotoxin (ng/m³)	Respirable endotoxin (ng/m³)	Ammonia (p.p.m.)	Inhalable dust emission (mg/h per 500 kg)	Respirable dust emission (mg/h per 500 kg)	Ammonia emission (mg/h per 500 kg)
Laying hens in cages (n = 26)								
Denmark	1.6	0.23	85	13	6.1	642	82	2,160
England	1.5	0.21	430	48	11.9	872	161	9,316
Germany	1.0	0.03	21	2	1.6	633	24	602
Netherlands	0.8	0.09	20	2	5.9	398	46	1,624
Laying hens in percheries (n = 22)								
Denmark	4.9	0.92	148	19	25.2	3,131	637	10,892
England	2.2	0.35	1,978	139	8.3	1,771	467	7,392
Germany	–	–	–	–	–	–	–	–
Netherlands	8.9	1.26	268	15	29.6	4,340	682	9,455
Broilers (n = 33)								
Denmark	3.8	0.42	70	6	8.0	1,856	245	2,208
England	9.9	1.14	128	42	27.1	6,218	706	8,294
Germany	4.5	0.63	6,000	239	20.8	2,805	394	7,499
Netherlands	10.4	1.05	381	41	11.2	4,984	725	4,179

The age of the broilers was about 4 weeks.

Table 10.2. Microorganisms identified in the air of a small broiler house, in newly opened food, in clean wood-shavings and in bird lungs. (From Madelin and Wathes, 1989.)

Microorganism	Air[†]	Food	Wood-shavings	Lungs[‡]
Fungi				
Acremonium sp.	1	+	−	0
Acremonium strictum	0	+	−	0
Aspergillus spp. (total)	7	+	−	9
A. amstelodami	1	−	−	0
A. candidus	4	−	−	5
A. echinulatus	1	−	−	0
A. flavipes	1	−	−	0
A. flavus	7	+	−	0
A. fumigatus	3	−	−	1
A. glaucus	6	+	−	0
A. nidulans	1	−	−	0
A. repens	7	+	−	9
A. sydowii	1	−	−	0
A. terreus	1	−	−	1
Aureobasidium pullulans	1	−	−	0
Basidiomycetes with clamps	6	−	−	6
Botrytis cinerea	1	−	−	0
Candida fennica	5	−	−	0
Cladosporium spp.	5	−	−	3
Debaryomyces hansenii	7	−	−	2
Fusarium sp.	2	−	−	0
Malbranchea sulfurea	0	−	−	0
Mucor racemosus	3	+	−	0
Paecilomyces variotii	5	−	+	5
Penicillium spp. (total)	8	+	+	7
P. brevicompactum	8			
P. chrysogenum	8			
P. corylophilum	8	Not identified to species		
P. granulatum	0			
P. purpurogenum	3			
Rhizopus nigricans	3	−	+	1
Rhizopus rhisopodiformis	6	−	+	1
Rhodotorula sp.	2	+	−	2
Scopulariopsis brevicaulis	8	+	−	4
Syncephalastrum racemosum	5	−	−	0
Stemphylium sp.	1	−	−	0
Verticillium sp.	0	+	−	0
Actinomycetes				
Streptomycete (white)	1	+	−	1
Thermoactinomyces thalpophilus	4		Not tested	
Thermoactinomyces vulgaris	4		Not tested	
Bacteria				
Staphylococcus spp. (total)	7	+	−	9
S. hominis	7			5
S. saprophyticus	7	Not identified to species		8
S. xylosus	7			1

Table 10.2. *Continued.*

Microorganism	Air[†]	Food	Wood-shavings	Lungs[‡]
Micrococcus sp.	2	–	–	6
Bacillus sp. (Gram-positive)	0	–	+	0

[†]Number of weeks, out of a total of 8, in which the specified microorganism was recovered from the air; + = present; – = not present; [‡]number of birds, out of a total of 10 at 8 weeks of age, from which the specified microorganism was isolated from the lungs.

Airborne microorganisms constitute the third category of aerial pollutants in a broiler house, and these have been catalogued by several authors. Specific pathogens (for example, Newcastle disease virus) can be isolated from the air but only when there is a disease outbreak. Table 10.2 shows an example of this catalogue in terms of the long list of bacteria, viruses and fungi (in the form of fungal propagules) that have been recovered from the air of a broiler house (Madelin and Wathes, 1989). In this study, day-old cockerels were kept in duplicate groups of 250 on wood shavings litter. Although the rooms were small in comparison with a commercial broiler house, there is no reason to believe that the findings are atypical. Both qualitative and quantitative measurements of bioaerosol concentration and type were recorded weekly within the birds' breathing zone, i.e. approximately 10–15 cm above the floor. (A bioaerosol is an aerosol comprising particles of biological origin or activity, which may affect living things through its infectivity, allergenicity or toxicity, or through pharmacological or other processes. Particle sizes may range from aerodynamic diameters of approximately 0.5 to 100 μm (Cox and Wathes, 1995).) The numerical concentration of dust particles was measured with an optical particle counter. Airborne fungal and actinomycete propagules were collected with a six-stage Andersen sampler and airborne bacteria with a May three-stage liquid impinger. Dust particles were examined microscopically from samples collected with a May cascade impactor.

Concentrations of the common aerial pollutants were high in comparison with fresh air (Fig. 10.1), even though the rooms were ventilated at a minimum rate to maintain an optimum air temperature (rather than air quality). The concentration of respirable dust particles (diameter 0.5–5.0 μm) was between 10^7 and $10^8/m^3$. The majority of these particles were skin squames with a smaller proportion of down or feather fragments, food and faecal debris. The peak concentration of viable fungal propagules was approximately 10^6 colony-forming units per m^3, with an even higher concentration of viable respirable bacteria. The range of airborne fungal species was large (Table 10.1) and some fungal propagules were isolated from the lungs.

Thus, typical broiler chicks will be exposed continuously to a dense but invisible cloud of gases, dust and microorganisms. Assuming a minute volume of 760 ml (D. McKeegan, personal communication) and the concentrations of aerial pollutants shown in Fig. 10.2, a 1.6 kg chicken in a typical English broiler house will inhale each day a substantial burden of aerial pollutants. The equivalent dose inhaled by a woodland chicken would be considerably less. Furthermore, the actual quantities inhaled by the housed chicken may be an underestimate since it is exposed to higher

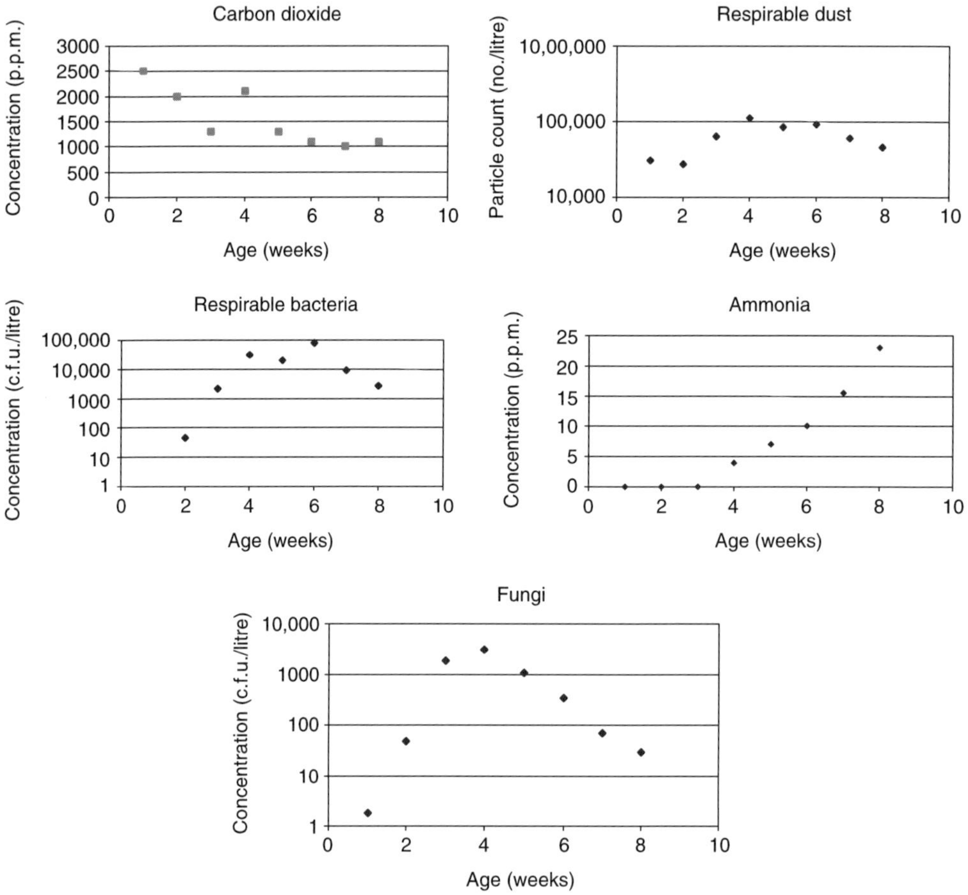

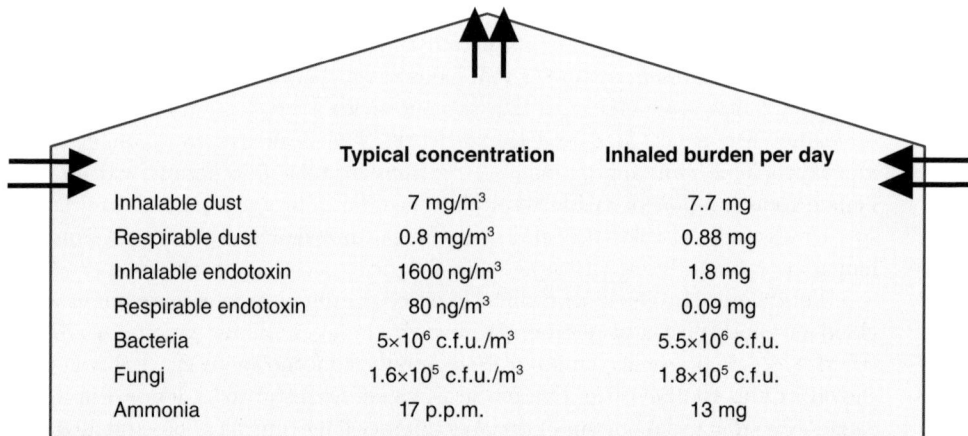

Fig. 10.1. Concentrations of ammonia and other aerial pollutants in a broiler house alter as the birds grow. (From Madelin and Wathes, 1989.)

	Typical concentration	Inhaled burden per day
Inhalable dust	7 mg/m³	7.7 mg
Respirable dust	0.8 mg/m³	0.88 mg
Inhalable endotoxin	1600 ng/m³	1.8 mg
Respirable endotoxin	80 ng/m³	0.09 mg
Bacteria	5×10⁶ c.f.u./m³	5.5×10⁶ c.f.u.
Fungi	1.6×10⁵ c.f.u./m³	1.8×10⁵ c.f.u.
Ammonia	17 p.p.m.	13 mg

Fig. 10.2. Typical concentrations and burdens of aerial pollutants in a broiler house, calculated assuming a minute volume of 760 ml for a 1.6 kg broiler chicken.

concentrations of pollutants than those recorded when it is scratching about in the litter. Few scientists have had the patience to analyse the composition of poultry house dust in terms of its physical, chemical and microbiological properties or to identify the numerous odorants that are present (Hartung, 1988). The above description of the inhaled burden of aerial pollutants on a broiler chicken is therefore necessarily crude and incomplete.

Impact of Aerial Pollutants on the Welfare of Broiler Chickens

There are potentially three means by which exposure to aerial pollutants in a broiler house could affect the welfare of broiler chickens adversely. First, veterinarians and farmers have long suspected that chronic exposure to certain pollutants increases the incidence and severity of multifactorial respiratory disease. Secondly, aerial pollutants could affect olfaction, a sense that is involved in conspecific recognition, food selection and the detection of alarm and other pheromones (Jones and Roper, 1997). Finally, chronic exposure may affect various organs and systems in the chicken by mechanisms that are unknown. Nevertheless, any adverse effects will be integrated by the chicken and may be manifested in its behaviour in the presence of aerial pollutants, either immediately or shortly after exposure has ceased.

The evidence that chronic exposure of poultry to aerial pollutants affects their health is based upon studies mainly carried out in the 1960s and 1970s with ammonia. Anderson *et al.* (1964) showed that exposure to relatively high concentrations of ammonia (20 p.p.m. for 72 h or 50 p.p.m. for 48 h) increased susceptibility to respiratory infection in chickens with Newcastle disease virus. However, no effect of ammonia on susceptibility was found for Marek's disease in chickens (Brewer and Koon, 1973) and air sacculitis in turkeys (Wolfe *et al.*, 1968). The physiological mechanisms by which ammonia could affect susceptibility are by damage to pulmonary ultrastructure (Oyetunde *et al.*, 1978) and/or a direct effect upon microbial virulence, as has been shown by Hamilton *et al.* (1998) for *Pasteurella multocida*, which is a pathogen involved in progressive atrophic rhinitis in pigs. The evidence reviewed by Kristensen and Wathes (2000) suggests that ammonia exposure (60–70 p.p.m.) may cause keratoconjunctivitis, i.e. inflammation of the cornea and conjunctiva (Valentine, 1964; Lille, 1970; Quarles and Kling, 1974). Clearly, there is a need to repeat these early studies with modern strains of poultry using representative concentrations of aerial pollutants. This can be best achieved by both epidemiological and laboratory studies to demonstrate, once and for all, whether or not aerial pollutants are involved in the aetiology of respiratory disease in poultry.

Until comparatively recently, doubt was expressed about the importance of olfaction in birds. Jones and Roper (1997) concluded that the domestic fowl has a powerful sense of olfaction, which is involved in environmental familiarization, elicitation and fear responses, feeding and drinking and the avoidance of noxious substances. Using electrophysiological techniques, McKeegan *et al.* (2002a,b) have quantified the stimulus–response characteristics of both olfactory bulb neurones and nasal trigeminal receptors for ammonia in the fowl. The median response threshold for olfactory bulb neurones was 3.75 p.p.m.; that is, at the lower end of the concentrations found in broiler houses (Table 10.1). The threshold was much higher for the trigeminal nociceptors (median = 2320 p.p.m.), which corresponds with the function of this organ in detecting noxious stimuli. The corollary of these studies is that broiler chickens can readily sense ammonia in broiler houses but it is probably not painful; their olfactory acuity for other odorants is unknown but likely to be significant.

There are few studies of the effects of aerial pollutants on olfaction in poultry. Acute exposure to certain pollutants could mask the perception of biologically relevant odorants while chronic exposure could damage olfactory receptors, thereby affecting olfactory perception. These are the olfactory equivalent of an animal seeing a dimly lit world through dark glasses. Recently, McKeegan and colleagues (D.E.F. McKeegan, T.G.M. Demmers, R.B. Jones, C.M. Wathes and M.J. Gentle, unpublished) tested the latter hypothesis in hens exposed to 20 p.p.m. ammonia for 12 days before their olfactory acuity for ammonia was measured physiologically. Ammonia exposure increased the detection threshold for ammonia. The effects on olfactory thresholds for other odorants are not known. No other studies of pollutants and poultry olfaction have been reported in the literature, but a lack of acute and chronic effects, with one exception, was shown with pigs in a set of experiments by Jones *et al.* (2000, 2001) and Kristensen *et al.* (2001).

The broiler chicken's behaviour in the presence of aerial pollutants can provide an insight into the combined effects of various physiological and pathological mechanisms. Morrison *et al.* (1993) studied the aversion of pigs and fowls to atmospheric ammonia (up to 60 p.p.m.) in acute (30 minutes to 12 hours) preference tests. They concluded that 'the level of ammonia commonly present in commercial buildings appears to be of no great consequence to the animals' perceived well-being'. Subsequently, a series of experiments on ammonia aversion was undertaken at Silsoe Research Institute with both pigs and fowls, and has been summarized by Wathes *et al.* (2002). Broiler chickens or adult laying hens were placed in a large preference chamber (Figs 10.3 and 10.4) and given a choice of atmospheres polluted with different concentrations of ammonia with a nominal maximum of 40 and 45 p.p.m. respectively; the control was fresh air with a nominal concentration of approximately 0 p.p.m. ammonia. Full details are given in the original reports (Kristensen *et al.*, 2000; Jones, 2002). Fresh air was significantly preferred to ammoniated atmospheres and both the duration and frequency of visits to ammoniated atmospheres declined with increasing ammonia concentration (Table 10.3). The most surprising observation was that the aversion to ammonia was not immediate but delayed, avoidance behaviour taking at least 15 min. This cannot be because ammonia was not detected at these concentrations (McKeegan *et al.*, 2002a). Jones (2002) subsequently showed that the aversion might be explained in terms of ammonia absorption through the respiratory tract and its effect on the acid–base balance of the blood.

Ammonia is an abundant pollutant in broiler houses and the only one whose impact on poultry welfare has been investigated in any detail; the topic was reviewed by Kristensen and Wathes in 2000. The 'five freedoms' of the Farm Animal Welfare Council provide a summary framework to assess the effects of ammonia on poultry welfare (Table 10.4). The majority of studies included in this review involved exposure to concentrations of ammonia exceeding 25 p.p.m., which had gross effects on disease and performance. The mean concentrations of ammonia ranged between 8 and 27 p.p.m. in the four European countries surveyed by Groot Koerkamp *et al.* (1998) (Table 10.1), but there is a need to study responses to lower concentrations

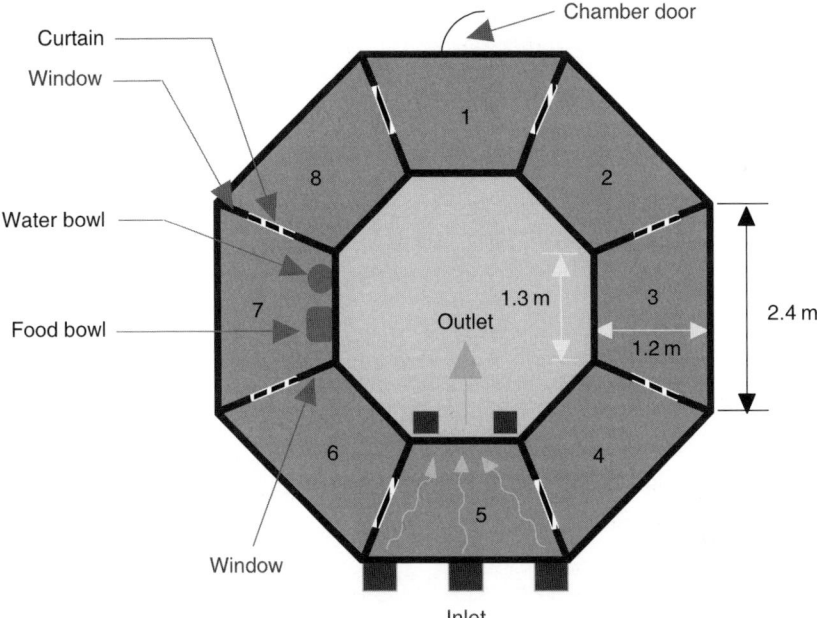

Fig. 10.3. Plan view of the environmental preference chamber.

of ammonia in the range 5–25 p.p.m. since this may be more representative of conditions in modern broiler houses. On the basis of the review by Kristensen and Wathes (2000) and the evidence given above, chronic exposure of broiler chickens to ammonia has an adverse effect on chicken welfare. There is insufficient evidence to draw a definitive conclusion about the other aerial pollutants in a broiler house.

Air Quality Standards in Broiler Houses

There are three criteria that may be used to set the standard for air quality in broiler houses but only one that relates directly to the birds' welfare as reviewed above. The second criterion concerns stockperson health. Epidemiological studies of the health of pig and poultry farmers have now established clear guidelines for occupational exposure to aerial pollutants (see review by Donham *et al.*, 2002). In the UK, the Health and Safety Commission sets statutory limits to control an employee's exposure to hazardous substances. The occupational exposure standard (OES) is set at a level at which there is no indication of risk to worker health. However, OES's are set for industry in general and no special hazard is recognized for poultry farmers, despite the reported prevalence of respiratory impairment of 8% in stockpeople working with laying hens (Whyte *et al.*, 1998). If this finding can be generalized to broiler stockpeople, it implies that the OES's are too high.

Table 10.5 shows the current guidelines for air quality in livestock houses based on human health; the recommendations of Donham *et al.* (2002) for swine health are also shown. For the former criterion, Donham concludes that the current limits are

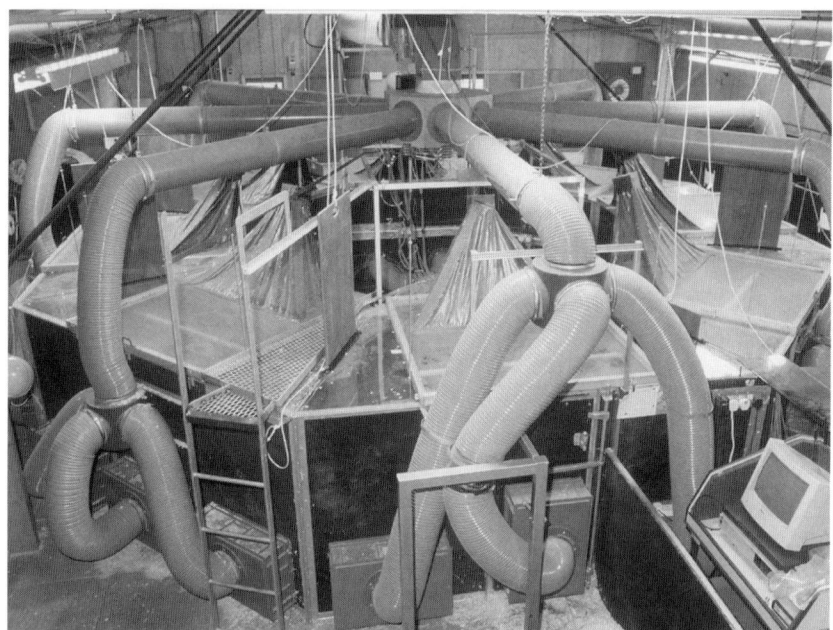

Fig. 10.4.   The environmental preference chamber at Silsoe Research Institute.

Table 10.3.   Mean (back-transformed) visit duration and occupancy by broiler chickens in ammoniated atmospheres. (From Jones, 2002.)

	Nominal ammonia concentration (p.p.m.)			
	0	10	20	40
Mean visit duration (min)	37	28	20	15
Mean occupancy (%)				
Bright light	37	41	15	6
Dim light	65	25	9	2

The lighting schedule was 12 h bright (102 lux), 4 h dim (11 lux), 4 h dark, 4 h dim.

set too high by a factor of between 3 and 4. His recommendations on the basis of swine health are similar to those for human health, which is a convenient marriage of interests.

The third criterion concerns the environmental impact of the emissions of aerial pollutants upon the countryside. This is demonstrated in various government regulations; for example, the EC Integrated Pollution, Prevention and Control Directive (IPPC), which requires large broiler farms (40,000 birds) to limit pollutant emissions by the best available technology not entailing excessive costs. Ammonia, methane and nitrous oxide are the major gaseous pollutants that are emitted during broiler production. Ammonia is hazardous to natural habitats, such as heaths and bogs, because of eutrophication and/or acidification. Ammonia and ammonium compounds formed in the atmosphere are deposited via dry and wet processes and can also be transported over long distances within the UK and to continental

Table 10.4. Evidence for the effects of ammonia on the Farm Animal Welfare Council's Five Freedoms. (From Kristensen and Wathes, 2000, with minor modifications.)

Freedom	Evidence from original papers reviewed by Kristensen and Wathes (2000)
Freedom from hunger, thirst and malnutrition	Ammonia may reduce food intake and cause weight loss. Effects on thirst, feeding and drinking behaviour unknown
Freedom from discomfort	Ammonia may irritate mucous membranes
Freedom from pain, injury and disease	Ammonia causes air sac lesions and keratoconjunctivitis and may increase susceptibility to certain diseases
Freedom to express normal behaviour	Ammonia affects foraging, preening and resting behaviour
Freedom from fear and distress	Not studied

Table 10.5. Exposure limits for air quality in livestock buildings.

	Occupational exposure limits for human health		Exposure limits for swine health
	Eight-hour occupational exposure standard in UK; HSE, 2002	Recommendation by Donham *et al.*, 2002	Recommendation by Donham *et al.*, 2002
Total inhalable dust (mg/m³)	10	2.5	3.7
Respirable dust (mg/m³)	4	0.23	0.23
Endotoxin (EU/m³)	NA	100	150
Carbon dioxide (p.p.m.)	5000	1540	1540
Ammonia (p.p.m.)	25	7.0	11.0
Total microbes (c.f.u./m³)	NA	4.3×10^5	4.3×10^5

NA = not available.

Europe. Methane and nitrous oxide are greenhouse gases that contribute to climate change via global warming, while nitrous oxide is also harmful to the earth's ozone layer. A target of 297 kilotonnes/year in 2010 has been set for the total emission of ammonia in the UK (DEFRA, 2002b). Livestock production is the major source of ammonia, and poultry account for 14% of the total emission. This target is only 85% of the emissions in 1999 and DEFRA estimates that it can be achieved under 'business as usual' conditions. Furthermore, there are no plans at present to monitor or control emissions from individual farms. A standard for air quality in broiler houses cannot therefore be set on this criterion.

There is therefore a body of evidence upon which to set standards of air quality in broiler houses. Arguably, occupational exposure standards for poultry stockpeople are too high and should be lowered to the limits approaching those suggested by Donham *et al.* (2002). However, the direct evidence for poor chicken welfare arising from chronic exposure to aerial pollutants is slim and limited mainly to studies with ammonia. On the one hand, exposure of fowls to ammonia at concentrations generally exceeding those found in modern poultry houses affects susceptibility to some respiratory diseases: whether a subdued or subclinical response occurs at lower concentrations is unknown. On the other hand, broiler chickens (and adult hens) are averse to ammonia at concentrations of 20 p.p.m. and higher, indicating that various physiological systems and processes are adversely affected. DEFRA (2002a) recommends a maximum ammonia concentration of 20 p.p.m. on the grounds of welfare, and this is consistent with the scientific evidence. If a cautionary approach is taken, then a suggested upper guideline for the ammonia concentration in broiler houses is 10 p.p.m. The human OES for total inhalable dust is 10 mg/m^3 for an 8-hour exposure. This could be used as the guideline for broiler welfare, though exposure of the birds is continuous and, on a time-weighted basis, the guideline concentration could be reduced to about 3.3 mg/m^3, which is similar to that proposed by Donham *et al.* (2002) for swine health. Again, if a cautionary approach is taken a guideline for the concentration of total inhalable dust is 3.7 mg/m^3.

A further question arises about the standard for air quality for other aerial pollutants apart from ammonia. Can ammonia or another pollutant serve as a general index of air quality? The concentration of an aerial pollutant is determined by the balance between the strengths of its sources and the rate at which it is removed from the air by various physical means (Wathes *et al.*, 1983). Non-reactive gases like carbon dioxide are only removed by ventilation, whereas a reactive gas like ammonia may interact chemically; for example, by absorption on wet building surfaces or dust particles. In addition to dilution by ventilation, particulate matter may, according to its size, sediment from the air or impact on surfaces, while airborne microbes are also effectively removed from the air by the further process of microbial death. These removal mechanisms of dilution by ventilation, chemical reaction, sedimentation and impaction, and microbial death are additive and determine the overall concentration of an aerial pollutant. All pollutants will be affected by dilution by ventilation, which implies that non-reactive gases can, in the first analysis, serve as an overall marker of air quality. Carbon dioxide is a metabolic by-product of both broiler chickens and litter processes and could be a convenient marker, given that sensors for the concentration of carbon dioxide in the air are available. However, its dynamic relationship with other pollutants has not been analysed in detail.

The minimum ventilation rate needed in a broiler house to maintain concentrations of ammonia of 20 p.p.m. and (total) inhalable dust of 10 mg/m^3 can be calculated assuming steady-state production (Table 10.1) and removal, a negligible concentration of pollutant entering the building, and uniform mixing. To a first approximation and using the average emission rates in Table 10.1, the rates of production of ammonia and inhalable dust are 5500 and 4000 mg/h per 500 kg live weight, giving minimum ventilation rates of 360 and 400 m^3/h per 500 kg live weight of broilers respectively. These rates are greater than the current recommended minimum of 300 m^3/h/500 kg (D.R. Charles, personal communication), which was based on experiments carried out in the 1970s and subsequent commercial experience. Provided the above assumptions are met, the current recommended concentrations of ammonia (20 p.p.m.) and inhalable dust (10 mg/m^3) will be exceeded in broiler houses with 4-week-old broilers when the current minimum ventilation rate is used. Of course, if the ventilation rate is higher than the minimum (in summer, for example), the pollutant concentrations will be lower *pro rata*. However, if the lower guideline concentrations are adopted, then faster minimum ventilation rates will be needed. Use of such a rapid minimum ventilation rate to clear ammonia and dust (in winter) would mean that the target for air temperature of approximately 21°C would be missed. It is therefore unlikely that targets for both air hygiene and air temperature can be met in a broiler house unless litter management, nutrient supply and environmental management are of the highest standard. Adoption of the new technology of precision livestock farming (Wathes *et al.*, 2001) is one means by which these standards can be achieved.

Conclusions

Air hygiene in a broiler house is poor by comparison with the quality of air experienced by a chicken outdoors. It is remarkable that the modern broiler chicken tolerates the high burden of aerial pollutants, and yet there are grounds for concern that its welfare may be compromised by chronic exposure over its brief life. There is strong evidence for the adverse effects of ammonia on broiler welfare and the current recommended maximum concentration is justified. However, a cautionary approach implies that the current guidelines for air quality should be revised and lower limits considered.

References

Anderson, D.P., Beard, G.W. and Hanson, R.P. (1964) The adverse effects of ammonia on chickens including resistance to infection with Newcastle disease virus. *Avian Diseases* 8, 369–379.

Brewer, R.N. and Koon, J. (1973) Effects of various levels of ammonia on Marek's disease in young chickens. *Avian Diseases* 17, 851–854.

Cox, C.S. and Wathes, C.M. (eds) (1995) *Bioaerosols Handbook.* Lewis Publishers, Boca Raton, Florida.

DEFRA (2002a) Code of recommendations for the welfare of livestock: meat chickens and breeding chickens. Department for Environment, Food and Rural Affairs, London.

DEFRA (2002b) On the implementation in the United Kingdom of Directive 2001/81/EC on national emission ceilings for certain atmospheric pollutants. Department for Environment, Food and Rural Affairs, London.

Donham, K.J., Thorne, P.S., Breuer, G.M., Powers, W., Marquez, S. and Reynolds, S.J. (2002) Exposure limits related to air quality and risk assessment. In: *Iowa Concentrated Animal Feeding Operations Air Quality Study.* Iowa State University and The University of Iowa Study Group, http://www.public-health.uiowa.edu/ehsrc/CAFOstudy.htm, pp. 164–183.

Groot Koerkamp, P.W.G., Metz, J.H.M., Uenk, G.H., Phillips, V.R., Holden, M.R., Sneath, R.W., Short, J.L., White, R.P., Hartung, J., Seedorf, J., Schröder, M., Linkert, K.H., Pedersen, S., Takai, S.H., Johnsen, J.O. and Wathes, C.M. (1998) Concentrations and emissions of ammonia in northern Europe. *Journal of Agricultural Engineering Research* 70, 79–95.

Hamilton, T.C.D., Roe, J.M., Hayes, C.M. and Webster, A.J.F. (1998) Effects of ammonia inhalation and acetic acid pre-treatment on the colonisation kinetics of toxigenic *Pasteurella multocida* within the upper respiratory tract of swine. *Journal of Clinical Microbiology* 36, 1260–1265.

Hartung, J. (1988) Zur Einschätzung der biologischen Wirkung von Spurengasen der Stalluft mit Hilfe von zwei bakteriellen Kurzzeittests. [On the estimation of the biological effect of trace gases present in animal house air with the aid of two bacterial short-term tests.] *Fortschritt-Berichte VDI* 15 N0. 56. VDI-Verlag, Düsseldorf.

Health and Safety Executive (2002) Occupational exposure limits. EH40/2002. Health and Safety Executive, London.

Jones, E.K.M. (2002) Behavioural responses of broiler fowl to atmospheric ammonia. PhD Thesis, University of Bristol, UK.

Jones, J.B., Carmichael, N.L., Wathes, C.M., White, R.P. and Jones, R.B. (2000) The effects of acute simultaneous exposure to ammonia on the detection of buried odourised food by pigs. *Applied Animal Behaviour Science* 65, 305–319.

Jones, J.B., Wathes, C.M., Persaud, K.C., White, R.P. and Jones, R.B. (2001) Acute and chronic exposure to ammonia and olfactory acuity for n-butanol in the pig. *Applied Animal Behaviour Science* 71, 13–28.

Jones, R.B. and Roper, T.J. (1997) Olfaction in the domestic fowl: a critical review. *Physiology and Behaviour* 62, 1009–1018.

Kristensen, H.H. and Wathes, C.M. (2000) Ammonia and poultry welfare: a review. *World's Poultry Science Journal* 56, 235–245.

Kristensen, H.H., Burgess, L.R., Demmers, T.G.M. and Wathes, C.M. (2000) The preferences of laying hens for different concentrations of atmospheric ammonia. *Applied Animal Behaviour Science* 68, 307–318.

Kristensen, H.H., Jones, R.B., Schofield, C.P., White, R.P. and Wathes, C.M. (2001) The use of olfactory and other cues for social recognition by juvenile pigs. *Applied Animal Behaviour Science* 72, 321–333.

Lillie, R.J. (1970) Air pollutants affecting the performance of domestic animals. Agricultural Handbook 380, Agricultural Research Service, US Department of Agriculture, Washington, DC.

Madelin, T.M. and Wathes, C.M. (1989) Air hygiene in a broiler house: comparison of deep litter with raised netting floors. *British Poultry Science* 30, 23–37.

McKeegan, D.E.F., Demmers, T.G.M., Wathes, C.M., Jones, R.B. and Gentle, M.J. (2002a) Stimulus–response functions of single avian olfactory bulb neurones. *Brain Research* 953, 101–111.

McKeegan, D.E.F., Demmers, T.G.M., Wathes, C.M., Jones, R.B. and Gentle, M.J. (2002b) Response characteristics of nasal trigeminal nociceptors in *Gallus domesticus*. *Neuroreport* 13, 1–3.

Morrison, W.D., Pirie, P.D., Perkins, S., Braithwaite, L.A., Smith, J.H., Waterfall, D. and Doucett, C.M. (1993) Gases and respirable dust in confinement buildings and the response of animals to such airborne

contaminants. In: *Fourth International Symposium on Livestock Environment, American Society of Agricultural Engineers*, pp. 734–742.

Oyetunde, O.O.F., Thomson, R.G. and Carlson, H.C. (1978) Aerosol exposure of ammonia, dust and *Escherichia coli* in broiler chickens. *Canadian Veterinary Journal* 19, 187–193.

Phillips, V.R. and Pain, B.F. (1998) Gaseous emissions from the different stages of European livestock farming. In: Matsunaka, T. (ed.) *Environmentally Friendly Management of Livestock Waste*. Rakuno Gakuen University, Ebetsu, Sapporo, Japan, pp. 59–75.

Quarles, C.L. and Kling, H.F. (1974) Evaluation of ammonia and infectious bronchitis vaccination stress on broiler performance and carcass quality. *Poultry Science* 53, 1592–1596.

Seedorf, J., Hartung, J., Schröder, M., Linkert, K.H., Phillips, V.R., Holden, M.R., Sneath, R.W., Short, J.L., White, R.P., Pedersen, S., Takai, H., Johnsen, J.O., Metz, J.H.M., Groot Koerkamp, P.W.G., Uenk, G.H. and Wathes, C.M. (1998) Concentrations and emissions of airborne endotoxins and microorganisms in livestock buildings in northern Europe. *Journal of Agricultural Engineering Research* 70, 97–109.

Takai, H., Pedersen, S., Johnsen, J.O., Metz, J.H.M., Groot Koerkamp, P.W.G., Uenk, G.H., Phillips, V.R., Holden, M.R., Sneath, R.W., Short, J.L., White, R.P., Hartung, J., Seedorf, J., Schröder, M., Linkert, K.H. and Wathes, C.M. (1998) Concentrations and emissions of airborne dust in livestock buildings in northern Europe. *Journal of Agricultural Engineering Research* 70, 59–77.

Valentine, H. (1964) A study of the effects of different ventilation rates on the ammonia concentrations in the atmosphere of broiler houses. *British Poultry Science* 5, 149–159.

Wathes, C.M. (1995) Bioaerosols in animal houses. In: Cox, C.S. and Wathes, C.M. (eds) *Bioaerosols Handbook*. Lewis Publishers, Boca Raton, Florida, pp. 547–577.

Wathes, C.M. (1998) Aerial emissions from poultry production. *World's Poultry Science Journal* 54, 241–251.

Wathes, C.M., Jones, C.D.R. and Webster, A.J.F. (1983) Ventilation, air hygiene and animal health. *Veterinary Record* 113, 554–559.

Wathes, C.M., Frost, A.R., Gordon, F. and Wood, J. (eds) (2001) *Integrated Management Systems for Livestock*. Occasional Publication No. 28, British Society of Animal Science, Edinburgh.

Wathes, C.M., Jones, J.B., Kristensen, H.H., Jones, E.K.M. and Webster, A.J.F. (2002) Aversion of pigs and domestic fowl to atmospheric ammonia. *Transactions of the American Society of Agricultural Engineers* 45, 1605–1610.

Whyte, R.T., Williamson, P.A.M and Lacey, J. (1998) Comparison of the aerial environments in intensive and semi-intensive poultry laying houses and implications for stockman respiratory health. In: *Spring Conference of the International Egg Commission, London, 2 March 1998*, 10 pp.

Wolfe, R.R., Anderson, D.P., Cherms, F.L. and Roper, W.E. (1968) Effect of dust and ammonia air contamination on turkey response. *Transactions of the American Society of Agricultural Engineers* 11, 515–518.

11 Stocking Density

W. BESSEI

Applied Farm Animal Ethology and Small Animal Production, Institute for Animal Breeding and Husbandry, University of Hohenheim, Stuttgart, Germany

Introduction

High stocking density is a major welfare issue in intensive livestock production systems. It is usually discussed in the context of negative experiences of the birds, such as crowding, social strife, lack of physical space for locomotor activity and exploration and physiological stress (Zayan, 1985). Hence, it is assumed that reduction of stocking density increases well-being, and attempts are being made to restrict stocking density to an acceptable level by laws and regulations. The perception of acceptability, however, differs widely according to the country, region and organization concerned, and there is no evidence that the existing rules are based on scientific knowledge. The economic aspects of stocking density have been studied extensively, and it is generally known that increasing the stocking density increases the net profit of broiler production, even though growth rate and mortality are affected negatively by extreme conditions. This leads to an obvious conflict between animal welfare and profitability. In the highly competitive international poultry meat trade, local limitations on stocking density can distort the marketing system (EU, 2000). Therefore, the aim should be to restrict stocking density internationally, so as to avoid economic disadvantages for farmers producing under animal-friendly conditions. Concurrently with the regulations, it is important to establish a control system that can be operated under practical conditions.

This chapter will first review the main existing regulations on stocking density for broilers in EU countries. Secondly, the interrelationships between stocking density and characteristics related to welfare will be treated, and finally proposals for measuring and assessing broiler welfare, in the context of stocking density, will be made.

Regulations and Recommendations for Stocking Density in EU Countries

The most important regulations on stocking density for broilers in EU countries are shown in Table 11.1.

The 'Recommendations Concerning Domestic Fowl (*Gallus gallus*)' of the Standing Committee of the European Convention for the Protection of Animals kept for Farming Purposes of the Council of Europe (Council of Europe, 1995) considers the problem of stocking density. The provisions are essentially as follows:

- Stocking density should be such that:
    - All birds are able to exercise and perform normal patterns of behaviour.
    - All birds are able to reach food and water easily.
    - Any bird wishing to move from a crowded area to a more open space is able to do so.
    - The birds shall have access to litter in order to peck, scratch and dust-bath.
- In intensive systems, feeders and drinkers should be arranged so that no bird has to move more than 3 metres in order to feed and drink; *at high stocking density* it is necessary to reduce this distance. [Italics added.]
- Management measures should prevent the occurrence of leg problems (e.g. by low-energy diets for the first 3 weeks of the birds' lives; providing daylight from the first hours of life, providing perches, *lowering stocking densities*, improving air circulation). [Italics added.]

The recommendations of the European Council are the basis for the development of a directive by the EU Commission, which will regulate more in detail the minimum standards for broilers kept in the EU. A scientific report, 'The Welfare of Broiler Chickens', has been prepared by an expert group of the Scientific Committee on Animal Health and Animal Welfare. This report (EU, 2000) provides the scientific background for specific recommendations concerning stocking density, light and other welfare requirements. As regards stocking density, it is stated that 'welfare problems are likely to occur when stocking density exceeds 30 kg/m^2'.

Table 11.1. Recommendations on stocking density in the EU.

EU (2000)	Slaughter age, weight and ventilation rate/climatic condition should be considered; it appears that welfare problems are likely to emerge when stocking rates exceed 30 kg/m^2
	Such stocking densities should only be allowed when the producer is able to maintain air and litter quality
A.V.E.C. (1997)	. . . stocking density depends on the housing capacity and quality of equipment and standard of managment
Danske Fjekraeraad (1997, 2000)	Maximum 40 kg/m^2
FAWC (2992)	34 kg maximum stocking density; should not be exceeded at any time
BML Germany (1993)	30–37 kg/m^2 depending on management conditions
Voluntary agreement	Maximum 35 kg/m^2
Austria	35 kg/m^2
Swiss law	20 birds/m^2 or 30 kg/m^2 maximum
Sweden	20–36 depending on management scoring

Special legal regulations for broilers exist in Switzerland. The law for animal protection limits stocking density in large groups to 20 birds or 30 kg live weight per m². In Austria, several states have established maximum stocking rates of 30 kg/m². In Germany, a report of the Federal Ministry of Agriculture recommends stocking densities between 30 and 37 kg/m², depending on the management conditions (Stellungnahme und Empfehlungen der Sachverständigengruppe des Bundesministeriums für Landwirtschaft und Forsten zur artgemässen und verhaltensgerechten Geflügelmast. Unpublished report submitted to the Federal Ministry of Agriculture (BML), Bonn, Germany, 1993). In addition, an agreement on minimum standards for broilers has been signed on a voluntary basis between the broiler producers' associations and the State ministries concerned (Bundeseinheitliche Eckwerte für eine freiwillige Vereinbarung zur Haltung von Jungmasthühnern (Broiler, Masthähnchen) und Mastputen, Unpublished document of the Federal Ministry of Agriculture BML 321-3545/2, Bonn, 2 September 1999), in which the maximum density allowed is 35 kg/m² the day before slaughter. Higher densities may be tolerated under certain conditions.

A more flexible regulation concerning stocking density has been established in Sweden. Considering the fact that the influence of stocking density on the welfare of the birds depends on related management factors such as ventilation rate, litter conditions, etc., a scoring system was developed which grades the management quality of the farms. Stocking densities up to 36 kg/m² are permitted under good management conditions. Farmers who do not participate in the scoring scheme are allowed a maximum stocking density of only 20 kg/m² (Berg, 1998). In Denmark, the current maximum stocking density is 43–44 kg/m², but a working group of the Ministry of Justice (Anonymous, 2000) has recommended the maximum stocking density be reduced gradually to 40 kg/m².

In many other countries there are no specialized rules on the welfare of broilers. Nevertheless, production and marketing programmes exist (for example, for premium chicken meat production EWG 1538/91, and the Label programmes), which fulfil and even exceed the standards of welfare mentioned above. With regard to the EU poultry meat marketing regulations, the maximum stocking density inside the broiler house varies from 25.0 kg to 27.5 kg/m² (extensive indoor and free range) and 40 kg/m² (traditional free range). Additional space of 1 and 2 m² per bird must be provided for free range and traditional free range systems respectively. In the Free Range Total Freedom scheme, unlimited range has to be provided. The maximum density under the Freedom Food scheme in the UK is 30 kg/m². In France there are two Label Rouge programmes in which maximum stocking densities indoors are 11 and 20 birds per m², and free range has to be provided.

The EU regulations on organic broiler production allow maximum stocking densities of 10 birds or 25 kg/m².

Effect of Stocking Density on Performance and Welfare-related Traits

Stocking density and performance traits

Numerous experiments have been performed on the influence of stocking density on growth rate, feed efficiency, mortality and carcass quality, covering a wide range of densities from less than 10 to over 80 kg/m². The highest stocking rates, of more than 80 kg/m², have been reported in caged broilers by Andrews (1972). In deep litter systems, stocking rates of 50 kg/m² have been tested by Shanawany (1988) and Grashorn and Kutritz (1991). Most other experiments covered densities from about 20 to 40 kg/m² (Scholtyssek and Gschwindt-Ensinger, 1983; Scherer, 1989; Cravener *et al.*, 1992; Wiedmer and Hadorn, 1998).

There is a general trend of reduced growth rate when stocking rate exceeds 30 kg/m² on deep litter. In some experiments the growth rate was depressed even at stocking rates below 30 kg/m². Since feed intake was also reduced with increasing stocking density, the access to the feeder was assumed to be a causal factor for growth depression. The experiments of Scholtyssek (1974), Scholtyssek and Gschwindt (1980) and Scholtyssek and Gschwindt-Ensinger (1983), in which feeder space per bird varied from 1.9 to 4.6 cm or from 1.9 to 3.2 cm, showed no significant effect on feed intake or growth rate.

It has to be considered that, in broilers, feeding activity takes only a small part of the bird's time budget (Bessei, 1993), which, under a continuous light programme, is regularly distributed over the time of day. Hence, access to feed is not likely to be a causal factor for reduced growth rate under commercial production conditions. This finding is in contrast to the assumption of the European Council cited above. There is, however, evidence that heat stress is the main reason for the decline in performance with increasing stocking density. Grashorn and Kutritz (1991) tested stocking densities from 30 to 50 kg/m² under high and low ventilation rates. It was shown in this experiment that the depression of growth rate started later at a high ventilation rate than at a low rate (Fig. 11.1). Comparing different stocking densities in cages and on deep litter, Scholtyssek and Gschwindt-Ensinger (1983) found that depression of growth rate started earlier on deep litter than in cages. Perforated floor systems in combination with under-floor ventilation increased growth rate compared with conventional deep litter systems at the same stocking rate (Arkenau *et al.*, 1997). These results led to the assumption that heat stress may be the main underlying factor for growth depression at high stocking rates. This was confirmed by a study

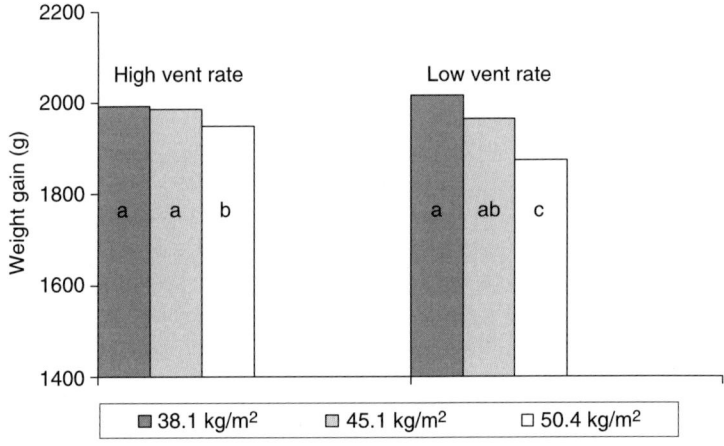

Fig. 11.1. Influence of high and low ventilation rate and stocking density on body weight gain in 6-week-old broilers. (From Grashorn and Kutritz, 1991.)

with stocking densities of 20–40 kg/m², in which the temperature was recorded at and under the surface of the deep litter layer and also between the birds (Reiter and Bessei, 2000). With increasing stocking rate, the temperature increased to more than 30°C under and on the surface of the litter, while a normal temperature (about 20°C) was recorded 1 m above the birds, independently of stocking rate (Fig. 11.2).

The influence of stocking rate was less evident on feed conversion than on growth rate. Improved feed conversion rate with increasing stocking density has been found in various experiments (Scholtyssek, 1974; Scholtyssek and Gschwindt, 1980; Shanawany, 1988; Grashorn and Kutritz, 1991; Cravener *et al.*, 1992). In some experiments, there was no effect of stocking density on feed conversion (Waldroup *et al.*, 1992; Scholtyssek and Gschwindt-Ensinger, 1983), and a reduced feed conversion was found by Scholtyssek (1974). Shanawany (1988) reported an increased mortality rate as stocking density increased. In most other experiments, however, there was no clear relationship between mortality and stocking rate (Proudfoot *et al.*, 1979; Cravener *et al.*, 1992; Grashorn, 1993a; Wiedmer and Hadorn, 1998).

It can be concluded that reduced feed intake and growth rate constitute a sensitive indicator of poor welfare associated with high stocking density in deep litter systems. The reactions of feed efficiency and mortality to increasing stocking rates are less consistent and may serve only as indicators of poor welfare under extreme conditions.

Stocking density and weight development

It is generally assumed that higher stocking rates, in kg/m², can be tolerated by heavier birds. It is argued that the size of the bird increases in both the horizontal and

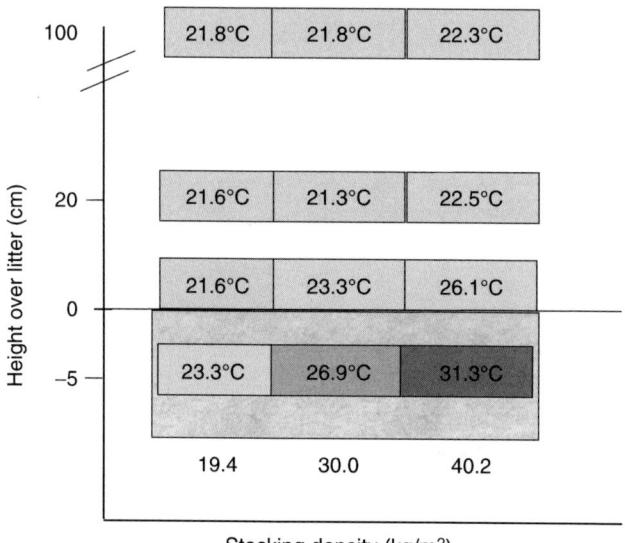

Fig. 11.2. Influence of stocking density on the temperature at different sites of the broiler house. (From Reiter and Bessei, 2000.)

the vertical planes, and the latter direction is not a limiting factor. However, the dissipation of metabolic heat depends on body weight relative to the body surface as well as on the insulation capacity of the feather cover. Both factors make heat dissipation more difficult as the bird's weight and age increase. Thus, in contrast to current practice, it may be concluded that older and heavier birds may in fact need more space in terms of kg/m^2.

Stocking density and behaviour

The behaviour of broilers has been studied over a wide range of stocking density. There was a decrease of locomotor activity and scratching behaviour when the stocking rate increased from less than 10 birds per m^2 in small groups to higher densities and larger groups (Blokhuis and van der Haar, 1990; Lewis and Hurnik, 1990; Bessei and Reiter, 1993). In other experiments using larger groups and stocking rates from 10 to 25 birds per m^2, there was no significant difference in locomotor and other behavioural activities, such as feeding, drinking, scratching and resting, as stocking rate varied (Scherer, 1989; Bessei, 1995) (Figs 11.3 and 11.4). Preston and Murphy (1989) observed individually marked broilers under commercial conditions (15 birds per m^2). They reported that the broilers moved throughout the total available area, and density was obviously not a physical problem for locomotion. There were, however, frequent disturbances of resting behaviour at a stocking density of about 28 kg/m^2. Sixty per cent of the resting periods were shorter than 1 minute (Murphy and Preston, 1988). This was explained by the birds stepping over their lying pen-mates when moving to the feeder and drinker. The short periods of lying down may also have been caused by the high temperature of the litter.

It can be concluded that significant effects of stocking density on behavioural activities, including disturbance of rest, occur at stocking densities that are far below those found under commercial conditions. There is a sharp decline in most behavioural activities within the first weeks of life, and time spent resting increases

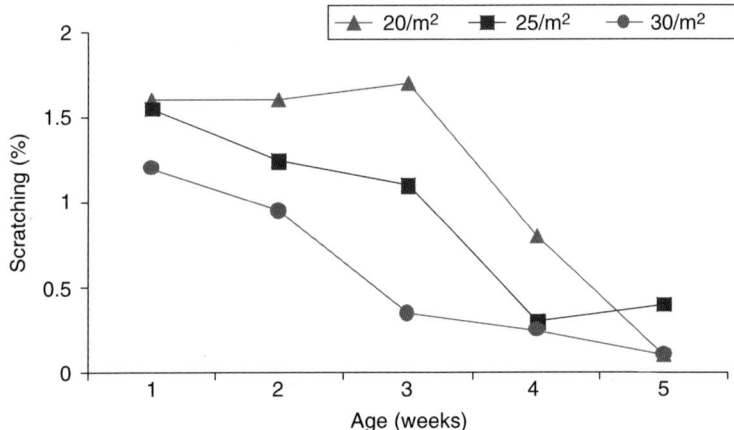

Fig. 11.3. Influence of stocking density (birds/m^2) on scratching behaviour from 1 to 5 weeks of age. (From Bessei, 1992.)

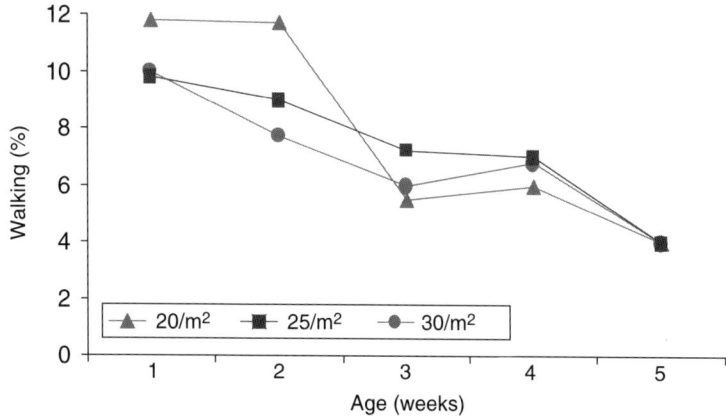

Fig. 11.4. Influence of stocking density (birds/m²) on locomotor behaviour of broilers from 1 to 5 weeks of age. (From Bessei, 1992.)

accordingly. The effects of stocking density on behavioural traits are small, especially at the end of the growing period, when physical space is at a minimum. The frequency of disturbance of resting birds and the short periods of resting are indicators of impaired welfare.

Stocking density and morphological changes

Morphological changes such as acute and chronic dermatitis, breast blisters, leg weakness and soiled plumage have been reported as responses to stocking density. Most experiments have demonstrated that the incidence of these problems increases with increasing stocking density (Weaver *et al.*, 1973; Proudfoot *et al.*, 1979; Cravener *et al.*, 1992; Gordon, 1992). Grashorn and Kutritz (1991), however, did not find such changes in response to stocking density. It seems that factors related to stocking density, rather than stocking density itself, cause these problems. Algers and Svedberg (1989) found that wet litter and ammonia concentration were clearly related to the incidence of dermatitis, leg problems and soiled plumage, but they found no correlation of these kinds of damage with stocking density, which varied from 20 to 35 birds per m². Wet litter and ammonia have been found to cause breast blisters and skin lesions by Harms *et al.* (1977), Proudfoot *et al.* (1979), Weaver and Meijerhof (1991) and Grashorn (1993b). High stocking density has also been considered the cause of scabby hip syndrome (Proudfoot and Hulan, 1985; Frankenhuis *et al.*, 1991). It was assumed that at high stocking density birds step on their penmates and thus injure the skin, especially in the area of the back and the legs. Since scabby hips have been reported only in a few experiments with high stocking density, it is assumed that other factors must act concurrently with high stocking density to produce this syndrome.

In conclusion, morphological changes indicate that there are welfare problems that are related to stocking density.

Monitoring Stocking Density under Practical Conditions

Stocking density in broiler houses is usually expressed as the number of birds or live weight per m² floor space at the end of the fattening period. As broiler production in Europe is highly standardized, body weight at a determined slaughter age is predictable and stocking density can easily be calculated on the basis of number of birds and the size of the poultry house. In practice, however, there are many factors which make any statement on stocking density difficult.

Counting the birds in a broiler house of commercial size is not feasible. Therefore, the number of birds has to be calculated on the basis of the number of chicks delivered by the hatchery and records of mortality. The number of the chicks declared on the hatchery bill is often not reliable and a surplus of chicks is supplied to compensate for early mortality. This number may not appear on the bill. The mortality records are also often not complete; especially during the first weeks of age, dead chicks may not be found and recorded. There is also variation in growth rate under standardized conditions, which depends on chick quality and climatic conditions. In addition, the date of slaughter may be delayed for technical or economic reasons.

While each individual factor does normally not lead to severe mistakes in the estimation of stocking density, the concurrent occurrence of a delivery of surplus chicks, low mortality and excellent growth rate can produce excessive densities.

The most accurate figures for stocking density may be obtained from records of bird number and weight from the slaughter plant. The use of this post-mortem information may be criticized because it does not enable measures to be taken to reduce the density when the maximum stocking density was exceeded. However, it may be useful to identify those farms which regularly produce under high stocking densities, and to take measures to avoid this in subsequent batches.

The amount of feed consumed may also provide information on the live weight produced per square metre of poultry house. With a given energy content of the feed and assuming that the feed conversion ratio will be as expected, the amount of live weight can be estimated without having records of the number of birds. Variation in feed conversion rate caused by food spillage, ambient temperature and digestive problems can, however, bias the estimate. It is also not easy to measure precisely the total feed consumption of a batch of broilers when there is no special equipment to weigh the feed left at the end of the fattening period. If reliable data on feed consumption are available, they may be used to verify the estimates of live weight. Feed consumption alone may not be a useful tool for predicting stocking density.

The percentage of floor space that is covered by the birds has been measured in broilers and turkeys in relation to their age and body weight (Ellerbrock, 2000; unpublished report by Petermann, S. and Roming, L. Untersuchungen zur Masthähnchenhaltung im Regierungsbezirk Weser-Ems. Teil I: Tierschutzrelevante Aspekte. Hannover, Germany, 1993). The regression of percentage floor coverage on body weight per square metre could serve as a means of estimating stocking density (Fig. 11.5). The problem with this approach is that the birds do not spread evenly

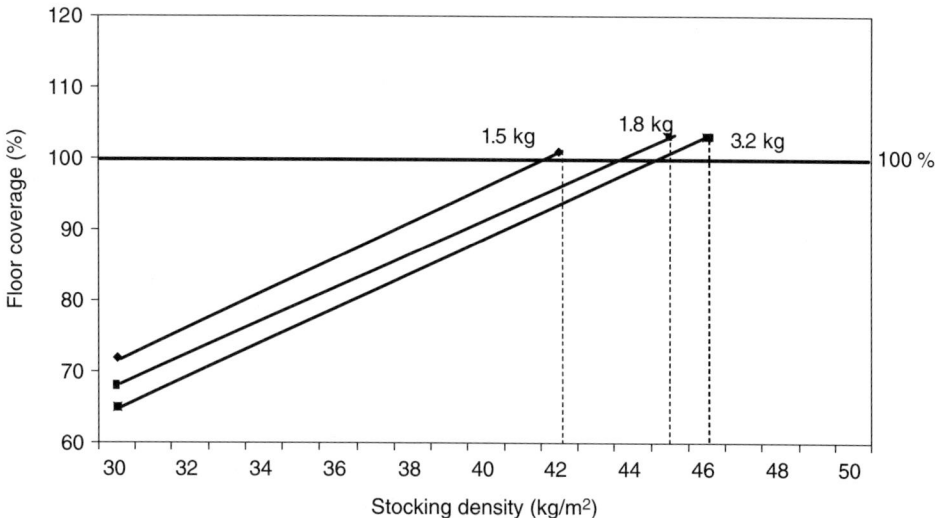

Fig. 11.5. Percentage of floor coverage in fattening broilers in response to different live weights at slaughter and stocking density. (After data from unpublished report by Petermann, S. and Roming, L. Untersuchungen zur Masthähnchenhaltung im Regierungsbezirk Weser-Ems. Teil I: Tierschutzrelevante Aspekte. Hannover, Germany, 1993.)

throughout the available floor space. During inspection, in particular by unfamiliar persons, the birds tend to crowd in certain areas, which makes objective assessment of floor coverage impossible. This method could be further developed by videotaping randomly selected areas throughout the broiler house. At present there exists no such method that could be used for regular inspection.

Given these problems of measuring or estimating stocking density, and given the finding that welfare problems arise not from physical stocking density as such, but from factors related to it, the assessment of welfare should focus on the indicators that result from excessive stocking density rather than on physical space. Among the indicators discussed in the previous paragraphs, the following are considered important and easy to record by on-farm inspectors:

- Condition of the deep litter, scored using one of the established scoring systems, such as those of Ekstrand *et al.* (1997) and Weaver and Meijerhof (1991).
- Temperature above and under the surface of the litter (sampled in defined areas of the broiler house).
- Foot and skin conditions (e.g. hock burn, foot pad lesions, breast blisters).

On-farm records of chick placement, mortality, feed consumption and growth rate, and records from the slaughterhouse, such as number of birds, live weight and causes of downgrading, may be used (as far as they are available and reliable) as supplementary information.

General Conclusions

At present most welfare regulations focus on the control of minimum requirements of space and other environmental factors. It is assumed that welfare (as expressed by behavioural, morphological and physiological indicators) will not be compromised when the maximum stocking density is not exceeded. However, the upper limit of stocking density is likely to be exceeded when good management practice allows a high growth rate and low mortality. It is therefore recommended that the inspection of broiler farms should focus on indicators that compromise the birds' welfare rather than on physical space. This gives the farmer a wider variety of options with which to develop his or her management system so as to improve the animals' welfare.

References

Algers, B. and Svedberg, J. (1989) Effects of atmospheric ammonia and litter status on broiler health. In: *Proceedings of the Third European Symposium on Poultry Welfare, Tours, France*, pp. 237–241.

Andrews, L.D. (1972) Cage rearing of broilers. *Poultry Science* 51, 1194–1197.

Anonymous (2000) Slagtekyllinger – Delrapport afgivet af Justietsministeriets arbejdsgruppe om slagtefjerkrae. Copenhagen, Denmark.

Arkenau, E.F., Macke, H. and Van den Weghe, H. (1997) Einfluss der Bodenbelüftung in Broilermastställen auf Tierleistung und Tierverluste. *Züchtungskunde* 69, 307–313.

Berg, C.C. (1998) Foot-pad dermatitis in broilers and turkeys: prevalence, risk factors and prevention. Swedish University of Agricultural Science, Uppsala.

Bessei, W. (1995) Das Verhalten von Broilern. In: *Jahrbuch für die Geflügelwirtschaft 1995*, pp. 27–30.

Bessei, W. and Reiter, K. (1993) Der Einfluss der Besatzdichte auf das Verhalten von Broilern. In: KTBL-Schrift 361, Aktuelle Arbeiten zur artgemässen Tierhaltung 1992, Darmstadt, Germany, pp. 203–212.

Blokhuis, H.J. and van der Haar, J.W. (1990) The effect of the stocking density on the behaviour of broilers. *Archiv für Geflügelkunde* 54, 74–77.

Council of Europe (1995) Recommendations concerning domestic fowl (*Gallus gallus*). Standing Committee of the Council of Europe, Strasbourg.

Cravener, T.L., Roush, W.B. and Mashaly, M.M. (1992) Broiler production under varying population density. *Poultry Science* 71, 427–433.

Ekstrand, C., Algers, B. and Svedberg, J. (1997) Rearing conditions and foot pad dermatitis in Swedish broiler chickens. *Preventive Veterinary Medicine* 31, 167–174.

Ellerbrock, S. (2000) Beurteilung verschiedener Besatzdichten in der intensiven Putenmast unter besonderer Berücksichtigung ethologischer und gesundheitlicher Aspekte. PhD thesis, Tierärztliche Hochschule Hannover, Germany.

EU (2000) The welfare of chickens kept for meat production. Health and Consumer Protection Directorate-General, European Commission, Brussels.

Frankenhuis, M.T., Vertommen, M.H. and Hemminga, H. (1991) Influence of claw clipping, stocking densities and feeding space on incidence of scabby hips in broilers. *Poultry Science* 32, 227–230.

Gordon, S.H. (1992) The effect of broiler stocking density on bird welfare and performance. *British Poultry Science* 33, 1120–1121.

Grashorn, M.A. (1993a) Untersuchungen zur Ätiologie und Pathogenese des Plötzlichen Herztods bei Masthühnern. *Hohenheimer Arbeiten* 123, Verlag Eugen Ulmer, Stuttgart, Germany.

Grashorn, M.A. (1993b) High stocking densities and incidence of leg disorders in broiler chickens. In: Whitehead, C.C. (ed.) *Bone Biology and Skeletal Disorders in Poultry*. Carfax, Abingdon, UK, p. 352.

Grashorn, M. and Kutritz, B. (1991) Der Einfluss der Besatzdichte auf die Leistung moderner

Broilerherkünfte. *Archiv für Geflügelkunde* 55, 84–90.

Harms, R.H., Damron, B.L. and Simpson, C.F. (1977) Effect of wet litter and biotin and/or whey on the production of foot pad dermatitis in broilers. *Poultry Science* 56, 291–296.

Lewis, N.J. and Hurnik, J.F. (1990) Locomotion of broiler chickens in floor pens. *Poultry Science* 69, 1087–1093.

Murphy, L.B. and Preston, A.P. (1988) Time-budgeting in meat chickens grown commercially. *British Poultry Science* 29, 571–580.

Preston, A.P. and Murphy, L.B. (1989) Movement of broiler chickens reared in commercial conditions. *Poultry Science* 30, 519–532.

Proudfoot, F.G. and Hulan, H.W. (1985) Effects of stocking density on the incidence of scabby hip syndrome among broiler chickens. *Poultry Science* 64, 2001–2003.

Proudfoot, F.G., Hulan, H.W. and Ramey, D.M. (1979) The effect of four stocking densities on broiler carcass grade, the incidence of breast blisters, and other performance traits. *Poultry Science* 58, 791–793.

Reiter, K. and Bessei, W. (2000) Einfluss der Besatzdichte bei Broilern auf die Temperatur in der Einstreu und zwischen den Tieren. *Archiv für Geflügelkunde* 64, 1–3.

Scherer, P.J. (1989) Einfluss unterschiedlicher Haltungsbedingungen auf das Verhalten von Broilern unter Berücksichtigung von Leistungsdaten. Dissertation, ETH Zürich, Switzerland.

Scholtyssek, S. (1974) Die Bedeutung des Futterplatzes in unterschiedlich besetzten Mastabteilen. *Archiv für Geflügelkunde* 38, 41–45.

Scholtyssek, S. and Gschwindt, B. (1980) Untersuchungen zur Besatzdichte und Futterplatz in der Bodenmast. *Archiv für Geflügelkunde* 44, 220–224.

Scholtyssek, S. and Gschwindt-Ensinger, B. (1983) Leistungsvermögen einschliesslich Befiederung und Belastbarkeit von Broilern bei unterschiedlicher Besatzdichte in Bodenhaltung. *Archiv für Geflügelkunde* 47, 3–8.

Shanawany, M.M. (1988) Broiler performance under high stocking densities. *British Poultry Science* 29, 43–52.

Waldroup, A.L., Skinner, J.T., Hierholzer, J.M., Kopek, M. and Waldroup, W.P. (1992) Effects of bird density on Salmonella contamination of prechill carcasses. *Poultry Science* 71, 844–849.

Weaver, W.D. and Meijerhof, R. (1991) The effect of different levels of relative humidity and air movement on litter conditions, ammonia levels, growth and carcass quality for broiler chicks. *Poultry Science* 70, 746–755.

Weaver, W.D., Beane, W.L. and Siegel, P.B. (1973) Methods of rearing sexes and stocking densities on broiler performance: an experiment conducted by a poultry science curriculum club. *Poultry Science* 52, 2100–2101.

Wiedmer, H. and Hadorn, R. (1998) Revision Tierschutzverordnung: Kurzmast-Besatzdichten lassen. *Schweizerische Geflügelzeitung* 2, 10–15.

Zayan, R. (1985) Spacing patterns of laying hens kept at different densities in battery cages. In: Zayan, R. (ed.) *Social Space for Domestic Animals*. Martinus Nijhoff, Dordrecht, The Netherlands.

12 Transport and Handling

M.A. MITCHELL[1] AND P.J. KETTLEWELL[2]

[1]Roslin Institute, Roslin, Midlothian, UK; [2]Silsoe Research Institute, Bedford, UK

Introduction

Auditing of all the procedures and processes associated with meat animal production and processing is becoming more common, and has resulted from both public and political pressure to establish full traceability of animals entering the food chain and to ensure the highest standards of food hygiene and safety (Fallon, 2001; Pettitt, 2001). Auditing procedures and associated quality assurance (QA) schemes are intimately linked to the implementation of hazard analysis and critical control point (HACCP) identification, which is focused specifically upon the reduction of biological, chemical and physical contamination of meat foodstuffs (Quinn and Marriott, 2002; DeGraft-Hanson, 2003). It is apparent that, as public demand for meat produced in systems of high welfare standards increases, then an important development is the creation of QA schemes including quality control based upon independent welfare auditing (Webster, 2001; Main *et al.*, 2001). Ethical or welfare audits could be imposed at every stage of production, including standards upon the farm, during collection and transport, and at slaughter (Barton-Gade, 2002). Several welfare audit systems have been implemented in North America by retail and food outlet companies for the production and slaughter of cattle, pigs and poultry (Livestock Handling Quality Assurance; Welfare of Pigs during Transport; available at www.grandin. com). Crucially, these system are designed around the identification of critical control points at which the welfare of the animals is most likely to be compromised (Poultry Slaughter Plant Audit: Critical Control Points for Bird Welfare: Critical Control Points (CCPs) of Humane Slaughter and Handling; available at www. grandin.com). As yet these approaches have not been applied to the catching, handling and transport of poultry. This chapter will, none the less, examine all currently available scientific information relating to these procedures and identify the critical control points (CCPs) that may impose the greatest stress upon broiler birds and thus constitute the major threats to their welfare. The methods by which stress and welfare may be quantified will be addressed. In turn, possible improvements in procedures, practices and transport vehicles will be described. Finally, the sound scientific basis for welfare auditing schemes based on CCPs and knowledge of measurable bird responses and behaviours will be proposed.

Broiler Production, Handling and Transport: Background

The catching and transport of poultry probably constitute the largest commercial handling and translocation operation for a single class of livestock in the world. Despite regional economic pressures, worldwide poultry production continues to expand. In the year 2002 the number of broiler chick placings worldwide was approximately 44 billion (FAOSTAT, 2003).

At slaughter age, intensively produced broilers are caught (harvested) and loaded on to vehicles for transport from their geographically dispersed sites of rearing to centralized processing plants. The catching and handling procedures and sub-sequent 'livehaul' with their associated problems are major determinants of the efficiency and profitability of large-scale commercial broiler production (Nilipour, 1996). The welfare of poultry during catching and on journeys has received attention (e.g. Gerrits et al., 1985; Kettlewell and Turner, 1985; Bayliss and Hinton, 1990; Metheringham and Hubrecht, 1996). Concern about broiler welfare in transit has resulted in pressure upon producers and governments, particularly in European Union member states, to improve conditions, standards and regulations. European broiler production is increasingly focused on fewer, larger farms. Coupled with a reduction in the number of slaughter plants owing to the imposition of improved hygiene and welfare standards, birds may often be transported over large distances with accompanying extended journey durations.

The logistics of catching, handling, transport, lairage and slaughter may result in broilers being held for extended periods in the transport containers. Together with unpredictable external climatic conditions and environments, these practices consti-tute significant sources of compromised bird welfare, reduced production efficiency and decreased product quality. For example, Warriss et al. (1990), in a survey of 19.3 million broilers in the UK, found the typical total marketing time for broilers (from loading of the first bird to unloading of the last bird) was 3.6 hours, with a range between 2.2 and 5.5 hours. The maximum recorded total marketing time was 12.8 hours. This considerable variation highlights the challenges facing the industry in scheduling the collection and delivery of birds to the processing plant.

The effect of the time interval between loading birds on farm and processing has been studied in relation to carcass quality (Carlyle et al., 1997). The study considered a total of 156,000 broilers with total journey times ranging between 3 and 7 hours. The data showed a positive correlation between increasing time and increased breast bruising but a negative correlation between journey time and wing bruising. While this was a preliminary study, it concluded that birds confined to transport containers for prolonged periods had a higher level of breast bruising as they were restricted to sternal recumbency. Clearly, the effects of handling and transport can have marked effects on the welfare of the birds, and on subsequent product quality, before the birds reach the processing plant.

Catching or Harvesting of Broilers

The handling and collection of birds has been reviewed by a number of authors (Kettlewell and Turner, 1985; Gerrits *et al.*, 1985; Bayliss and Hinton, 1990; Scott, 1993; Kettlewell, 2000). The predominant practice throughout Europe until recently has been manual catching. To achieve the desired commercial catching rates of broilers from litter floors, birds are picked up by the legs in handfuls (four to eight birds per operative). While picking up birds gently by their sides is an ideal in respect of bird welfare, it is not commercially viable. The most recent welfare based guidelines issued in the UK by DEFRA (2003) state that 'Chickens should be caught individually by grasping both legs, just above the feet. If carried in groups, care must be taken to ensure birds can be held comfortably without distress or injury, and carrying distances must be kept to a minimum. No more than three birds should be carried in one hand.' Similar statements within the code specify the requirements of the catching crew personnel, the designation of a responsible person within the team, measures to be taken to prevent chickens crowding together on the house floor and the maximum stocking density within the transport drawer (RSPCA, 1999).

Not only is the task of catching and loading chickens very demanding physically, but also the conditions within the house can be very unpleasant. During catching, house lights are often turned off (or significantly dimmed) so the operators are working in near-dark conditions close to the floor, where dust levels can be very high. Coupled with this hostile work environment, catchers are expected to lift at least 1000 birds per hour, which, for 2 kg birds over an 8-hour shift, equates to lifting a total weight of 16 tonnes. There can be a high incidence of work-related back problems amongst the catchers and, because of the nature of the work, it fails to attract highly motivated personnel. These factors clearly influence the attention paid to bird welfare during catching and handling.

Whilst there is considerable scope for physical injury to the birds during loading, this is generally kept to a minimum by the application of guidelines, which are enforced by farmers, catching supervisors and processors. It is very difficult to assess injury (except gross insults) to birds prior to loading as the nature of the operation precludes easy inspection of the birds at the farm after they have been loaded. It is sometimes possible to ascribe retrospectively any injury or downgrading which might have occurred at the farm, but this is detected at unloading rather than on-farm.

The practice, efficiency and effects upon the birds of the catching process are largely determined by the transport containers employed, the methods of loading and unloading, and vehicle design. These interactive factors are often defined by the overall processing system that an integrated broiler producer/processor chooses to use.

Loose crates and fixed crate systems have now largely been superseded by modular systems on grounds of improved efficiency and animal welfare. There are two types of modules: the loose drawer and the coop dump.

The loose drawer system is typified throughout Europe by the Easyload handling system (Anglia Autoflow Ltd, Diss, UK), which now accounts for the majority of UK broiler production (there are currently 85 Easyload installations in the UK and 195 worldwide). Each module comprises a number of plastic drawers located within a metal framework. The specific number of drawers depends on the type of poultry to be carried. The drawers are arranged in three columns, the module width being equivalent to the full width of the transport vehicle (2.4 metres). Individual drawer dimensions are 2.4 m long, 1.2 m wide and, for broilers, 0.22 m high, with a design capacity of 50 kg. In a typical 12-drawer broiler module each drawer would hold a maximum of 25 birds, each weighing approximately 2 kg. Stocking density can be adjusted to suit the live weight of the birds and the climatic conditions.

By moving the transport container to the birds, catching rates can be increased and a typical manual collection rate of 1000–1500 birds per man-hour is expected. The large open top of each drawer ensures minimal injury to the birds during loading, and by partially closing the drawer during loading the possibility of birds escaping is greatly reduced.

Automated modular handling at the factory allows the loaded drawers to be moved through a covered conveyor system where the birds are presented directly under the shackling line. This is the first time since loading that the top of the drawer is exposed, and use of this arrangement reduces bird activity before shackling.

The 'coop dump' modular system is so named because the original method of unloading containers was simply to tip the birds on to a conveyor belt supplying the shackling line. The modular cages incorporate a hinged front door, which is opened to allow loading of each compartment. The loaded container is removed from the shed and placed on to the transport vehicle. In some cases, to assist loading, the empty container is tipped backwards to allow birds to slide gently to the back of each compartment. At the factory, modules are removed from the lorry and placed on a tipping system, which empties the birds by tipping the module forwards. Early versions of this system relied on the weight of the birds to burst open the cage door, resulting in a mass of birds falling on to a cross conveyor for transfer to the shackling line. More recent designs have incorporated a series of conveyors that align with each compartment level so that, as the module is tipped, birds from each level are conveyed to the transfer conveyor.

The concept of mechanized harvesting or catching is not new. Early proposals for such systems were discussed by Nelson (1982, 1984), Bingham (1986) and De Koning *et al.* (1987). A number of ideas have been proposed, and some patented, for improved live bird collection to address the requirement to move birds out of the house to the transport vehicle. These have included the use of multiple conveyors (Anonymous, 1979a), pneumatic conveying systems (Nesheim *et al.*, 1979), collection mats (Gerrits and de Koning, 1980) and even overhead shackles (Anonymous, 1979b). An alternative type of crateless vehicle was also reported (Manbeck, 1974; Reed, 1974), where a series of horizontal conveyors were mounted on to the vehicle itself. Birds carried from the house were then loaded on to the rear of each belt in turn, as the belts advanced slowly during loading.

One of the major obstacles in the development and uptake of commercially viable broiler harvesters has been the requirement for them to operate in existing broiler houses. Many of the houses in the UK are over 20 years old and reflect the standard designs that prevailed at the time they were installed. The majority of houses feature queen posts, which serve to support the roof structure. There are still very few clearspan houses. A typical house may be 60–100 m long by 10–20 m wide, the roof being supported by two lines of posts. Access is further compromised by the lack of headroom at the eaves, which is sometimes as low as 1.5 m. Any harvester that is to match current commercial hand-catching must be capable of working in these conditions.

Berry *et al.* (1990) describe a broiler harvester developed in the UK by the Silsoe Research Institute which operates on a 'sweeping' principle, gathering birds on to an angled conveyor using slowly rotating rubber fingers mounted on a set of near-vertical rotors. As the collection head is advanced on to the birds, they are picked up by their sides and transferred on to the angled conveyor belt, which conveys them to the rear of the machine and into transport containers. This development was fully patented.

The only commercially available harvester currently in use within the UK is that developed by Anglia Autoflow (the Easyload Automatic Harvesting System). There are two systems in use, each harvesting 35,000 birds per day. In addition, there are five systems in use in the USA (each harvesting 50,000 birds per day) and one system in use in Australia (harvesting 35,000 birds per day). This new system extends the Easyload concept of live bird handling and affords automated handling from the broiler shed to the processing line. This machine uses a similar configuration of soft rubber fingers to the Silsoe machine, but the fingers are mounted on a horizontal rotor. The rotors pass over the top of the birds and draw them on to an inclined conveyor belt. The automatic catcher picks up and holds around 200 birds and then transfers them to the packing unit, which remains in the doorway of the house. The packing unit conveys and counts the birds into modular drawers, which are held within a module on the packing unit. Filled modules are then placed on to the transport vehicle and replaced with an empty module. The system is capable of operation at speeds of 5000 birds per hour and can be moved rapidly from site to site, requiring only two operators to catch and load the broilers.

The only published data comparing the effects of mechanical and manual collection of broilers are based on a prototype collection head developed at Silsoe Research Institute (Duncan *et al.*, 1986). The heart rate and duration of tonic immobility of small groups of birds were assessed immediately after either manual catching or lifting from the floor by the mechanized pick-up head. The data showed that both procedures acted as short-term stressors, but that the effects were more pronounced after manual catching than after mechanical collection. It was concluded that stress could be reduced and welfare improved in the birds by mechanical collection with carefully designed lifting machines.

However, the study only considered lifting birds off the floor and did not address the subsequent placement of birds into transport containers. A complete comparison was not possible at that time, as there were no machines available for testing which effected the complete transfer of birds from floor to container without manual intervention. It is surprising that further physiological and behavioural experiments have not been undertaken to evaluate and compare the effects of mechanical versus manual catching/harvesting since Duncan's (1986) discussion of the stressfulness of the procedures. Mechanical harvesting is being encouraged and supported by industry and welfare bodies in both Europe and North America on the grounds of lower costs, improved welfare, reduced bird injury, product damage and rejections, and improved working conditions for poultry operatives (Lacy and Czarick, 1998). The evidence for such claims remains equivocal and a recent study suggests that transport mortalities may be higher in some machine-caught flocks than in their manual equivalents, and that bruising may also be elevated in these birds (Ekstrand, 1998).

Auditing of Welfare in Catching and Handling

It is clear that substantial stress may be imposed during broiler catching and that significant numbers of injuries are caused by inconsiderate handling procedures. These injuries may severely compromise the birds' welfare and the problems may be exacerbated by the additional stressors imposed during the subsequent journey to the slaughterhouse. With existing knowledge it is possible to identify the critical control points of the process.

First, the method of picking from the floor is crucial, as are the number of birds lifted and the method of holding. Secondly, the care with which birds are placed into the transport container and the way in which this container is closed will both influence the well-being of broilers.

The available evidence suggests that mechanical harvesting systems may reduce risks to the birds from each of these.

The following assessments could therefore provide a basis for auditing (these assessments may be undertaken at both the farm and the processing plant):

- At the *farm*, the catching process could be assessed by characterizing the type of container and/or module employed; the use of manual or mechanized catching procedures (different assessment?); the numbers of broilers that are picked up manually at each take; and the number of birds that are carried by each operative and in each hand. Further points would include whether handling is visibly careful and considerate; the length of time taken to load each container and vehicle; whether the containers are of a drawer type or have a gate opening, which may involve placing birds in the furthest portion of the container by manual means; how the containers and drawers are closed and what procedures are in place to ensure that injuries are not caused by injudicious closing of drawers or gates. It may also be possible to attribute and quantify some injuries per load at the farm.
- At the *factory*, records of mortality, morbidity and injury can be examined. Whilst birds that are dead on arrival (DOA) or, more strictly, 'dead on shackling', will include transport and lairage mortalities, it should be possible to assess more accurately the extent of catching injuries in the form of leg and wing fractures, dislocations and head-crush in drawer systems. If the analysis of DOAs and carcass rejections is thorough and supported by the appropriate inspection of birds on the line for injuries, then the extent of catching damage can be audited and the welfare of the birds assessed. All such inspections should be supplemented by concurrent examinations of stress and welfare of the birds during transport, as outlined below.

Road Transport of Broilers

Prior to and during transit, birds may be exposed to a variety of potential stressors, including the thermal demands of the transport microenvironment, acceleration, vibration, motion, impacts, fasting, withdrawal of water, social disruption and noise (Nicol and Scott, 1990; Mitchell *et al.*, 1992; Mitchell and Kettlewell, 1993, 1998). Each of these factors and their various combinations may impose stress upon the birds, but it is well recognized that thermal challenges and in particular heat stress constitute the major threat to animal well-being and productivity (Mitchell and Kettlewell, 1998; Mitchell *et al.*, 2000, 2001; Weeks and Nicol, 2000; Nilipour, 2002). The imposition of thermal loads upon the birds in transit will result in moderate to severe thermal stress and consequent reduced welfare (Mitchell *et al.*, 1992, 2001; Mitchell and Kettlewell, 1998), increased mortality due to either heat or cold stress (Hunter *et al.*, 2001), and induced pathology, including muscle damage and associated changes in product quality (Gregory, 1998; Mitchell, 1999).

These issues can be addressed by means of a multidisciplinary approach involving the following: (i) the full characterization of the thermal microenvironment on commercial vehicles; (ii) understanding the factors that determine the micro-environment; (iii) the birds' physiological requirements: optimum conditions and limits; and (iv) strategies and solutions: improved ventilation. These will now be discussed in turn.

Full characterization of the thermal microenvironment on commercial vehicles

The distribution of temperature and humidity within the load of chickens is not uniform (Kettlewell and Mitchell, 1993; Mitchell and Kettlewell, 1998). On typical closed transporters, temperature lifts of 10–20°C in the thermal core may be encountered. Thermal core temperatures of 25–26°C accompanied by water vapour densities of 5–17 g per m^3 may be observed regularly on commercial transporters in the UK during winter transport, when external temperatures may be 5–8°C. On such vehicles, an 18°C gradient between the thermal core and air inlets will exist, with a parallel humidity gradient. Thus, the complex nature of the thermal microenvironment on broiler vehicles may result in a risk of severe heat stress for some birds in relatively mild weather. Core temperatures above 30°C and vapour densities greater than 20 g per m^3 have been reported in the UK (Kettlewell and Mitchell, 1993). Knezacek *et al.* (2000) measured a 60°C temperature lift in one container in a broiler transporter operating with minimal passive ventilation when the external temperature was −28°C in central Canada. Clearly, thermal stress is even more likely with elevated external temperatures in southern Europe, and in countries whose climate is subtropical or tropical. Heat stress will increase evaporative heat loss and thus potentially body weight loss (Mitchell *et al.*, 2003). High levels of preslaughter stress, including heat stress, may affect meat quality, including meat colour (Kannan *et al.*, 1997). Transport *per se* can increase the incidence of haemorrhages (Kranen *et al.*, 2000). Heat stress during transport may be compounded by holding the birds at the processing plant prior to slaughter. Studies have demonstrated that lairage or holding microenvironments may be poorly controlled (Quinn *et al.*, 1998); in consequence hyperthermia or heat stress may occur during this period, resulting in acid–base balance disturbances, muscle damage, further reductions in bird welfare (Hunter *et al.*, 1998; Sandercock *et al.*, 1999), and depletion of liver glycogen and meat quality alterations (Warriss *et al.*, 1999). It is now recommended that time in lairage should be minimized and kept below 1 hour.

Understanding the factors that determine the microenvironment

The internal thermal microenvironment in the transport containers is the product of the inlet air temperature and humidity, airflow rate and the heat and moisture production of the birds (Mitchell *et al.*, 2000). The passive ventilation regimes of most commercial broiler transport vehicles result in low rates and heterogeneous distribution of airflow within the bioload. Studies have characterized the pressure profiles over the surface of and within commercial broiler vehicles (Baker *et al.*, 1996; Dalley *et al.*, 1996; Hoxey *et al.*, 1996). It is these pressures that drive passive ventilation within the vehicle. A central feature is the tendency for air to move in the same direction as the motion of the vehicle; thus air tends to enter at the rear and move forward over the birds, exiting towards the front. This pattern accounts for the distribution of temperatures and humidities observed on commercial vehicles, the existence of the thermal core, the ingress of water spray and bird wetting, and the pattern of dead-on-arrivals and thermal stress found within the load (Hunter *et al.*, 1997, 2001; Mitchell *et al.*, 1997, 1998a). When vehicles are stationary there is no external force driving the ventilation; thus heat and moisture removal is then dependent upon free convection. Problems of heat stress may be markedly exacerbated even on open or semi-open vehicles, particularly when stationary in hot and humid weather. Solid-floored containers restrict the flow of air. Any practical solution to these problems must involve modification and improvement of the ventilation regime.

The birds' physiological requirements: optimum conditions and limits

The degree of physiological stress imposed upon slaughter-weight broilers by a range of temperature and humidity combinations has been determined in transport simulation studies (Mitchell and Kettlewell, 1998; Mitchell *et al.*, 2000, 2001). Thermal loads were expressed as apparent equivalent temperatures (AET) derived from dry bulb temperature, water vapour pressure and the corrected psychometric constant (Mitchell *et al.*, 1996; Mitchell and Kettlewell, 1998). This approach allowed definition of thermal comfort zones, optimum transport conditions and acceptable limits for temperature and humidity for broilers in transport crates under commercial transport conditions (Fig. 12.1). In essence it is recommended that temperature–humidity combinations yielding an AET greater than 65°C in transport containers should be avoided at all times as these thermal loads would be associated with severe thermal stress (danger zone) and increased mortality. In practice this equates to maintaining the in-crate dry bulb temperature at 26–27°C or less, because the water vapour densities in commercial transport containers are typically 70–80%. Temperatures and humidities equating with AET values of 40–65°C may impose mild to moderate physiological stress (alert) and those less than 40°C may be considered to be associated with minimal stress and are thus safe (Mitchell and Kettlewell, 1998; Cockram and Mitchell, 1999). The physiological modelling procedure provides the

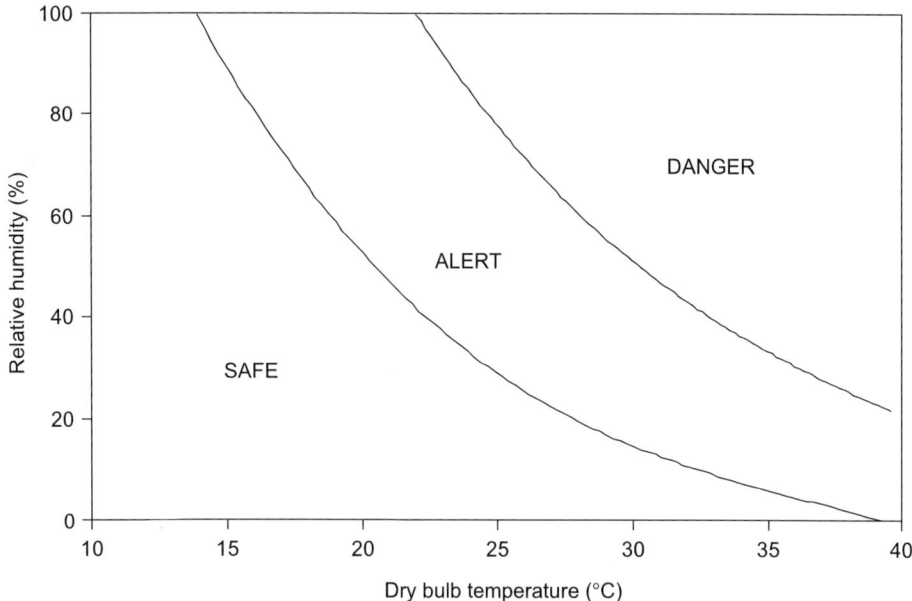

Fig. 12.1. Thermal comfort zones for broiler transport. Safe limit AET (apparent equivalent temperature) = 40°C; danger limit AET = 65°C or greater.

target conditions to be achieved by modification and improvement of vehicle ventilation.

Strategies and solutions: improved ventilation

A number of modifications to existing passively ventilated vehicles may be made in order to improve air flow through the load space. The success of these strategies is limited, particularly in the hostile climatic conditions encountered in many major broiler-producing regions. If current transport numbers per vehicle are to be maintained to ensure economic efficiency, then forced or mechanical ventilation may be the only feasible option. In order to define the specifications of such a system, it is necessary to determine three factors:

- The ranges and limits for temperature and humidity within the transport containers and the optimum thermal environment for transport.
- The acceptable temperature and humidity lifts above ambient for a range of external conditions.
- The heat and moisture loads for dissipation by the ventilation system.

Physiological response modelling has provided the data necessary for factors 1 and 2. Total heat and moisture production of slaughter-weight broiler chickens were measured in a fan-ventilated commercial broiler transporter that was instrumented to function as a direct calorimeter (Mitchell *et al.*, 1998b; Kettlewell *et al.*, 2000, 2001a). The lorry contained birds weighing 1.75 kg. Ambient temperature (T_e)

during the experiments ranged from 9 to 17°C. Two fan flow rates (2.8 and 4.9 m³ per second) were employed. The mean total heat production on the lorry was 26.1 ± 3.0 kW, which could be partitioned into 63% sensible and 37% latent components. This value represents 9.5 ± 1.0 W per bird. On the vehicle water loss was 1.43 mg per bird per second, and it may thus be calculated that to maintain a specified temperature lift within the transport container the ventilation system must be capable of dissipating loads of at least 5.41 W per kg and 0.82 mg per kg per second. In practice, removal of larger heat and water vapour loads may be required. On the basis of these findings, it is proposed that the effective ventilation rate for vehicles in this configuration, fully loaded and operating at environmental temperatures of up to 20°C, should be 0.6 m³ per second per tonne live weight (Kettlewell *et al.*, 2001). Several fans should be installed on the vehicle, giving a maximum ventilation rate which will exceed this value. To achieve maximum fan efficiency they should be used to extract air at sites of low pressure, where air will naturally tend to leave the vehicle (Baker *et al.*, 1996; Dalley *et al.*, 1996; Hoxey *et al.*, 1996). A vehicle based on these design principles (Kettlewell and Mitchell 2001a,b) is now operating commercially within the UK (Fig. 12.2).

Auditing Welfare in Broiler Transport

As explained above, the thermal microenvironment is often the major source of stress and reduced welfare of broilers in transit. Heat and cold stress can add to problems caused by catching injury, existing pathology, food and water withdrawal, vibrations, impacts and acceleration and possible pollutants. In order to assess these variables

Fig. 12.2.   Concept 2000, a mechanically ventilated broiler transport vehicle with automated control, now in commercial service.

and to audit welfare it is necessary to first consider the type of transport containers and vehicles employed. Are the containers of an appropriate nature for adequate ventilation, are they free of protrusions which may cause injury and are they well maintained and cleaned? Are the vehicles passively or mechanically ventilated? If they are, what system is employed? What are the prevailing weather conditions? What was the duration of the journey and were there any journey interruptions or hold-ups? How quickly were the birds unloaded and what was the duration of the lairage time? What was the condition of the birds upon arrival, were there obvious mortalities, and, if so, where in the load did these occur? Were the birds wet or dry and did they appear dirty or clean? As the thermal microenvironment poses the biggest threat to welfare, what was the body temperature of the birds? Were they panting or shivering upon arrival? All such observations can retrospectively be correlated with dead-on-arrivals, down-gradings and meat quality assessments. Incorporation of these assessments can therefore form the basis of a sound, practicable welfare audit system for broilers in transit. In future, an audit system may also contain a requirement for automatic meaningful recording of the on-board thermal microenvironment so that any adverse conditions can be identified and the duration of exposure determined. The recorded temperatures and humidities would be compared with the recommended thermal envelope, and poor environmental and ventilation control could then be identified. Such auditing, with the appropriate penalties for failure to conform and an advisory support system to improve practices, procedures and vehicle design and operation, would help maximize the welfare of broilers during transport.

Summary

Modern broiler production systems involve the intensive rearing of large numbers of birds on geographically dispersed sites and their subsequent harvesting/catching at slaughter weight and transport on purpose-built vehicles to centralized processing plants. Both catching and transport represent potential risks to bird welfare and to product quality. Catching and transport systems are often interdependent and integrated, and their design and operation are major determinants of bird welfare and productivity. Transport container and vehicle design must take due account of the climatic conditions and operational requirements and constraints in different regions of the world. The imposition of stress and accidental injury during handling and in transit are issues of major concern to producers, the veterinary profession, welfare organizations, governments and the public. Fuller understanding of the origins of handling and transport stress and characterization of the birds' biological requirements allow the design and development of improved methods of catching and transport. This review examines the available catching methods, including mechanical harvesting, and reports the current position as regards the assessment of the efficacy of such systems. The role of the transport thermal microenvironment in the aetiology of physiological stress in transit is also considered. Fan-ventilated broiler transport vehicles may improve broiler welfare in transit. Simple and practical ways to audit welfare during catching and transport are suggested.

Conclusions

- Catching, handling and transport constitute potential major sources of stress and reduced welfare in broilers.
- Manual catching can pose a threat to bird welfare, and improved practices should be introduced, monitored and assessed as part of any future auditing system. Possible methods for this process are presented in this chapter.
- Mechanical catching may represent an opportunity for improving welfare standards of broiler catching, but the implementation of this new technology should be closely monitored and assessed and the process audited with as much vigour as is applied to manual procedures.
- Handling and catching may be audited through measurements and observations made at both the farm and processing/slaughter plant.
- The thermal microenvironments encountered in broiler transport can constitute a major potential stressor and thus a primary threat to the birds' welfare.
- Models from research have established the acceptable ranges and limits for temperature and humidity in transit.
- In practice, effects of the thermal conditions can be simply and routinely assessed from:
  - body temperature and environmental conditions;
  - appearance and behaviour (panting, shivering, mucous membranes, etc.);
  - carcass characteristics.
- Vehicle and ambient environments should be routinely monitored and recorded.
- Vehicle environments should be controlled wherever possible.
- Birds should be routinely examined on arrival and behaviour, their appearance noted and mortality distributions determined (where possible).
- Carcass characteristics should be correlated with transport and handling conditions.
- Methods for measuring and auditing the status of the birds are available.
- Research findings have facilitated improvements in procedures and environments (and legislation, codes of practice and education).
- On this basis, simple approaches for assessing and improving welfare under commercial conditions can be developed.

Acknowledgements

This research programme was funded by the Ministry of Agriculture, Fisheries and Food (DEFRA), a Clean Technology Grant from the Biotechnology and Biological Research Council, and commercial funding from the UK broiler industry. The assistance of many colleagues at both research institutes is acknowledged, but particular thanks are given to the contributions from Ailsa Carlisle, Richard Hunter, Dale Sandercock, Chris Hampson, Nigel Green, Nick Teer and Roger Hoxey.

References

Anonymous (1979a) *Poultry Industry* 1979 *Disease Directory*. Agricultural Press Ltd, Sutton, UK, pp. 12–13.

Anonymous (1979b) *Poultry World* 1979 *Disease Directory*. Agricultural Press Ltd, Sutton, UK, pp. 21–23.

Anonymous (1999) *Poultry World* 1999 *Disease Directory*. Agricultural Press Ltd, Sutton, UK, pp. 13–25.

Baker, C.J., Dalley, S., Yang, X., Kettlewell, P.J. and Hoxey, R. (1996) An investigation of the aerodynamic and ventilation characteristics of poultry transport vehicles: Part 2, Wind tunnel experiments. *Journal of Agricultural Engineering Research* 65, 97–113.

Barton-Gade, P. (2002) Welfare of animal production in intensive and organic systems with special reference to Danish organic pig production. *Meat Science* 62, 353–358.

Bayliss, P.A. and Hinton, M.H. (1990) Transportation of broilers with special reference to mortality rates. *Applied Animal Behaviour Science* 28, 93–118.

Berry, P.S., Kettlewell, P.J. and Moran, P. (1990) The AFRC Mark I experimental broiler harvester. *Journal of Agricultural Engineering Research* 47, 153–163.

Bingham, A.N. (1986) Harvesting broilers. *British Poultry Science* 27, 150–150.

Carlyle, W.W.H., Guise, H.J. and Cook, P. (1997) Effect of time between farm loading and processing on carcass quality of broiler chickens. *Veterinary Record* 141, 364.

Cockram, M.S. and Mitchell, M.A. (1999) Role of research in the formulation of rules to protect the welfare of farm animals during road transportation. In: Russel, A.J.F., Morgan, C.A., Savory, C.J., Appleby, M.C. and Lawrence, T.L.J. (eds) *Farm Animal Welfare – Who Writes the Rules?* Occasional Publication No. 23, British Society of Animal Science, pp. 43–64.

Dalley, S., Baker, C.J., Yang, X., Kettlewell, P.J. and Hoxey, R. (1996) An investigation of the aerodynamic and ventilation characterisitics of poultry transport vehicles: Part 3. Internal flow field calculations. *Journal of Agricultural Engineering Research* 65, 115–127.

DEFRA (2003) Codes of recommendation for meat chickens and breeding chickens. Department for Environment, Food and Rural Affairs, London.

DeGraft-Hanson, J. (2003) Hazard analysis critical control point system and its impact on the meat and poultry industry. *Avian and Poultry Biology Reviews* 14, 79.

De Koning, K., Gerrits, A.R. and Migchels, A. (1987) Mechanized harvesting and transport of broilers. *Journal of Agricultural Engineering Research* 38, 105–111.

Duncan, I.J.H. (1986) Some thoughts on the stressfulness of harvesting broilers. *Applied Animal Behaviour Science* 16, 97.

Duncan, I.J.H., Slee, G.S., Kettlewell, P.J., Berry, P.S. and Carlisle, A.J. (1986) Comparison of stressfulness of harvesting broiler chickens by machine and by hand. *British Poultry Science* 27, 109–114.

Ekstrand, C. (1998) An observational cohort study of the effects of catching method on carcass rejection rates in broilers. *Animal Welfare* 7, 87–96.

Fallon, M. (2001) Traceability of poultry and poultry products. *Revue Scientifique et Technique de l'Office International des Epizooites* 20, 538–546.

FAOSTAT (2003) FAOSTAT Database. Available at: http://apps.fao.org

Gerrits, A.R. and de Koning, K. (1980) Van hok naar haak. *Pluimveehouderij* 10 (43), 86–95.

Gerrits, A.R., de Koning, K. and Migchels, A. (1985) Catching poultry. *Poultry* 1985 (July), 20–23.

Gregory, N.G. (1998) Livestock presentation and welfare before slaughter. In: Gregory, N.G. (ed.) *Animal Welfare and Meat Science*. CAB International, Wallingford, UK, pp. 15–41.

Hoxey, R.P., Kettlewell, P.J., Meehan, A.M., Baker, C.J. and Yang, X. (1996) An investigation of the aerodynamic and ventilation characteristics of poultry transport vehicles. Part I. Full scale measurements. *Journal of Agricultural Engineering Research* 65, 77–83.

Hunter, R.R., Mitchell, M.A. and Matheu, C. (1997) Distribution of 'dead on arrivals' within the bio-load on commercial broiler transporters: correlation with climate conditions and ventilation regimen. *British Poultry Science* 38 Supplement, S7–S9.

Hunter, R.R., Mitchell, M.A., Carlisle, A.J., Quinn, A.D., Kettlewell, P.J., Knowles, T.G. and Warriss, P.D. (1998) Physiological responses of broilers to pre-slaughter lairage: effects of the thermal micro-environment? *British Poultry Science* 39, S53–S54.

Hunter, R.R., Mitchell, M.A. and Matheu, C. (2001) Mortality of broiler chickens in transit – correlation with the thermal micro-environment. In: Stowell, R.R., Bucklin, R. and Bottcher, R.W. (eds) *Proceedings of the 6th International Livestock Environment Symposium, Louisville, Kentucky, USA, 21st–23rd May 2001*, pp. 542–549.

Kannan, G., Heath, J.L., Wabeck, C.J., Souza, M.C.P., Howe, J.C. and Mench, J.A. (1997) Effects of crating and transport on stress and meat quality characteristics in broilers. *Poultry Science* 76, 523–529.

Kettlewell, P.J. and Turner, M.J.B. (1985) A review of broiler chicken catching and transport systems. *Journal of Agricultural Engineering Research* 31, 93–114.

Kettlewell, P.J. and Mitchell, M.A. (1993) The thermal environment on poultry transport vehicles. In: Collins, E. and Boon, C. (eds) *Livestock Environment IV. Proceedings of the Fourth International Symposium*. American Society of Agricultural Engineers, St Joseph, Michigan, pp. 552–559.

Kettlewell, P.J. and Mitchell, M.A. (2001a) Comfortable ride: concept 2000 provides climate control during poultry transport. *Resource: Engineering and Technology for a Sustainable World* 8, 13–14.

Kettlewell, P.J. and Mitchell, M.A. (2001b) Mechanical ventilation: improving the welfare of broiler chickens in transit. *Journal of the Royal Agricultural Society* 162, 175–184.

Kettlewell, P.J., Mitchell, M.A. and Meehan, A. (1993) The distribution of thermal loads within poultry transport vehicles. *Agricultural Engineer* 48, 26–30.

Kettlewell, P.J., Hoxey, R.P. and Mitchell, M.A. (2000) Heat produced by broiler chickens in a commercial transport vehicle. *Journal of Agricultural Engineering Research* 75, 315–326.

Kettlewell, P.J., Hampson, C.J., Green, N.R., Teer, N.J., Veale, B.M. and Mitchell, M.A. (2001) Heat and moisture generation of livestock during transportation. In: Stowell, R.R., Bucklin, R. and Bottcher, R.W. (eds) *Proceedings of the 6th International Livestock Environment Symposium, Louisville, Kentucky, USA, 21st–23rd May, 2001*, pp. 519–526.

Kettlewell, P.J., Hampson, C.J., Berry, P.S., Green, N.R. and Mitchell, M.A. (2000) New developments in bird harvesting, live haul and unloading in the United Kingdom. In: *World Poultry Congress, Montreal, Canada, 22nd–24th August, 2000*. S3.13.01 CD-ROM. Events International Meeting Planners, Montreal, Canada, pp. 1–20.

Knezacek, T.D., Audren, G.F., Mitchell, M.A., Kettlewell, P.J., Hunter, R.R., Classen, H.L., Stephens, S., Olkowski, A.A., Barber, E.M. and Crowe, T.G. (2000) Temperature heterogeneity and moisture accumulations in trailers transporting broilers under Canadian winter conditions. *Poultry Science* 79, Supplement 1, 31.

Kranen, R.W., Lambooij, E., Veerkamp, C.H., van Kuppevelt, T.H. and Veerkamp, J.H. (2000) Haemorrhages in muscles of broiler chickens. *World's Poultry Science Journal* 56, 93–126.

Lacy, M.P. and Czarick, M. (1998) Mechanical harvesting of broilers. *Poultry Science* 77, 1794–1797.

Main, D.C.J., Webster, A.J.F. and Green, L.E. (2001) Animal welfare assessment in quality assurance schemes. *Acta Agriculturae Scandanavica (Section A)* Supplement 30, 108–113.

Manbeck, H.B. (1974) A mechanised handling system for handling broilers in bulk. *Transactions of the American Society of Agricultural Engineers* 17, 1191–1194.

Metheringham, J. and Hubrecht, R. (1996) Poultry in transit – a cause for concern. *British Veterinary Journal* 152, 247–249.

Ministry of Agriculture, Fisheries and Food (1999) *Meat Hygiene Enforcement Report*. Issue 32, December 1999.

Mitchell, M.A. (1999) Muscle abnormalities – pathophysiological mechanisms. In: Richardson, R.I. and Mead, G.C. (eds) *Poultry Meat Science. Poultry Science Symposium Series, Volume 25*. CAB International, Wallingford, UK, pp. 65–98.

Mitchell, M.A. and Kettlewell, P.J. (1993) Catching and transport of broiler chickens. In: Savory, C.J. and Hughes, B.O. (eds) *Proceedings of the Fourth European Symposium on Poultry Welfare.* Universities Federation for Animal Welfare, Wheathampstead, UK, pp. 219–229.

Mitchell, M.A. and Kettlewell, P.J. (1998) Physiological stress and welfare of broiler chickens in transit: solutions not problems! *Poultry Science* 77, 1803–1814.

Mitchell, M.A., Kettlewell, P.J. and Maxwell, M.H. (1992) Indicators of physiological stress in broiler chickens during road transportation. *Animal Welfare* 1, 91–103.

Mitchell, M.A., Kettlewell, P.J., Carlisle, A.J. and Matheu, C. (1996) The use of apparent equivalent temperature (AET) to define the optimum thermal environment for broilers in transit. *Poultry Science* 75 (Supplement 1), 18.

Mitchell, M.A., Carlisle, A.J., Hunter, R.R. and Kettlewell, P.J. (1997) Welfare of broilers during transportation: cold stress in winter – causes and solutions. In: Koene, P. and Blokhuis, H.J. (eds) *Proceedings of the Fifth European Symposium on Poultry Welfare.* WAPS, University of Wageningen and Institute of Animal Science and Health, Wageningen, Netherlands, pp. 49–52.

Mitchell, M.A., Hunter, R.R., Kettlewell, P.J. and Carlisle, A.J. (1998a) Body temperature responses to the thermal micro-environment in transit from farm to slaughter. *Poultry Science* 77 (Supplement 1), 110.

Mitchell, M.A., Kettlewell, P.J., Hoxey, R.P. and Macleod, M.G. (1998b) Heat and moisture production of broilers during transportation. A whole vehicle direct calorimeter. *Poultry Science* 77 (Supplement 1), 4.

Mitchell, M.A., Carlisle, A.J., Hunter, R.R. and Kettlewell, P.J. (2000) The responses of birds to transportation. In: *Proceedings of the World Poultry Congress, Montreal, Canada, 22nd–24th August 2000.* S3.13.05 CD-ROM, Events International Meeting Planners, Montreal, Canada, pp. 1–14.

Mitchell, M.A., Kettlewell, P.J., Hunter, R.R. and Carlisle, A.J. (2001) Physiological stress response modeling – application to the broiler transport thermal environment. In: Stowell, R.R., Bucklin, R. and Bottcher, R.W. (eds) *Proceedings of the 6th International Livestock Environment Symposium, Louisville, Kentucky, USA, 21st–23rd May 2001,* pp. 550–555.

Mitchell, M.A., Carlisle, A.J., Hunter, R.R. and Kettlewell, P.L. (2003) Weight loss in transit: an important issue in broiler transportation. *Poultry Science* 82 (Supplement 1), 101, S52.

Nelson, G.S. (1982) A mechanized system for harvesting broilers. *Arkansas Farm Research* 31, 4.

Nelson, G.S. (1984) A mechanized system for harvesting broilers. *Transactions of the ASAE* 27, 41–44.

Nesheim, M.C., Austic, R.E. and Card, L.E. (1979) *Poultry Production.* Lea and Febiger, Philadelphia, Pennsylvania, pp. 381–384.

Nicol, C.J. and Scott, G.B. (1990) Pre-slaughter handling and transport of broiler chickens. *Applied Animal Behaviour Science* 28, 57–73.

Nilipour, A.H. (1996) Live haul, making or losing money? *World Poultry* 12 (12), 49–53.

Nilipour, A.H. (2002) Poultry in transit are a cause for concern. *World Poultry* 18 (2), 30–33.

Pettitt, R.G. (2001) Traceability in the food animal industry and supermarket chains. *Revue Scientifique et Technique de l'Office International des Epizooites* 20, 584–597.

Poultry World (2001) EU facts and forecasts. *Poultry World* 2001 (September), 14–15.

Quinn, A.D., Kettlewell, P.J., Mitchell, M.A. and Knowles, T.G. (1998) Air movement and the thermal microclimates observed in poultry lairages. *British Poultry Science* 39, 469–476.

Quinn, B.P. and Marriott, N.G. (2002) HACCP plan development and assessment: a review. *Journal of Muscle Foods* 13, 313–330.

Reed, M.J. (1974) Mechanical harvesting of broiler chickens. *Transactions of the American Society of Agricultural Engineers* 17, 74–81.

Royal Society for the Prevention of Cruelty to Animals (1999) *Welfare Standards for Chickens.* Freedom Food Limited, RSPCA, Causeway, UK.

Sandercock, D.A., Hunter, R.R., Nute, G.R., Mitchell, M.A. and Hocking, P.M. (2001) Acute heat stress-induced alterations in blood acid– base status and skeletal muscle membrane integrity in broiler chickens at two ages: implications for meat quality. *Poultry Science* 80, 418–425.

Scott, G.B. (1993) Poultry handling – a review of mechanical devices and their effect on bird welfare. *World's Poultry Science Journal* 49, 44–57.

Warriss, P.D., Bevis, E.A. and Brown, S.N. (1990) Time spent by broiler chickens in transit to processing plants. *Veterinary Record* 127, 617–619.

Warriss, P.D., Knowles, T.G., Brown, S.N., Edwards, J.E., Kettlewell, P.J., Mitchell, M.A. and Baxter, C.A. (1999) Effects of lairage time on body temperature and glycogen reserves of broiler chickens held in transport modules. *Veterinary Record* 145, 218–222.

Webster, A.J.F. (2001) Farm animal welfare: the five freedoms and the free market. *Veterinary Journal* 161, 229–237.

Weeks, C. and Nicol, C. (2000) Poultry handling and transport. In: Grandin, T. (ed.) *Livestock Handling and Transport*, 2nd edn. CAB International, Wallingford, UK, pp. 363–384.

Welfare of Animals (Transport) Order (1997) Statutory Instrument 1997 No. 1480. Ministry of Agriculture, Fisheries and Food, London.

Xin, H., Sell, J.L. and Ahn, D.U. (1996) Effects of light and darkness on heat and moisture production of broilers. *Transactions of the American Agricultural Society* 39, 2255–2258.

13 Primary Processing of Poultry

S. WOTTON AND L.J. WILKINS

School of Veterinary Science, University of Bristol, Langford, Bristol, UK

Billions of broilers are processed for food each year in centralized, highly automated plants. Experimental and commercial trials carried out over a number of years have led to improvements in the design and operation of the primary processing area for poultry. This chapter combines the results from a number of publications together with unpublished data to provide an insight into the optimum design for a primary processing system based upon electrical waterbath stunning.

Hang-on

The logistics of the hang-on area should be carefully examined to ensure that the welfare of the bird is optimized. The birds should be presented to the operatives in a way that will minimize the physical effort required to remove them from the transport containers and hang them on the shackle-line. Mishandling or rough handling at hang-on will result in a compromise of bird welfare with subsequent downgrading of the carcass. Carcass downgrading will be expressed as bruises to the legs and/or an increase in the occurrence of wing damage, particularly the prevalence of red wingtips as a result of wing flapping. The likely causes of wing flapping at hang-on are discussed in the following paragraphs.

- *Pain associated with shackling.* This could occur for various reasons:
  - Compression of the periosteum in the leg. During shackling, the leg of the bird is compressed to the extent that the periosteum is squeezed against the tibia. Experiments at Langford have shown that birds find the shackling process painful. Analysis of the information available suggests that the main pain component is short-lasting, but the intensity of pain increases as the forces applied are increased. These results were confirmed by Gentle and Tilson (2000), who concluded that shackling is likely to be a very painful procedure.

- Leg size. The pain experienced during shackling can also be affected by the size or cross-sectional diameter of the broiler's leg. It has been shown that male broilers have consistently thicker legs than female broilers, which would increase the forces applied to the shanks during shackling. Genetic development of poultry has also resulted in birds with larger shank diameters, yet shackle design has changed little to account for the new, heavier strains of broiler. If shackle design is changed to account for the larger leg size, other provisions must be considered to ensure both the physical restraint of the bird during processing and the electrical contact between the shank and the shackle.
- Leg weakness and trauma. A bird suffering from leg weakness or catching damage (e.g. dislocation of the femur) would be subject to chronic pain during shackling and conveyance to the waterbath stunner.
- *Bird inversion.* This has been demonstrated to increase the levels of stress that poultry are subjected to during the shackling process.
- *Rough shackling.* This can contribute significantly to wing flapping. It can occur as a result of:
  - Excessive force being used to hang the birds.
  - Inexperienced shacklers.
  - Fast line speed and relatively small shackling teams. This will increase the incidence of birds being miss-hung; e.g. by one leg only, or because two legs have been forced into one side of the shackle.
  - Shackle spacing. Tight shackle spacing when combined with fast line speeds can result in the operative at the last hang-on position hanging birds without being able to see the shackle and having to move adjacent shackled birds aside in order to hang birds on the shackle line.
- *Lighting levels.* Low lighting has a quietening effect on poultry and many processing plants employ blue lighting in this area for this purpose. The MAFF code of practice (MAFF, 1991), soon to be superseded, referred to a legislative requirement that, where ante-mortem health inspection takes place at a slaughterhouse, the lighting must be capable of attaining not less than 540 lux.
- *Noise in the shackling bay.* There are two types of noise: general background noise and impulse sounds caused, for example, by the hiss of air vented from a pneumatic system. The latter sounds are likely to be particularly frightening to birds but both background and impulse sounds may cause disturbance to birds and therefore can compromise their welfare by causing them to wing flap.

It has been reported that when shacklers run their hands down the legs and body of the birds, the incidence of wing flapping is reduced (Gregory and Bell, 1987). Keeping hold of the bird's legs for 1 or 2 seconds after shackling has a similar effect. This calming effect has led to the development of breast comforting aprons, which are designed to maintain physical contact with the breasts of birds as they are conveyed to the waterbath stunner. It is important to ensure that the plastic sheet used for this purpose commences before hang-on and that it is continuous up to, and into, the waterbath. In addition, it is recommended that the bottom of the comforter should be below the level of the bird's heads to reduce incident light and visual distraction. The material used should be rigid and fixed so that heavier birds cannot distort the apron and adjacent smaller birds lose contact. Joints between different sections of the breast comforter should be smooth to prevent bird disturbance.

It is important that the people who are employed at hang-on should have an affinity for birds, and the supervisor of the hang-on team should receive formal training (e.g. the Poultry Welfare Officer training course, University of Bristol). The supervisor should be responsible for in-house training of operatives and special attention should be paid to any downgrading that is a direct result of the rough handling of birds.

Shackle Line

It has been suggested that broilers should be allowed to settle on the shackles for a minimum of 12 seconds and turkeys 25 seconds before entering the waterbath. Gregory and Bell (1987) found that after these times the majority of birds had stopped flapping. The current UK legislation [Welfare of Animals (Slaughter or Killing) Regulations (HMSO, 1995)] requires that no bird be suspended for more than 6 minutes in the case of a turkey or 3 minutes in other cases before being stunned or killed. Therefore, in the event of a broiler line breakdown, processors have a maximum time of 3 minutes. This period commences not at the moment of the breakdown but from the time the broiler that is about to enter the waterbath stunner was hung on the line. Processing plants with shackle lines that are about 12 seconds between the last hang-on point and waterbath entry will extend their potential down-time before action is required to either remove or despatch shackled birds. In addition, an extension of this time will: (i) increase the number of birds that require attention in the event of a significant breakdown; and (ii) increase the time that birds are exposed to a potentially painful procedure (shackling) before they are stunned. A proposed amendment to the legislation in the UK will reduce the maximum hanging times prior to stunning to 3 minutes for turkeys and 2 minutes for other poultry species.

In the event of a breakdown, the birds can be removed and reshackled when the line recommences. However, birds that have been shackled initially are likely to have received physical damage (trauma) to their legs. When the legs are shackled for a second time, this damage is likely to result in increased levels of pain. In addition, access to birds may be limited; for example, when the shackle lines are high. Alternatively, birds could be killed before removal from the shackle. Legislation (HMSO, 1995) stipulates that, for birds only, decapitation or dislocation of the neck is a permitted method of killing to prevent pain or distress. Research at Langford has demonstrated that there are doubts about the welfare aspects of decapitation and neck dislocation (Gregory and Wotton, 1990) and this has led to the development of an alternative stunning system for use in the casualty slaughter of poultry (Hewitt, 2000). A pneumatically powered percussive device has been developed and is commercially available (Accles & Shelvoke, Birmingham, UK) for use either as a back-up to the killer or for the despatch of shackled birds in the event of a line breakdown. It is hoped that both neck dislocation and decapitation will be gradually phased out.

Ante-mortem wing flapping on the shackle line has been shown to result in an increased incidence of red wing tips (Table 13.1).

Table 13.1.   Ante-mortem wing flapping and red wing tips. (Adapted from Gregory *et al.*, 1989.)

	Birds with red wing tips (%)	
	Did not flap	Did flap
Chickens (*n* = 11, 993)	4%	23%
Turkeys (*n* = 1,616)	7%	42%

There are various other factors that can contribute to an increase in wing flapping in shackled birds between hang-on and entrance to the waterbath stunner.

- *Bends or unevenness in the line.* Any bends (especially those that are particularly tight) in the line and swinging shackles can initiate wing flapping. Tight bends may also cause temporary loss of physical contact with the breast comforter or loss of visual contact between neighbouring birds, which has been shown to increase wing flapping. Loss of visual contact can also result when turkeys are hung every other shackle. Unevenness of the line can occur where welded joints in the overhead line are not ground down sufficiently smooth and the shackle trolleys consequently jolt over the rough area.
- *Obstructions.* Obstructions, such as apertures in the wall that are insufficient to allow birds to pass through with extended wings, crate lids, stanchions, etc., can increase the incidence of flapping and allow birds to come into physical contact with obstructions if they flap. Birds must be unable to reach any obstructions with outstretched wings.
- *Lighting.* Low levels of lighting are generally present between hang-on and the waterbath stunner. However, shafts of sunlight or light from a brightly lit processing area when a door is left ajar have been shown to cause an increase in the incidence of wing flapping.
- *The presence of personnel.* Plant personnel walking underneath the shackle line or close to the birds can initiate isolated cases of wing flapping. Humans are perceived by poultry as their top predator and bird agitation has been known to result from the close proximity of operatives. The area surrounding the shackle line should not be a thoroughfare for staff to access the lairage or the stunning area.
- *Water spray.* Legislation (HMSO, 1995) requires that appropriate measures are taken to ensure that the electrical stunning current passes efficiently; in particular that there are good electrical contacts and the shackle-to-leg contact is kept wet. Recent research (Perez-Palacios, 2003) has demonstrated that, provided the shackles are wet prior to hang-on, i.e. through the positioning of the shackle washer or through the use of a shackle spray prior to hang-on, there is no added benefit to be gained through the use of a shackle/leg spray. The use of a shackle/leg spray could stimulate birds to flap and/or result in a wet bird, and could thus have implications for effective electrical stunning. Wet birds require significantly higher currents than dry birds to produce an effective stun.
- *Pre-stun shocks.* A pre-stun shock as the bird enters the waterbath is highly likely to stimulate wing flapping immediately prior to immersion and possibly result in the bird 'flying' the waterbath stunner.

Pre-stun Shocks

Pre-stun shocks occur when the bird makes electrical contact with the 'live' water-bath before it is stunned. It is generally accepted that for stunning to be immediate, as defined in the legislation (HMSO, 1995), the bird must be rendered unconscious before it is able to feel any pain associated with the stunning method. The time necessary for a bird to perceive a painful stimulation is 100–150 ms. Therefore, the head of the bird must be immersed in the live waterbath within about 100 ms of first electrical contact to ensure that it does not receive an electrical shock before it is stunned (HMSO, 1995).

The various factors that will affect the incidence of pre-stun shocks in poultry are as follows:

- An entry ramp that is wet and in physical contact with the live waterbath. Birds that are drawn up such an uninsulated entry ramp can obtain a pre-stun shock through this wet contact.
- Very slow line speeds. These make it even more important to provide an insulated entry ramp to guide birds cleanly and quickly into the waterbath stunner (Wotton and Gregory, 1991).
- A poorly designed entry ramp. This, when combined with a relatively slow line speed, can allow contact to be made between the bird and the live water before the head is immersed (Wotton and Gregory, 1991).
- The use of a dipped shackle line, particularly with turkeys. The anatomy of turkeys results in the wings hanging lower than the head when the bird is inverted and shackled. The wings of the turkey are likely to be immersed significantly before the head if the bird is lowered into the waterbath.

The angle of the entry ramp can have an effect on the incidence of pre-stun shocks. A steeply inclined entry ramp that extends over the water is an effective way of ensuring that both the head and the wings enter the water together. However, some birds react badly to being pulled up a steep incline and may start to flap their wings, which increases the likelihood of pre-stun shocks. Therefore, the ramp or entry flume should be constructed to form a continuation of a breast-rubbing apron. Turkeys can be drawn up the flume (Fig. 13.1) on their breasts and when they are conveyed over the lip of the flume they enter the water in one swift movement. Consequently, the head and the wings enter the water at the same time. However, this device cannot be implemented in plants where the shackle line dips into the stunner, because, for the flume to be effective, a horizontal shackle-line must be used.

Research at Bristol has produced a turkey waterbath entrance ramp design that is based on an extended breast-rubbing apron, which is twisted to form an entry flume. The angle of the flume is quite shallow to allow fast line speeds and the lip of the flume is of a sufficient length (depending on bird size and line speed) to hold back each bird, to enable a clean entry into the live water. The shackle-stabilizing chains control the degree of restraint that can be imposed on each bird before it is released and swung down into the water. The modified breast comforter and flume (Fig. 13.1) has been fitted and tested in the bleeding area of a major processor (Fig. 13.3) to produce the following recommended dimensions. The overall shape and size of the entry flume prior to forming are shown in Fig. 13.2. The desired shape was obtained by fixing the top half in the vertical plane, so as to form a continuation of the breast-rubbing apron, and twisting the front extended portion of the cut-away lip to a height similar to that of the top half (Figs 13.1, 13.2 and 13.3).

The operation of the entry flume was adjusted when in position within the bleeding area of a processing line and the presentation of the birds from the flume was videotaped from two angles for detailed analysis. The results led to several recommendations regarding the construction of a production version of the modified entry flume based on the following data points (Figs 13.1 and 13.2):

- A concave-shaped lip at the top of the ramp produced optimum release of all sizes of bird.
- $x = 48$ cm was the optimum width for all sizes of bird.

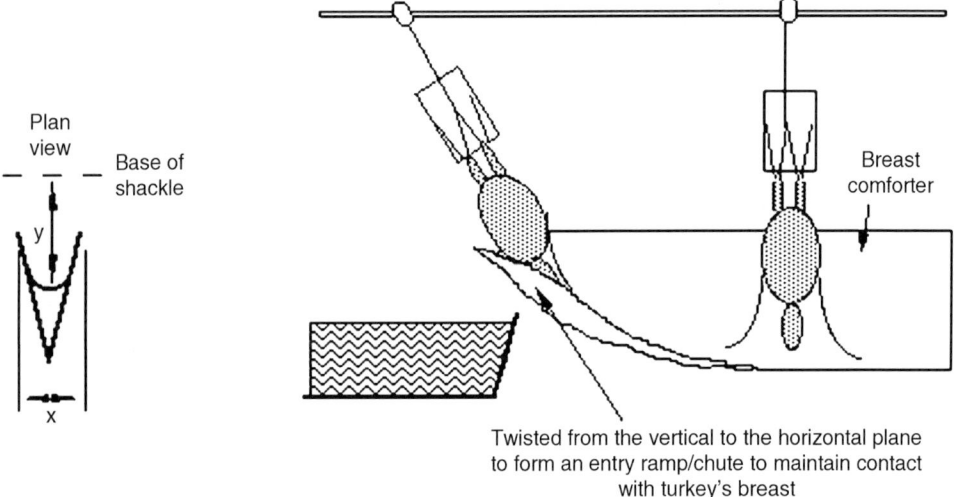

Fig. 13.1.   Schematic diagram of entry flume.

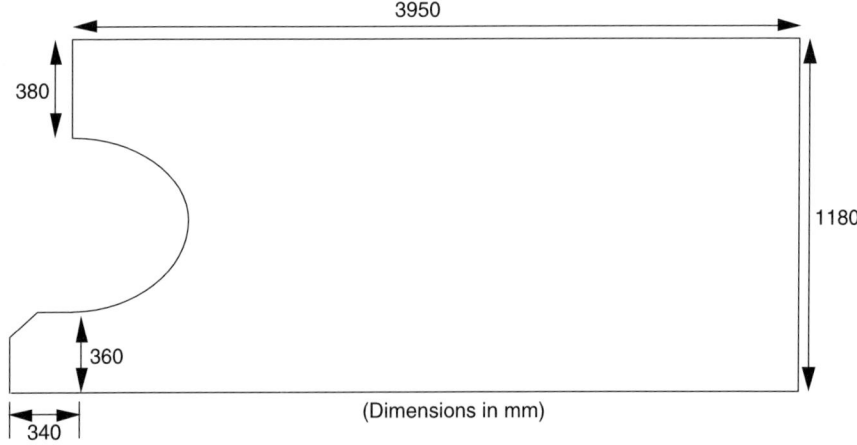

Fig. 13.2.   Shape and dimensions of turkey waterbath entry flume prior to forming.

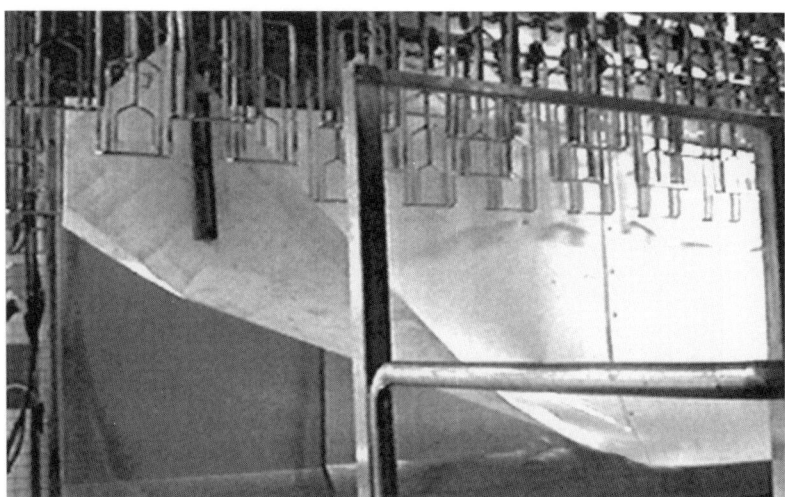

Fig. 13.3. Turkey entry flume mounted in bleeding area for testing.

- y = 25 cm (from shackle base to flume lip) was the optimum height of the lip of the ramp for the smallest birds (8 kg) under trial.
- y = 35 cm (from shackle base to flume lip) was the optimum height of the lip of the ramp for the largest birds (20 kg) under trial.

The range of adjustments necessary to cater for turkeys weighing from 8 to 20 kg is from a minimum of 25 cm (from shackle base to flume lip) to a maximum of 35 cm. The width can remain fixed at 48 cm. The front edge of the ramp should be cut away in a concave shape, as detailed in Fig 13.3. This solution to the potentially painful problem of pre-stun electric shocks in turkey plants requires the adoption of horizontally conveyed primary processing lines. The choice of dipped lines by equipment manufacturers has exacerbated the problem of pre-stun shocks. The entry flume can be mounted with the waterbath stunner such that adjustments to waterbath height when different sizes of bird are processed will adjust the height of the entry flume simultaneously. We recommend that entry flumes should be initially mounted separately from the waterbath to enable fine tuning of the final position to take place under normal processing conditions. Scaled down versions of this flume may be used for broilers.

Waterbath Design

The adoption of an entry flume necessitates the implementation of a new design of waterbath. The wide, open-sided waterbath shown in Fig. 13.4 will enable birds to enter the water cleanly without interruption; for example, a bird with wings outstretched could contact the sides of a traditional enclosed, narrow waterbath and delay or obstruct its entry. The lack of sides enables staff to witness bird entry to determine the extent of pre-stun shocks.

The health and safety of staff can be protected by encasing the waterbath in a weldmesh cage that will prevent accidental electrocution but still allow inspection

Fig. 13.4.   Open-sided waterbath design.

of the bird throughout the whole stunning process. Given a waterbath length requirement of 4000 mm, the weight of the filled bath would be approximately 1080 metric tonnes and therefore a hydraulic lifting system is required that can raise or lower a filled waterbath quickly during processing.

Pre-stun shocks not only affect bird welfare but can also influence the prevalence of carcass downgrading. It has been reported that turkeys that undergo preslaughter excitement and struggling exhibit a higher degree of redness in the breast muscle. Pre-stun shocks have also been shown to affect the number of turkeys with red wingtips and haemorrhaging in the musculature. Carcass downgrading conditions have also been observed in broilers, where struggling on the killing line, indicative of a reaction to an electric shock, resulted in a larger number of broilers exhibiting haemorrhaging in the thigh muscles.

Stunning

Waterbath stunning systems are at present of constant voltage by design. A constant voltage is applied between a live waterbath electrode and an earthed rail that makes contact with the shackles. On average, more than ten birds are immersed in the waterbath and subjected to the stunning voltage at any one time.

The major concern regarding the calibration of waterbath stunners for poultry is the large variation in impedance (resistance to the flow of electrical current) between birds within the stunner. Unpublished experimental investigations at this laboratory have demonstrated that the major variation in bird impedance is caused by the interface between the bird's leg and the shackle. Variations in the size of the legs of birds and the development of the leg scales will produce large variations in contact impedance. Large variation in contact impedance was recorded at a UK plant (unpublished data) that processed broilers and hens on the same line separately but with the same shackle fill. The voltage applied to the stunner was about 160 V, which produced a current of 1.3–1.8 A when broilers were processed. When the production was changed to hens, the voltage required to deliver the same current was 380 V. The lower impedance for broilers can be accounted for firstly by a larger diameter leg that fits more tightly in the fixed-diameter shackle. Secondly, broilers are very young birds (about 39 days old) with soft-skinned legs, which make relatively good contact with the shackle, whereas hens are much older (about 72 weeks) and have thin, dry, scaly legs that make poor contact. Legislation (HMSO, 1995) requires that 'appropriate measures are taken to ensure that the current passes efficiently, in particular that there are good electrical contacts and the shackle-to-leg contact is kept wet'. This legislative requirement recognizes the contribution the shackle-to-leg contact makes in introducing variation into the system.

Impedance varies both between species and between birds of the same species. Sparrey *et al.* (1992) proposed a range in impedance for broilers from 1000 to 2000 Ω. Wooley *et al.* (1986a,b) used stainless steel electrodes applied directly to birds and found a similar range in impedance for broilers, namely 1000–2600 Ω. Gregory and Wotton (1987) found a mean impedance for broilers to sinusoidal alternating current (50 Hz) of 1590 ± 318 (standard deviation) Ω per bird (range 1153–1665 Ω). This range in impedance was less than the range (1000–2600 Ω), reported by Schuett-Abraham and Wormuth (1991) and Schuett-Abraham *et al.* (1987) (Table 13.2). This disparity could be accounted for by the method used to calculate the average bird impedance. Gregory and Wotton (1987) recorded: (i) the total current flow over a 20-minute period during normal processing; and (ii) the voltage applied under bird load in a commercial waterbath stunner. The average total current flow was measured and the total impedance to current flow was calculated by Ohm's law. The average impedance per bird was subsequently calculated by dividing the total impedance by the number of birds in the waterbath at any one time. This method does not take into account any impedance attributable to: (i) the internal impedance of the stunner, (ii) the impedance of the water or brine, or (iii) the impedance of the contact between the shackle and earth potential.

Research has shown that it is the amplitude of current flow through individual birds that is important in determining whether a bird is adequately stunned. Gregory

Table 13.2.   Range in bird impedances. (From Schuett-Abraham and Wormuth, 1991; Schuett-Abraham *et al.*, 1987.)

Species/type	Bird impedance (Ω)
Broilers	1000–2600
Hens	1900–7000
Turkeys	800–5700
Ducks	1100–2400
Geese	1200–4100

and Wotton (1990) demonstrated that 105 mA per bird would render broilers insensible when a 50 Hz sinusoidal alternating current (AC) was applied for 4 to 5 seconds. Wilkins and others (1998) examined the effectiveness of a variety of waveforms and frequencies on broilers and showed that they were equally effective at inducing a stun at 105 mA. More recent research (Wotton and Wilkins, 1999) examined the effect of using a unipolar pulsed direct current at 550 Hz. The effectiveness of the stun was assessed from the birds' behaviour, and all birds were judged to be stunned effectively when more than 15 mA true RMS was applied for either 1 or 10 s. A number of other electrical frequencies and waveforms are used for stunning broilers in poultry processing plants, but their minimum effective currents are not known (Gregory, 1989). Gregory and Wotton (1994) demonstrated that hens required a similar current to broilers (105 mA per bird) but that recovery from the stun was more rapid, so the period of unconsciousness was shorter.

The problem of current control in a multi-bird waterbath stunner is illustrated in Figs 13.5 and 13.6. In the four-bird waterbath shown in Fig. 13.5, all four birds have the same impedance to current flow, equal to 1560 Ω. The total impedance can be calculated as follows:

$$\frac{1}{I_{total}} = \frac{1}{1560} + \frac{1}{1560} + \frac{1}{1560} + \frac{1}{1560} = 390 \ \Omega \qquad \text{(Equation 1)}$$

Applying Ohm's law (voltage = current × resistance) gives 164 = current × 390. Therefore, total current flow is 420 mA and, because current flow per bird is additive, the average current flow per bird is 105 mA.

However, in reality the resistance of each individual bird in a multi-bird waterbath stunner would always be different and the situation described in Fig. 13.6, where there is a range of bird resistances from 1000 to 2500 Ω, is more representative of what is seen under commercial conditions.

In this case:

$$\frac{1}{I_{total}} = \frac{1}{1000} + \frac{1}{1500} + \frac{1}{2000} + \frac{1}{2500} = 390 \ \Omega \qquad \text{(Equation 2)}$$

Therefore, the total current flow will also be 420 mA. However, because the birds have different resistances, and because they are arranged in parallel, they would receive different amounts of current, as follows:

(1) The 1000 Ω bird would receive $\dfrac{164 \text{ volts}}{1000 \ \Omega} = 164 \text{ mA}$ \qquad (Equation 3)

(2) The 1500 Ω bird would receive $\dfrac{164 \text{ volts}}{1500 \ \Omega} = 109 \text{ mA}$

(3) The 2000 Ω bird would receive $\dfrac{164 \text{ volts}}{2000 \ \Omega} = 82 \text{ mA}$

(4) The 2500 Ω bird would receive $\dfrac{164 \text{ volts}}{2500 \ \Omega} = 65 \text{ mA}$

This range of applied currents would result in the bird with a low resistance (1000 Ω) being exposed to a current level (164 mA) that would result in an effective

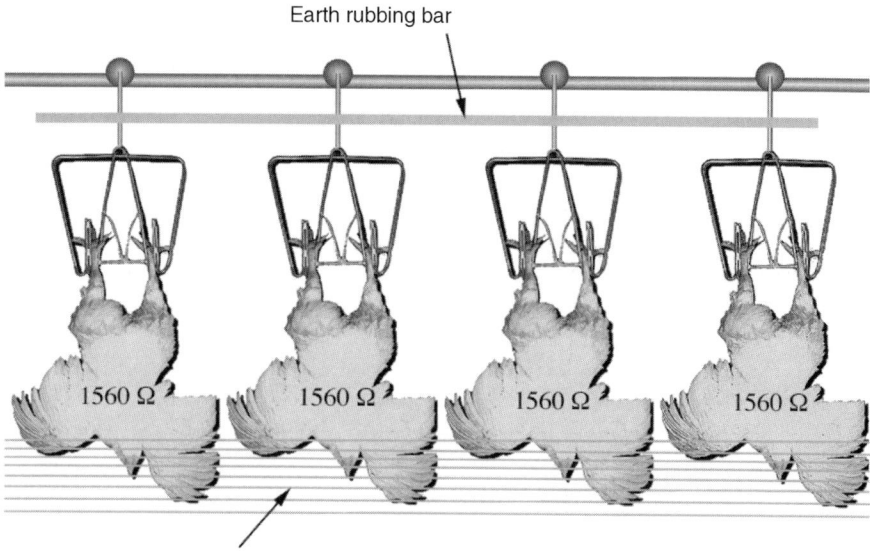

Fig. 13.5. Four birds in a waterbath stunner with the same impedance.

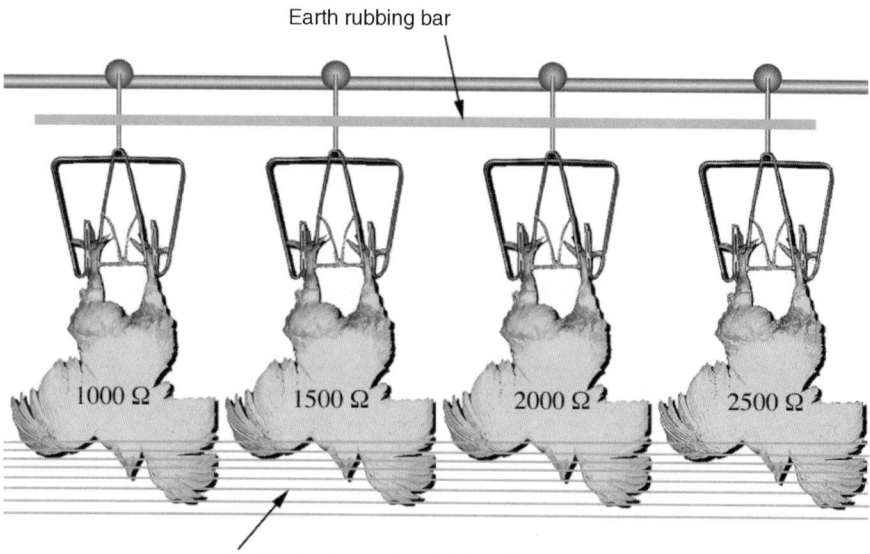

Fig. 13.6. Four birds in a waterbath stunner with different impedances.

stun but would probably produce a significant reduction in meat quality (Gregory and Wilkins, 1989a). The bird at the opposite end of the scale, with a high resistance (2500 Ω), would receive little current (65 mA) and would probably not be stunned but would produce acceptable carcass and meat quality. Thus, the processing plant management is under pressure to reduce the overall current levels that are applied in waterbath stunners in order to control the amount of downgrading in birds of low resistance, whilst the official veterinary surgeon is concerned with the number of birds that remain unstunned.

The solution to the problem produced by the range of bird resistances illustrated in Fig. 13.6 is to control the voltage applied to individual birds, as shown in Fig. 13.7. The amplitude of the voltage applied could be electronically controlled so that a constant current could be maintained through individual birds.

The concept of constant current control was developed at Silsoe Research Institute (Sparrey *et al.*, 1993) but the equipment manufacturing industry has yet to realize the commercial potential. Not only would the welfare of individual birds within a multi-bird waterbath stunner be protected by ensuring that all birds receive the required current to stun, but the level of downgrading incurred when birds are exposed to stunning currents above the required level would also be reduced.

The MAFF Code of Practice covering poultry slaughter (MAFF, 1991) advises that stunners should be fitted with easily visible ammeters so that it is possible to check that birds are receiving sufficient current. Although this Code of Practice has not been updated following the introduction of the 1995 legislation (HMSO, 1995), it remains a useful guidance document. Therefore, waterbath stunners should display the constant voltage applied and the applied current. Most waterbath stunner manufacturers use analogue ammeters but these are difficult to read in practice. Birds constantly enter and leave the waterbath, which creates large variations in total resistance within the system. This is reflected in constant movement of the analogue meter needle. The use of digital meters offers little advantage because averaging circuitry would be required to enable an average current value to be obtained and displayed. This averaging circuitry would damp down the responsiveness of the meter to a level that would affect its accuracy. Therefore, experimental work was undertaken to develop a robust current-recording device able to withstand the

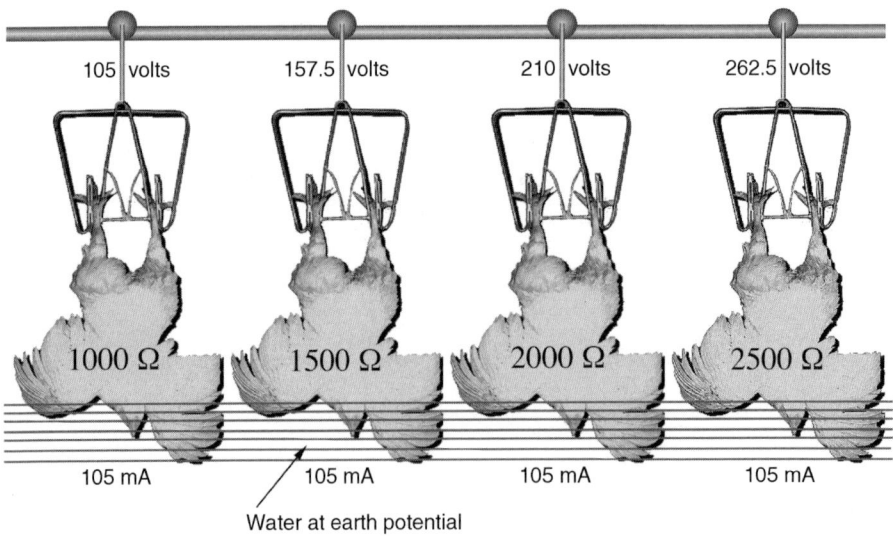

Fig. 13.7. Current flow and the associated voltages needed to generate the same current in four birds differing in resistance in a constant-current waterbath stunner.

rigours of a short journey along a poultry processing line through a waterbath stunner (Wotton, 2000). The unit would then display the average current that it monitored during the time in the waterbath and the duration of contact with the live water.

A commercial prototype (AGL Consultancy, Cobham, UK) has been developed from an experimental device and is shown in Fig. 13.8. The case is constructed of plastic and has stainless steel hangers that are articulated by ball-joints. The display window can be backlit by the operation of one of four large push-buttons mounted on the front panel. The other push-buttons function as switch-on, switch-off and reset. The waterbath electrode is fashioned from armour-protected cable and is connected to the Poultry Stun Monitor (PSM) unit via a 3 mm plug. This electrode terminates in a round, smooth stainless steel seal that prevents the lead from snagging on the entry ramp or during conveyance through the waterbath. Two electrode lengths are provided so that the device can operate with all sizes of poultry. These are 200 and 500 mm long respectively; when these dimensions are added to the length of the PSM from hangers to electrode socket, the overall length is 500 or 800 mm from the shackle hangers to the lead end.

The commercialization of the prototype PSM has resulted in a device that will function accurately (within 3%) across all the known waveforms and frequencies that are currently used worldwide to stun birds. The PSM reflects the potential difference between the shackle and the live water and can be used either to set up the stunner prior to hanging the birds or during normal processing. The PSM is robust and

Fig. 13.8. Poultry Stun Monitor in commercial operation.

waterproof (IP65); during commercial trials one of them was accidentally immersed in an electrically live turkey waterbath for about 30 minutes without damage or malfunction. The stainless steel hangers are simple to use and the length of the T-piece ensures that the PSM is both simple to hang and to remove. The lack of space between poultry stunners and automatic neck cutters does make removal from the line critical. However, the detachable neck permits the neck lead to be removed before the shackle line passes the neck cutter. Another method that was found to overcome this potential difficulty was to pull or push the PSM out of reach of the blade/s as it passed the killer.

The provision of a detachable flexible electrical bath probe means that, by changing the length of probe selected, one device can be used with all sizes of birds that are processed commercially. The construction of the probe was chosen to prevent the probe from snagging during passage through a waterbath stunner and the choice of connector enables quick removal of the probe from the body of the instrument. An additional function was added as an optional extra to the PSM to produce a data-logging model. The microcontroller used in the device contains a time/date function. This function can be used for verification purposes, such that the data recorded by the PSM can be downloaded to a PC via an RS232 data lead at a convenient time, using AGL Consulting's own acquisition and report software package. This additional function could prove useful to either poultry processors (in order to meet the specifications of their retailers) or those who have to enforce legislation (e.g. HMSO, 1995).

Effect of Waterbath Operation on Carcass and Meat Quality

The application of high voltages during stunning has been associated with poor bleeding, broken bones (Gregory and Wilkins, 1989a), exploded or damaged viscera, bruised wing joints and red wing tips (Heath, 1984), haemorrhages on the breast meat (Veerkamp and de Vries, 1983; Veerkamp, 1988) and split wishbones and separation of shoulder muscle tendons (Sams, 1996). There is a widespread belief within the poultry processing industry that the use of high stunning currents inevitably results in unacceptable levels of carcass damage (Bilgili, 1999). However, it has been demonstrated experimentally that stunning currents of up to 120 mA per bird are not responsible for any increases in the amount of carcass damage (Gregory and Wilkins, 1989a).

Blood Loss

Schedule 6 of UK legislation (HMSO, 1995) requires that any person engaged in the bleeding of any animal that has been stunned shall ensure that: (i) the bleeding is rapid, profuse and complete; (ii) the bleeding is completed before the animal regains consciousness; and (iii) the bleeding is carried out by severing at least one of the carotid arteries or the vessels from which they arise. However, some birds will be killed by the passage of the current during the stunning process and exsanguination will merely void the carcass of blood and have no bearing on bird welfare. In birds that are alive after stunning, the accuracy of the neck-cutting procedure is vital to their welfare. Gregory and Wotton (1986) demonstrated the importance of severing *both* carotid arteries to reduce the time to loss of brain responsiveness in birds that survive the stunning treatment. This is essential when higher frequencies are used in a bid to reduce carcass damage. Complete decapitation combined with a head macerator is an option recommended for good bird welfare.

Electrical waterbath stunning of poultry involves both the head and the body, with the heart included in the current pathway. Cardiac muscle is particularly sensitive to low frequency (50 Hz) alternating current, such that even at low currents of about 45 mA per bird (insufficient to induce an effective stun) ventricular fibrillation would be induced in about 10% of broilers (Wormuth *et al.*, 1981). Ventricular fibrillation is physiologically equivalent to a cardiac arrest (Heath, 1984). Therefore, at the recommended minimum current of 105 mA per bird about 90% of broilers would be killed in the stunner (Gregory and Wotton, 1987, 1990).

It has been demonstrated in all livestock species that a beating heart is not necessary for effective bleedout (Warriss and Wilkins, 1987; Gregory and Wilkins, 1989b). However, bleeding is slower after stunning with systems that cause cardiac arrest, particularly in poultry (Gregory and Wilkins, 1989b; Raj and Gregory, 1991). Poultry, in contrast to mammals, have unusually long clotting times. They rely on tissue thromboplastin for homeostasis rather than the generation of plasma thromboplastin, as in mammals (Bigland, 1964).

Wotton and Wilkins (1997) discussed the problem of poor bleeding and residual blood in turkeys. They suggested that the problem of poor bleeding could be exacerbated by good electrical stunning, which, at the recommended stunning current of 150 mA per bird, results in nearly 100% cardiac arrest (the turkeys are stun-killed). Poor bleedout was not caused by a lack of a beating heart. Rather, birds that were killed in the stunner were relaxed and did not show reflexes, whether of spinal or central origin, which, in a stunned but live bird, assisted bleedout through muscle pumping. Therefore, when stun-killed, turkeys that received a neck cut that severed at best one carotid artery and one jugular vein, and that in addition were given insufficient time to bleed fully before entry to scald, showed incomplete bleedout, and hence blood vessels that contained excessive residual blood. It was not the case that birds in which cardiac arrest had been induced would not bleed sufficiently given sufficient time, it was simply that poor neck-cutting in a relaxed carcass would extend the length of time necessary to bleed the carcass adequately. Wotton and Wilkins (1997) concluded that without a complete ventral neck-cut, which severed all the soft tissues, including the oesophagus, trachea and both carotid arteries and jugular veins, poor bleeding would always be a problem in turkeys. Thus it is possible to achieve good welfare with the recommended stunning current combined with efficient neck cutting, and at the same time avoid most meat quality issues.

The occurrence of red pygostyles and red feather tracts is almost exclusively a result of poor bleeding and has not been found to be significantly affected by stunning regimes, whether mechanical or electrical, although high-frequency waveforms may lessen the expression of these conditions. Among the factors that contribute to a difference in the rate or total amount of blood loss following a stunning method that induces cardiac arrest are the delay between killing and neck cutting, the particular blood vessels severed at neck cutting, and the length of time the birds are allowed to bleed before entry to scald (Raj, 1999).

Mouchoniere et al. (1999) evaluated the effect of increasing the stunning current frequency on the extent of blood loss in turkeys. Birds were stunned with 50, 300, 480, 550 and 600 Hz sinusoidal alternating current (150 mA per bird). The rate and extent of blood loss were measured over a 3-minute period following a unilateral neck cut that severed one carotid artery and one jugular vein. Both rate and extent of blood loss increased significantly as the current frequency increased, and the maximum difference in the rate of blood loss was evident during the first 40 seconds of bleeding. A reduced rate of bleeding was significantly correlated with the induction of cardiac arrest.

Bone Fractures

Direct electrical muscle stimulation in the waterbath stunner is thought to be the most important causal factor in the production of meat quality defects such as broken bones. Wilkins et al. (1998) found that pectoral bones that were broken as a direct result of electrical stunning would have a haemorrhage associated with the break. If there was no haemorrhage present, the break had occurred during carcass processing rather than at the point of stun. The occurrence of broken pectoral bones has been shown to be related to the amplitude of the stunning current at 50 Hz sinusoidal alternating current. Gregory and Wilkins (1989a) demonstrated that the number of broilers with one or more broken bones increased from 25 to 39% as the stunning current was raised from 74 to 269 mA per broiler. In turkeys, however, Gregory and Wilkins (1990) were unable to show any correlation between stunning current amplitude (75–250 mA) and the prevalence of broken bones. Gregory and Wilkins (1990a) demonstrated that 95% of frozen broiler carcasses had one or more broken bones but that only 3% of live birds had broken bones.

The use of stunning currents at frequencies higher than the conventional 50 Hz sine wave alternating current reduces the occurrence of broken pectoral bones in broilers and turkeys (Wilkins et al., 1998, 1999; Wotton and Wilkins, 1999). Wotton and Wilkins (1997) proposed that the advantage of higher frequencies was attributable to a reduction in direct muscle stimulation.

The use of higher frequencies has been shown to produce a significant reduction in the incidence of breast muscle haemorrhages, associated or not associated with a broken bone, in broilers (Wilkins et al., 1998). These authors also demonstrated that the use of high-frequency waveforms (1500 Hz) with broilers at the current levels recommended within the EU (100–105 mA/bird) could satisfy welfare requirements with regard to effectiveness of stunning as well as reducing carcass quality defects associated with the use of 50 Hz frequency waveforms at the same current level. Wilkins et al. (1999) demonstrated a more marked effect with turkeys (Fig. 13.9), in which increasing the stunning frequency from 50 or 100 Hz to 500 or 1500 Hz produced very significant improvements in breast muscle quality.

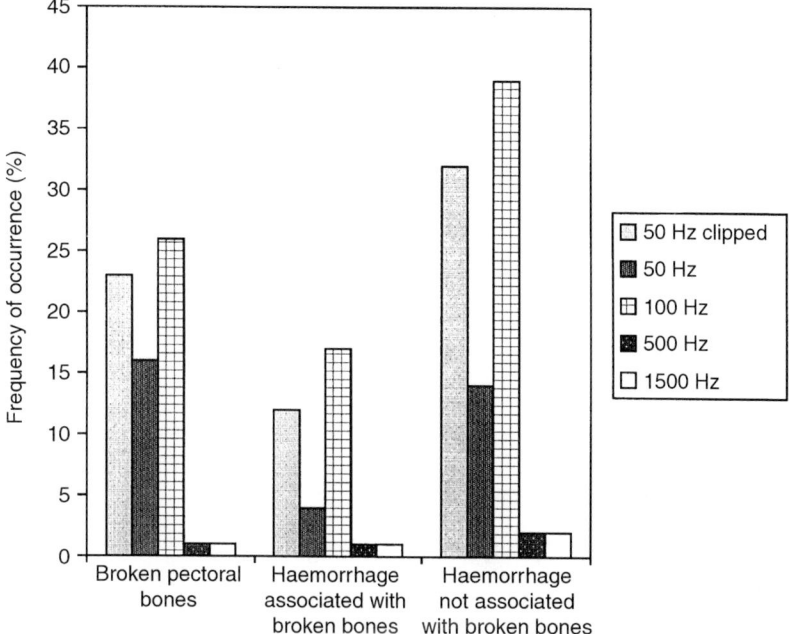

Fig. 13.9. The effect of stunning frequency and waveform at 150 mA per turkey on broken pectoral bones and breast muscle haemorrhages. (Adapted from Wilkins *et al.*, 1999.)

Breast muscle haemorrhages are particularly important when the fillets are intended for sale as prime portions after removal from the carcass. Given that birds are conventionally stunned using a constant voltage source in a multi-bird waterbath stunner, it is difficult to attribute carcass damage in individual birds to an overall waterbath current setting because the current flow through the individuals is unknown. A summary of the possible causes of carcass damage is given in Table 13.3. Most of these are attributable to factors associated with stunning.

The control of current levels through individual birds as described by Sparrey *et al.* (1993) offers huge advantages in terms of bird welfare when compared with constant voltage systems. The recommended current levels for alternating current systems are 105 mA per bird for broilers (Gregory and Wotton, 1990), 150 mA per bird for turkeys (Mouchoniere *et al.*, 1999) and 197 mA per bird for geese (Schuett-Abraham and Wormuth, 1988). Theoretically, control of the applied current should also significantly reduce the incidence of carcass defects, but in practice this is difficult to demonstrate experimentally (Wilkins *et al.*, 1999a).

Acknowledgements

The authors wish to thank the Department for Environment Food and Rural Affairs for funding the majority of this work.

Table 13.3. The likely causes of the common downgrading conditions of poultry.

Defect	Cause
Carcass quality defects	
Red wing tips	Ante-mortem flapping
	High stunning currents (110–150 mA)
	Residual blood from poor bleed-out
	Pre-stun shock
	Mechanical plucking
Red pygostyles	Poor bleeding
	High stunning currents (110–150 mA)
Red feather tracts	Poor bleeding
	High stunning currents (110–150 mA)
	Mechanical plucking
Engorged wing veins	Residual blood from poor bleed-out
Haemorrhage in wing and	High stunning currents (110–150 mA)
shoulder veins	Hard set plucker massaging residual blood out of broken vein
Meat quality defects	
Breast muscle haemorrhage	High stunning currents (130–190 mA)
	Contractile force of muscle spasm
Leg muscle haemorrhage	Leg catching and shackling pressure
	Postural position due to shackling
	High stunning currents (110–150 mA)
Broken bones	High stunning currents (110–150 mA)
	Hard set plucker
	Contractile force of muscle spasm
Blood smear on fillet	Residual blood from poor bleed-out

References

Bigland, C.H. (1964) Blood clotting times of five avian species. *Poultry Science* 43, 1035–1039.

Bilgili, S.F. (1999) Recent advances in electrical stunning. *Poultry Science* 78, 282–286.

Gentle, M.J. and Tilson, V.L. (2000) Nociceptors in the legs of poultry: implications for potential pain in pre-slaughter shackling. *Animal Welfare* 9, 227–236.

Gregory, N.G. (1989) Stunning and slaughter. In: Mead, G.C. (ed.) *Processing of Poultry*. Elsevier Applied Science, London, pp. 31–44.

Gregory, N.G. and Bell, J.C. (1987) The duration of wing flapping in chickens shackled before slaughter. *Veterinary Record* 121, 567–569.

Gregory, N.G. and Wilkins, L.J. (1989a) Effect of stunning current on carcass quality in chickens. *Veterinary Record* 124, 530–532.

Gregory, N.G. and Wilkins, L.J. (1989b) Effect of slaughter method on bleeding efficiency in chickens. *Journal of the Science of Food and Agriculture* 47, 13–20.

Gregory, N.G. and Wilkins, L.J. (1990) Effect of stunning current on downgrading in turkeys. *British Poultry Science* 30, 761–764.

Gregory, N.G. and Wotton, S.B. (1986) Effect of slaughter on the spontaneous and evoked activity of the brain. *British Poultry Science* 27, 195–205.

Gregory, N.G. and Wotton, S.B. (1987) Effect of electrical stunning on the electroencephalogram in chickens. *British Veterinary Journal* 143, 175–183.

Gregory, N.G. and Wotton, S.B. (1990) Comparison of neck dislocation and percussion of the head on visual evoked responses in the chicken's brain. *Veterinary Record* 126, 570–572.

Gregory, N.G. and Wotton, S.B. (1994) Effect of electrical stunning current on the duration of insensibility in hens. *British Poultry Science* 35, 463–465.

Gregory N.G., Austin, S.D. and Wilkins, L.J. (1989) Relationship between wing flapping at shackling and red wingtips in chicken carcasses. *Veterinary Record* 124, 62.

Heath, G.B.S. (1984) The slaughter of broiler chickens. *World's Poultry Science Journal* 40, 151–159.

Hewitt, L. (2000) The development of a novel device for humanely dispatching casualty slaughter. PhD thesis, University of Bristol.

HMSO (1995) *Welfare of Animals (Slaughter or Killing) Regulations*. Statutory Instrument No. 731. HMSO, London.

MAFF (1991) *The Welfare of Poultry at Slaughter*. MAFF code of practice, PB0733. HMSO, London.

Mouchoniere, M., Le Pottier, G. and Fernandez, X. (1999) The effect of current frequency during waterbath stunning on the physical recovery and rate and extent of bleed out in turkeys. *Poultry Science* 77, 485–489.

Perez-Palacios, S. (2003) Optimisation of poultry waterbath stunner electrical parameters. MSc dissertation, University of Bristol, UK.

Raj, A.B.M. (1999) Effects of stunning and slaughter methods on carcass and meat quality. In: Richardson, R.I. and Mead, G.C. (eds) *Poultry Meat Science*. CAB International, Wallingford, UK, pp. 231–254.

Raj, A.B.M. and Gregory, N.G. (1991) Efficiency of bleeding of broilers after gaseous or electrical stunning. *Veterinary Record* 128, 127–128.

Sams, A.R. (1996) Stunning basics. *Broiler Industry* 58, 36–38.

Schuett-Abraham, I. (1999) Humane stunning of poultry. *EC Seminar: Animal Welfare, August 24th to September 2nd, Dublin*.

Schuett-Abraham, I. and Wormuth, H.J. (1991) Anforderungen an eine Tierschutzgerechte elektrische Betaubung von Schlachtegeflugel. [Requirements for the humane electric stunning of slaughter poultry.] *Rundschau für Fleischhygiene und Lebensmittelüberwachung* 43, 7–8.

Schuett-Abraham, I., Wormuth, H.J. and Fessel, J. (1987) Vergleichende untersuchungen

zur Tierschutzgerechten elektrobetaubung verschiedener Schlachtgeflugelarten. [Comparative experiments on the humane electric stunning of different types of slaughter poultry.] *Berliner und Münchener Tierärztliche Wochenschrift* 100, 332–340.

Sparrey, J.M., Paice, M.E.R. and Kettlewell, P.J. (1992) Model of current pathways in electrical waterbath stunners used for poultry. *British Poultry Science* 33, 907–916.

Sparrey, J.M., Kettlewell, P.J., Paice, M.E.R. and Whetlor, W.C. (1993) Development of a constant current waterbath stunner for poultry processing. *Journal of Agricultural Engineering Research* 56, 267–274.

Veerkamp, C.H. (1988) What is the right current to stun and kill broilers? *Poultry Missel* 40, 30–31.

Veerkamp, C.H. and de Vries, A.W. (1983) Influence of electrical stunning on quality aspects of broilers. In: Eikelenboom, G. (ed.) *Stunning Animals for Slaughter*. Martinus Nijhoff, Boston, Massachusetts, pp. 197–212.

Warriss, P.D. and Wilkins, L.J. (1987) Exsanguination of meat animals. In: *Pre-slaughter Stunning of Food Animals. Proceedings of a Seminar Organised by the European Conference Group on the Protection of Farm Animals, Brussels*, pp. 150–158.

Wilkins, L.J., Gregory, N.G., Wotton, S.B. and Parkman, I.D. (1998) Effectiveness of electrical stunning using a variety of waveform-frequency combinations and consequences for carcass quality in broiler chickens. *British Poultry Science* 39, 511–518.

Wilkins, L.J., Gregory, N.G. and Wotton, S.B. (1999) Effectiveness of different electrical stunning regimens for turkeys and consequences for quality. *British Poultry Science* 40, 478–484.

Wooley, S.C., Borthwick, F.J.W. and Gentle, M.J. (1986a) Tissue resistivities and current pathways and their importance in pre-slaughter stunning of chickens. *British Poultry Science* 27, 301–306.

Wooley, S.C., Borthwick, F.J.W. and Gentle, M.J. (1986b) Flow routes of electric currents in domestic hens during pre-slaughter stunning. *British Poultry Science* 27, 403–408.

Wormuth, H.-J., Schuett-Abraham, I. and Fessel, J. (1981) Tierschutzgerechte elektrische Betaubung von Schlachtgeflügel. [Humane electrical stunning of poultry.] *Vet. Med. Berichte* 2. Dietrich Reimer, Berlin.

Wotton, S.B. (2000) The development of a poultry stun monitor for the measurement of current flow in poultry waterbath stunners. MSc thesis, University of Bristol, UK.

Wotton, S.B. and Gregory, N.G. (1991) How to prevent pre-stun electric shocks in waterbath stunners. *Turkeys* 39, 15 and 30.

Wotton, S.B. and Wilkins, L.J. (1997) Turkey processing: solving the problems of poor bleeding and the presence of residual blood in the carcass. *Poultry International* August, 36–40.

Wotton, S.B. and Wilkins, L.J. (1999) Effect of very low pulsed direct currents at high frequency on the return of neck tension in broilers. *Veterinary Record* 145, 393–396.

3 Welfare and Auditing Issues

14 Comparing Welfare in Different Systems

S.M. HASLAM AND S.C. KESTIN

School of Veterinary Science, University of Bristol, Langford, Bristol, UK

The degree of public interest in the welfare of food animals has led to the introduction of welfare assessment systems for several farmed species. Retail and restaurant groups have initiated the development of welfare standards to which producers are expected to adhere and to which they are audited. Additionally, accreditation schemes, run by producer groups and independent bodies, have burgeoned. Broiler chickens are perceived by the public to be one of the most intensively reared food animals and as such receive considerable public scrutiny (Stevenson, 1995; RSPCA, 2003). This has led the introduction of farm assurance schemes for broiler chickens by several UK retailers and to the development in the UK of the Assured Chicken Production scheme (ACP, Long Hanborough, UK, 2000).

Almost without exception, current (2003) welfare audit schemes are based on standards set around prescribing input measures, that is, prescribing resources available to the bird such as litter type or ventilation system capacity. Many authors consider that input measures are much less reliable as welfare assessment measures than output measures; that is, those based on the experience of the bird. There is some evidence that contemporary input-based schemes may not improve animal welfare (Main *et al.*, 2003a,b). As these schemes are based on pass/fail criteria, an overall comparison between two farms may not be easily made. Furthermore, for these schemes there is no weighting of individual welfare aspects, with the result that that some failures may be trivial and others very important to bird welfare. These input-based audit systems may not be applicable to all housing systems, where, for example, specifying a minimum ventilation fan capacity has little relevance in a free-range system. From this it is clear that, if improving welfare is an objective of a scheme, a more valid and flexible method is required.

A welfare assessment system based on output measures that are appropriately weighted for their importance to bird welfare could be a much more valid method of assessing and comparing bird welfare. This chapter briefly reviews a project in which a broiler welfare assessment system (Unitary Welfare Index (UWI)) for assessing and comparing broiler welfare on different farms and under different husbandry systems was developed and evaluated.

Essential Attributes of Welfare Assessment Measures

A number of authors have discussed the essential features and structure of practical welfare assessment systems (summarized in Sorensen and Sandoe, 2001). From this work it may be concluded that measures should be: (i) simple to measure; (ii) objective, in that different observers, observations and producer groups produce the same result; (iii) sensitive, in that they reflect small differences in welfare state; (iv) practicable to measure under audit conditions; and (v) a valid reflection of animal welfare.

Selection of Welfare Assessment Measures

For this exercise, welfare assessment measures relevant to broiler growing systems were identified empirically using the framework of the Five Freedoms (Farm Animal Welfare Council, 1993, cited in Webster, 1995). These measures were then classified as either resource (input) factors or bird-based (output) factors. The validity of the welfare assessment measures that were identified (that is, the relevance of each measure for broiler welfare) was assessed with reference to scientific literature in the field of broiler welfare science. Those measures that were considered to meet the essential criteria for a welfare assessment measure (discussed above) were selected for inclusion in the UWI. Output measures were used where possible and, where several measures were relevant to one aspect of welfare (that is, there was overlap in the area) the measure most closely meeting the selection criteria was chosen.

Thus, for example, welfare assessment measures generated from the Freedom from 'discomfort' included atmospheric ammonia level, litter quality, bird cleanliness and the level of contact dermatitis, all of which might be measured or assessed on a numerical rating scale and included in the UWI. However, from the relevant literature, it is clear that there is some overlap between these measures. Litter deterioration has been found to be associated with high levels of both hock and breast burn (Bruce *et al.*, 1990) and increased moisture content of the litter has been shown to increase levels of footpad dermatitis (Weaver and Meijerhof, 1991; Ekstrand *et al.*, 1997). Furthermore, ammonia levels are increased by high litter moisture levels (Carr *et al.*, 1990; Gustafsson and Martensson, 1990). However, ammonia levels may not be reliable as a welfare assessment measure as they have been shown to be directly related to house temperature and relative humidity, which may change from day to day (Carr *et al.*, 1990; Weaver and Meijerhof, 1991). Similarly, litter quality may be affected by specific husbandry practices, such as total litter replacement, which has been used to reduce levels of contact dermatitis (Menzies *et al.*, 1998).

In such circumstances, litter quality is a poor welfare assessment measure, as it may not reflect the welfare of the birds and so lacks validity. In this study, assessment of bird cleanliness was found to be difficult to standardize between assessors, rendering this welfare assessment measure less objective.

For these reasons, contact dermatitis level was selected to reflect bird comfort in terms of house ammonia levels, litter quality and bird cleanliness. Contact dermatitis levels are relatively simple and practicable to measure at the slaughterhouse, reliable and repeatable, valid as an output measure of bird welfare, and sensitive to small differences in welfare state.

As a result of such evaluation against essential criteria and elimination of overlap, the welfare assessment measures selected from the Five Freedoms were: (i) percentage mortality; (ii) percentage of birds with contact dermatitis; (iii) percentage of birds with severe leg weakness; (iv) stocking density; (v) enrichment and emergency provision; (vi) severity of thinning and feed restriction programmes; and (vii) degree of feather-pecking damage.

Weighting Welfare Assessment Measures

The welfare assessment measures selected may be classified into those which are characteristic of all broiler chicken flocks and those which are not always present. Thus, measures that apply to all flocks include mortality, leg weakness, contact dermatitis, stocking density and emergency and enrichment provision. Measures which are not experienced by all flocks include feather pecking, thinning and feed restriction programmes.

Measures that applied to all flocks formed the core of the index and were weighted using conjoint analysis of expert opinion (Haslam and Kestin, 2003). Briefly, an orthogonal set of 'plan cards', each of which represented a virtual broiler flock, described in terms of six of the core welfare assessment measures identified, were generated. Veterinary surgeons with postgraduate qualifications in poultry medicine and production or welfare science ethics and law and members of research groups registered with the UK Poultry Research Liaison Group were asked to score each house from 1 (worst welfare) to 10 (best welfare), taking account of all of the levels of each of the welfare assessment measures. The responses were analysed using the software SPSS Conjoint 8.0 to identify the utility, or weighting, that the individual experts had assigned to each welfare assessment measure. The average weightings were incorporated into the overall index. Thus, mortality and leg disorders were weighted highest (0.26 and 0.24, respectively) and emergency and enrichment provision lowest (both 0.10).

Welfare aspects not present in all flocks were included in the index as a penalty score, the level again being determined by expert opinion. The weighting and penalty scores for welfare assessment measures included in the index are summarized in Table 14.1.

Additionally, a requirement to highlight aspects where welfare may be poor in spite of a satisfactory overall score has been identified (Whay *et al.*, 2003). In order to meet this requirement, the level of each welfare assessment measure at which action should be taken to protect bird welfare was determined using expert opinion. A UWI scoresheet was created, and scores requiring action were denoted as shaded areas. The authors would be pleased to provide example UWI score sheets to interested parties.

Table 14.1.   The range and weightings of each welfare assessment measure incorporated in the UWI.

Welfare assessment measure	Weighting	Range	Welfare index
Mortality (%)	0.26	0.87–9.20	WI_{Mrt}
Foot burn (%)	0.16	0–100	WI_{Ftbn}
Leg Weakness (ELWS)	0.24	10–90	WI_{LW}
Stocking Density (kg/m²)	0.14	20–40	WI_{SD}
Enrichment Provision E (ErPES)	0.1	0–43	WI_{Enrich}
Emergency Provision (EmPES)	0.1	0–15	WI_{Emerg}
Feed Restriction Programme (severity on a NRS)	0; −0.03; −0.12; −0.18; −0.32	0–4	WI_{FR}
Thinning programme (no. of thinnings)	0; −0.12; −0.22; −0.31; −0.40	0–4	WI_{Thin}
Feather Damage (EFPS)	0; −0.07; −0.09; −0.22; −0.34	0–200	WI_{FP}

ELWS, Equivalent Leg Weakness Score; ErPES, Enrichment Provision Equivalent Score; EmPES, Emergency Provision Equivalent Score; EFPS, Equivalent Feather Pecking Score.

Conversion of Welfare Assessment Measures to Welfare Indices

In order to amalgamate the welfare assessment measures into an overall welfare score and to make each measure as sensitive as possible, all measures were converted to common units and scaled according to normal broiler production figures. For example, in the case of mortality level, the current distribution of flock mortality in the UK, supplied by the industry, is shown in Fig. 14.1.

To exclude outlying flocks, the normal range of flock mortality was taken to be from 0.87 to 9.2% (the 2.5 and 97.5 percentile values respectively) and the welfare assessment measure weighting (26 points) distributed evenly between these points. Thus, a flock with 9.2% mortality would score zero points for mortality while one with 0.87% mortality or less would score 26 points (26 being the weighting given by the experts in the conjoint analysis study). A flock with a mortality level of over 9.2% would fail the assessment. Levels of welfare assessment measures at which the flock fails the welfare assessment are denoted on the UWI score sheet as darkly shaded areas on the welfare index (WI) conversion tables. WIs for which a flock fails are recorded on the cover sheet of the UWI score sheet.

The ranges for each welfare assessment measure are included in Table 14.1.

The Unitary Welfare Index Structure

The overall structure of the UWI is represented by the following formula:

$$UWIscore = (WI_{Mrt} + WI_{Ftbn} + WI_{LW} + WI_{SD} + WI_{Enrich} + WI_{Emerg}) - (WI_{FR} + WI_{Thin} + WI_{FP})$$

A UWI audit document was developed, based on this structure, to facilitate the collection of the welfare assessment measures in chronological sequence, conversion

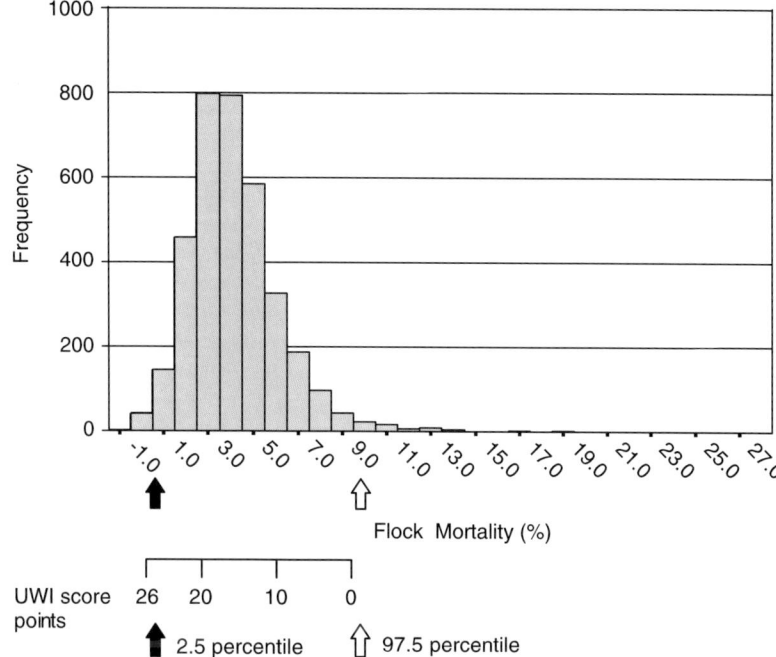

Fig. 14.1. Frequency distribution of flock mortality for flocks contributing to the Department of Agriculture Northern Ireland Database.

to UWI units, calculation of the UWI score and identification of welfare aspects for which action is required or when the flock fails the assessment.

Evaluation of the UWI Score as a Welfare Assessment System

The UWI scoring system as a welfare assessment tool was evaluated in a field trial on a series of farms belonging to one producer in the south-west of England.

Method

Ten broiler chicken farms were visited, representing three husbandry systems: traditional intensive, free-range and organic. For each flock visited, the UWI welfare measures were carefully collected. In parallel, a set of alternative welfare measures was made. The data were then examined to determine correlations between the UWI and individual WI scores and the alternative welfare evaluation measures. The alternative welfare measures collected for each flock are summarized in Table 14.2; each measure is annotated with references supporting it as a valid welfare assessment parameter.

Parent flock ages, mortality and culls by cause as well as number of birds placed and medication and light programmes were recorded from flock records. The bird

Table 14.2. Aspect of welfare reflected by each Welfare Assessment Measure used in the field evaluation study of the UWI developed.

Aspect of welfare assessed	Parameter measured	Reference
Dehydration	Packed cell volume (PCV), osmolality	Warriss *et al.*, 1993; Sommer *et al.*, 1999
Fear	Heterophil:lymphocyte ratio (H:L)	McFarlane and Curtis, 1989; Maxwell and Robertson, 1998
Stress	Corticosterone, non-esterified fatty acids (NEFA), triglycerides, glucose, total protein, globulin	Puvadolpirod and Thaxton, 2000
Bone remodelling	Alkaline phosphatase (ALKP)	Seregni *et al.*, 2001; Akalin *et al.*, 2002; Vergnoud *et al.*, 2002
Muscle damage	Creatine kinase (CK)	Grigor *et al.*, 1998; Hall *et al.*, 1999; Hocking *et al.*, 1996, 2001
Immune suppression	Thymus, liver, spleen and bursal weights	Puvadolpirod and Thaxton, 2000
Heart failure	Heart weight	Mirsalimi *et al.*, 1992; Wang and Hacker, 1993, 1996
Leg health	Leg bone deformation scores	Butterworth, 2002
Disease	DOAs and bird rejected at inspection	Yogaratnam, 1995; Olkowski, 1996; Elfadil, 1996; Ansondanquah, 1987
Trauma	Scratches and bruises	
Litter and air quality	Contact dermatitis levels	Bruce, 1990; Nairn and Watson, 1972; Harms and Simpson, 1975; Harms *et al.*, 1977; Greene, 1985; Hemminga and Vertommem, 1985; Martland, 1985; Bray and Lynn, 1986; McIlroy *et al.*, 1987; Weaver and Meijerhof, 1991; Ekstrand *et al.*, 1997

Akalin, A., Colak, O. and Alatus, O. (2002) Bone remodelling markers and serum cytokines in patients with hyperthyroidism. *Clinical Endocrinology* 57, 125–129.

Ansongdanquah, J. (1987) A survey of carcass condemnation at a poultry abattoir and its application to disease management. *Canadian Veterinary Journal – Revue Veerinaire Canadienne* 28(1–2), 53–56.

Bray, T.S. and Lynn, N.J. (1986) Effects of nutrition and drinker design on litter condition and broiler performance. *British Poultry Science* 27, 151 (abstr.).

Bruce, D.W., McIlroy, S.G. and Goodall, E.A. (1990) The epidemiology of a contact dermatitis of broilers. *Avian Pathology* 19, 523–537.

Butterworth, A. (2002) Infectious and skeletal components of lameness in broiler chickens *Gallus gallus domesticus* and their implications for welfare and hygiene. PhD thesis, University of Bristol, UK.

Ekstrand, C., Algers, B. and Svedberg, J. (1997) Rearing conditions and foot-pad dermatitis in Swedish broiler chickens. *Preventive Veterinary Medicine* 31, 167–174.

Elfadil, A.A., Vaillancourt, J.P., Meek, A.H., Julian, R.J. and Gyles, C.L. (1996) Description of cellulitis lesions and association between cellulitis and other catagories of condemnation. *Avian Disease* 40(3), 690–698.

Greene, J.A. (1985) A contact dermatitis of broilers – clinical and pathological findings. *Avian Pathology* 14, 31–63.

Grigor, P.N., Goddard, P.J. and Littlewood, C.A. (1998) The behavioural and physiological reactions of farmed deer to transport: effects of sex, group size, space allowance and vehicular motion. *Applied Animal Behaviour Science* 56, 281–295.

Hall, S.I.J., Broom, D.M. and Goode, J.A. (1999) Physiological responses of sheep during long road journeys involving ferry crossings. *Animal Science* 69, 19–27.

Table 14.2. *Continued.*

Harms, R.H. and Simpson, C.F. (1975) Biotin deficiency as a possible cause of swelling and ulceration of footpads. *Poultry Science* 54, 1711–1713.

Harms, R.H., Damron, B.L. and Simpson, C.F. (1977) Effect of wet litter and supplemental biotin and/or whey on the production of foot pad dermatitis in broilers. *Poultry Science* 56, 291–296.

Hemminga, H. and Vertommem, M.H. (1985) Faecal burn-spots upon the breast-skin of broiler chickens. *Pluimveehouderij* 15, 12–14.

Hocking, P.M., Maxwell, M.H. and Mitchell, M.A. (1996) Relationships between the degree of food restriction and welfare indices in broiler breeder females. *British Poultry Science* 37, 263–278.

Hocking, P.M., Maxwell, M.H. and Robertson, G.W. (2001) Welfare assessment of modified rearing programmes for broiler breeders. *British Poultry Science* 42, 424–432.

Martland, M.F. (1985) Ulcerative dermatitis in broiler chickens: the effects of wet litter. *Avian Pathology* 14, 353–364.

Maxwell, M.H. and Robertson, G.W. (1998) The avian heterophil:lymphocyte ratio: a review. *World's Poultry Science Journal* 54, 155–178.

McFarlane, J.M. and Curtis, S.E. (1989) Multiple concurrent stressors in chicks. 3. Effects on plasma corticosterone and the heterophil:lymphocyte ratio. *Poultry Science* 68, 510–521.

McIlroy, S.G., Goodall, E.A. and McMurray, C.H. (1987) A contact dermatitis of broilers – epidemiological findings. *Avian Pathology* 16, 93–105.

Mirsalimi, S.M., O'Brian, P.J. and Julian, R.J. (1992) Changes in erythrocyte deformability in NaCl-induced right-sided cardiac failure in broiler chickens. *American Journal of Poultry Research* 53, 2359–2363.

Nairn, M.E. and Watson, A.R.A. (1972) Leg weakness in broilers – a clinical and pathological characterisation. *Australian Veterinary Journal* 48, 645–656.

Olkowski, A.A., Kumor, L. and Classen, H.L. (1996) Changing epidemiology of ascites in broiler chickens. *Canadian Journal of Animal Science* 76(1), 135–140.

Puvadolpirod, S. and Thaxton, J.P. (2000a) Model of physiological stress in chickens 1. Response parameters. *Poultry Science* 79, 363–369.

Puvadolpirod, S. and Thaxton, J.P. (2000b) Model of physiological stress in chickens 2. Dosimetry of adrenocorticotropin. *Poultry Science* 79, 370–376.

Puvadolpirod, S. and Thaxton, J.P. (2000c) Model of physiological stress in chickens 3. Temporal patterns of response. *Poultry Science* 79, 383–390.

Seregni, E., Martinetti, A. and Ferrari, L. (2001) Clinical utility of biochemical marker of bone remodelling in patients with bone metastases of solid tumors. *Quarterly Journal of Nuclear Medicine* 45, 7–17.

Sommer, B., Leeb, C. and Troxler, J. (1999) Providing a practice-suitable method for judging housing systems of pregnant sows for proper keeping of animals in view of welfare using lesion pattern of the integument as an indicator. *Proceedings of the 1st International Workshop on the Assessment of Animal Welfare at Farm and Group Level.* Danish Institute of Agricultural Sciences and Royal Veterinary and Agricultural University.

Vergnoud, P., Lunt, M. and Scheidt-Nave, C. (2002) Is the predictive power of previous fractures for new spine and non spine fractures associated with biochemical evidence of altered bone remodelling. The EPOS study. *Clinica Chimica Acta* 322, 121–132.

Wang, J.Y. and Hacker, R.R. (1993) Effects of diaoxinxuekang on ascites in broilers. *Poultry Science* 72, 1467–1472.

Warriss, P., Kestin, S.C. and Brown, S.N. (1993) The depletion of glycogen stores and indices of dehydration in transported broilers. *British Veterinary Journal* 149, 391–398.

Weaver, W.D. and Meijerhof, R. (1991) The effect of different levels of relative humidity and air movement on litter conditions, ammonia levels, growth and carcass quality in broiler chickens. *Poultry Science* 70, 746–755.

Yogaratnam, V. (1995) Analysis of causes of high rates of carcass rejection at a poultry processing plant. *Veterinary Record* 137, 215–217.

carcasses were then subject to post-mortem examination during which hock, foot and breast burn scores, organ weights, leg bone deformation scores, and synovitis scores were recorded. Standardized assessment personnel recorded hock, foot and breast burn scores and percentage of birds scratched and bruised from a sub set of birds from the trial house at the slaughter plant. The percentage of birds Dead on Arrival (DOAs) and the percentage of birds rejected by the Meat Hygiene Service (MHI rejects) were collected and a histogram of bird weight distribution generated by the plant computer.

Results

The significant correlations between the UWI and individual WI scores and the other welfare evaluation data examined are presented in Tables 14.3 and 14.4. To simplify presentation and to avoid the inclusion of erroneous associations, only correlations that were significant at the 99% confidence level (when based on measurements on individual birds) with a correlation coefficient of at least ± 0.1 have been included. Correlations between welfare assessment measures based on house differences (thus involving fewer comparisons) are shown which were significant only at the 95% confidence level. These are indicated by a double asterisk. The total UWI score and individual WI scores show several supporting correlations with many of the individual bird welfare assessment measures.

In order to simplify the analysis, factor analysis was carried out using variables that were normally distributed, to condense the variability in the data to its principal components. Table 14.5 shows the contribution made to each of the principal factors by each variable included in the analysis. Factor 1 seems to be a measure of dehydration and factor 2 a measure of stress, whilst factor 3 is difficult to interpret but may be associated with an underlying variable other than welfare, specifically bird age, sex or growth rate. The correlations between principal factors and UWI and WI scores are presented in Table 14.6.

Conclusions

A considerable number of alternative welfare assessment measures were found to correlate with both the total UWI score and individual WI scores for the trial farms. The direction of the significant bivariate correlations identified largely supports the hypothesis that the UWI score and its constituent WI scores reflect bird welfare in the houses. The correlations tended to be weak, although with a high level of significance. This was unsurprising, as there are clearly many factors affecting bird welfare and it is unlikely that any of them individually will have a strong causative affect. Additionally, factor analysis of the evaluation data generated principal components that demonstrated correlations with all WI scores as well as the total UWI score. It may be concluded that this UWI produces a score that reflects the welfare state of broilers on the farm.

Table 14.3. UWI and bird-based welfare indices. Correlation between welfare assessment parameters and UWI scores (minus score for 'Emergency Provision') and bird-based welfare indices for Mortality, Leg Weakness and Foot burn.
**Denotes correlations based on house rather than individual bird measures.

Welfare index	Parameter	Spearman's rho	Significance
Total UWI Score	% Breast burn*	−0.77	0.01**
(minus WI for	% Mortality*	−0.86	0.01**
Emergency Provision)	Vocalization score*	0.60	0.05
	BdFoot burn*	−0.16	0.01
	Osmolality[†]	0.26	0.01
	PCV[†]	0.26	0.01
	Glucose[†]	0.21	0.01
	HtWtperBWt[‡]	0.15	0.01
	BursaWtperBWt[‡]	0.14	0.01
Mortality WI	% Breast burn*	−0.85	0.01**
	PCV[†]	0.25	0.01
	Osmolality[†]	0.22	0.01
	Glucose[†]	0.20	0.01
Leg Weakness WI	DOAs*	−0.73	0.01**
	% Total rejects*	−0.60	0.05**
	% Infected rejects*	−0.68	0.05**
	HouseFaecalCort*	−0.30	0.01**
	Total pathology*	−0.16	0.01
	ALKP*	−0.38	0.01
	NEFA*	−0.42	0.01
Foot burn WI	Feather score*	0.79	0.01**
	Av litter score*	−0.59	0.05**
	Av ammonia*	−0.66	0.05**
	Globulin*	−0.25	0.01
	TProtein*	−0.21	0.01
	NEFA[‡]	0.34	0.01
	HeartWtperBWt[‡]	0.20	0.01
	ALKP[‡]	0.47	0.01

*Supports UWI as welfare analogue; [†]may or may not support UWI as welfare analogue; [‡]does not support UWI as welfare analogue.

Utility of UWI for Comparison of Bird Welfare Between Flocks

The utility of the UWI for both comparing welfare between farms and highlighting areas in which action should be taken to improve welfare is exemplified by the UWI scores for two of the flocks visited for the field study described in this chapter and illustrated in Figs 14.2 and 14.3. Flock 10, an organic, free-range flock and flock 8, an intensive traditional flock of heavy cockerels, had very similar UWI scores: 57 and 59 respectively. On examination of individual WI scores for each flock, flock 10 failed for the mortality WI and scored poorly for emergency provision. This appeared to be

Table 14.4. Resource-based welfare indices. Correlation between welfare assessment parameters and resource-based welfare indices for Stocking Density and Enrichment. **Denotes correlations based on house rather than individual bird measures.

Welfare index	Parameter	Spearman's rho	Significance
Stocking Density WI	LiverWtperBWt*	−0.21	0.01
	HeartperBWt*	−0.15	0.01
	Total leg pathology*	−0.21	0.01
	ALKP*	−0.54	0.01
	PCV†	0.20	0.01
	CK†	0.28	0.01
	Globulin†	0.18	0.01
	TProtein†	0.18	0.01
	NEFA†	0.15	0.01
	BdFootbn‡	0.30	0.01
	InfRejects‡	0.68	0.05**
	SpleenWtperBWt‡	−0.20	0.01
Enrichment WI	% Hock burn*	−0.72	0.05**
	UWIGtScore*	0.59	0.05**
	AverageCort*	−0.61	0.05**
	HouseFaecalCort*	−0.143	0.07**
	Total Leg Pathology*	−0.15	0.01
	Bd Corticosterone*	−0.27	0.01
	ALKP*	−0.53	0.01
	BursaWtperBWt	0.13	0.01
	ThymusperBWt*	0.18	0.01
	HeartperBWt*	−0.15	0.01
	CK†	0.20	0.01
	Triglyceride‡	0.24	0.01
	Globulin‡	0.20	0.01
	Total Protein‡	0.23	0.01
	BdFoot burn‡	0.39	0.01

*Supports UWI as welfare analogue; †may or may not support UWI as welfare analogue; ‡does not support UWI as welfare analogue.

a result of poorly designed perimeter fencing and poor cover on range, with a resulting high level of mortality due to predation. The UWI score and bird welfare for flock 10 could be readily improved by providing fox-proof fencing and adequate cover on range; the UWI score system clearly demonstrated this to the producer.

Conclusions

The UWI was developed as a tool for integrating broiler welfare assessment measures so that bird welfare on different farms and production systems could be compared. It has proved to be practicable to apply on-farm and to correlate with alternative indices of welfare. The limited number of input and output parameters incorporated

Table 14.5. Structure matrix for components produced by factor analysis of evaluation welfare assessment parameters from individual birds.

	Component		
	1	2	3
LNOSMOL	0.704		
GLUCOSE	0.688		
PCV	0.654	0.370	
LNGlobulin	0.631		−0.418
LNHLratio	0.383	−0.356	
LNCsterone		0.596	0.379
LNThperWt		−0.545	
TRIGLYC		−0.498	
SPERWT		−0.448	0.338
LNAlkPhos			0.649
LNBPERWT			0.571
LNNEFA			−0.420
HTPERWT			0.397

Extraction method: principal component analysis. Rotation method: oblimin with Kaiser normalization.

Table 14.6. Significant correlations between the factors identified by factor analysis and the UWI and individual WI scores.

	Factor	Spearman's rho	Significance
UWI$_7$-E	1	0.27	0.001
UWI$_7$-E	2	0.15	0.001
UWI$_7$-E	3	−0.171	0.001
WIEnrich	2	−0.32	0.001
WIEnrich	3	0.23	0.001
WIFootbn	1	0.17	0.001
WIFootbn	2	0.20	0.001
WIMort	1	0.28	0.001
WIMort	2	0.12	0.01
WIMort	3	−0.134	0.01
WIGS	1	0.19	0.001
WIGS	3	−0.12	0.01
WISD	3	−0.37	0.001

in the index appears to give good discrimination between farms. As the weightings of WIs and the action levels were determined by expert opinion, the UWI assessment system will require adjustment as understanding of broiler welfare advances. The UWI should therefore be viewed as an evolving standard rather than a gold standard for broiler welfare assessment. Nevertheless, the UWI assessment system in its existing form may have a use in comparing bird welfare between farms and production systems and identifying areas where welfare needs to be improved.

Fig. 14.2.   House 8 – heavy cockerels in traditional intensive husbandry system.

Fig. 14.3.   House 10 – free range organic broilers.

References

Bruce, D.W., McIlroy, S.G. and Goodall, E.A. (1990) Epidemiology of a contact dermatitis in broilers. *Avian Pathology* 19, 523–537.

Carr, L.E., Wheaton, F.W. and Douglass, L.W. (1990) Empirical models to determine ammonia concentrations from broiler chicken litter. *Transactions of the ASAE* 33, 260–265.

Ekstrand, C., Algers, B. and Svedberg, J. (1997) Rearing conditions and foot-pad dermatitis in Swedish broiler chickens. *Preventive Veterinary Medicine* 31, 167–174.

Gustafsson, G. and Martensson, L. (1990) *Gases and Dust in Poultry Houses.* Swedish University of Agricultural Sciences, Department of Farm Buildings, Lund, Sweden.

Haslam, S. and Kestin, S. (2003) Use of conjoint analysis to weight welfare assessment measures for broiler chickens in UK husbandry systems. *Animal Welfare* 12, 669–675.

Main, D.C., Whay, H.R., Green, L.E. and Webster, A.J.F. (2003a) Effect of the RSPCA Freedom Food scheme on dairy cattle welfare. *Veterinary Record* 153, 227–231.

Main, D.C., Whay, H.R., Green, L.E. and Webster, A.J.F. (2003b) Preliminary investigation into the use of expert opinion to compare the overall welfare of dairy cattle farms in different farm assurance schemes. *Animal Welfare* 12, 565–569.

Menzies, F.D., Goodall, E.A., NcConaghy, D.A. and Alcorn, M.J. (1998) An update on the epidemiology of contact dermatitis in commercial broilers. *Avian Pathology* 27, 174–180.

RSPCA (2003) *Behind Closed Doors.* RSPCA, Horsham, West Sussex, UK.

Sorensen, J.T. and Sandoe, P. (2001) *Assessment of Animal Welfare at Farm and Group Level.* Taylor and Francis, Dublin.

Stevenson, P. (1995) The welfare of broiler chickens. Compassion in World Farming Trust, Petersfield, UK.

Weaver, W.D. and Meijerhof, R. (1991) The effect of different levels of relative humidity and air movement on litter conditions, ammonia levels, growth and carcass quality in broiler chickens. *Poultry Science* 70, 746–755.

Webster, J. (1995) *Animal Welfare: A Cool Eye Towards Eden.* Blackwell Science, Oxford, UK.

Whay, H.R., Main, D.C., Green, L.E. and Webster, A.J.F. (2003) Assessment of dairy cattle welfare using animal-based measurements. *Veterinary Record* 153, 197–202.

15 Human Factors Influencing Broiler Welfare

P.H. Hemsworth[1] and G.J. Coleman[2]

[1]Animal Welfare Centre, University of Melbourne and Department of Primary Industries, Victorian Institute of Animal Science, Werribee, Victoria, Australia; [2]Department of Psychology, Monash University, Caulfield Campus, Victoria, Australia

Introduction

The interactions between stockpeople and their animals have been shown in a number of livestock industries to influence the welfare and productivity of farm animals. While many of these interactions may appear mild and harmless to the animals, research in the livestock industries has shown that the frequent use of some routine behaviours by stockpeople can result in farm animals becoming highly fearful of humans. It is these high fear levels, through stress, that appear to limit animal welfare and productivity. This research has also shown that a major determinant of stockperson behaviour is the attitude of the stockperson towards interacting with his or her animals.

While not well quantified in the livestock industries, other important human characteristics that are likely to affect animal welfare include technical skills and knowledge, job motivation and commitment, job satisfaction and personality (Hemsworth and Coleman, 1998). Furthermore, human–animal interactions may influence a number of these job-related variables and consequently affect the work performance of the stockperson.

It is likely that, in the near future, the livestock industries, the general community and governments will place increasing emphasis on ensuring the competency of stockpeople who manage livestock (Hemsworth and Coleman, 1998). Furthermore, with the increasing reliance on audit procedures to safeguard animal welfare, the stockperson will take on the additional role of determining the effectiveness of these procedures in protecting animal welfare. The aim of this chapter is therefore to review some of the most recent research on human–animal interactions in order to consider their implications for poultry welfare and the role of the stockperson in the effectiveness of an audit procedure designed to protect animal welfare.

Fear and Animal Welfare

Fear is generally considered an undesirable emotional state of suffering in both humans and animals (Jones and Waddington, 1992). Studies in the broiler, dairy, egg and pig industries have consistently shown negative relationships between fear and productivity (for details of these studies see Hemsworth and Coleman, 1998): high levels of fear of humans (assessed from approach behaviour or, conversely, avoidance behaviour of animals with respect to humans) were correlated with reduced animal productivity. Handling studies in many farm animal species not only support the results of these studies in the livestock industries but also indicate that high levels of fear of humans may depress animal welfare (for details of these studies see Hemsworth and Coleman, 1998). For example, Gross and Siegel (1979, 1980, 1982) found that young chickens that received frequent human contact from an early age and of an apparent positive nature, such as gentle touching, talking and offering food on the hand, had improved growth rates and feed efficiency and were more resistant to infection than birds that either received minimal human contact or had been deliberately scared.

Role of Human Factors in Animal Fear and Animal Welfare

Studies in the dairy and pig industries have shown significant sequential relationships between the stockperson's attitudes and behaviour towards animals and the fear of humans and productivity of dairy cows and pigs (Coleman *et al.*, 1998; Hemsworth *et al.*, 1989, 2000; Breuer *et al.*, 2000). This research indicates that the attitude of the stockperson to interaction with his or her animals is a key antecedent of the behaviour of the stockperson and that the behaviour of the stockperson towards his or her animals is an important determinant of the animal's fear of humans. The existence and importance of these sequential relationships is demonstrated in intervention studies in the dairy and pig industries in which cognitive–behavioural training of stockpeople, designed to specifically improve these key attitudes and behaviours of stockpeople, decreased animal fear and improved animal productivity and welfare (Hemsworth *et al.*, 1994a, 2002; Coleman *et al.*, 2000).

Less research has been conducted in the broiler and egg industries but studies by Barnett *et al.* (1992), Hemsworth *et al.* (1994b, 1996) and Cransberg *et al.* (2000) found significant relationships between the behaviour of the stockperson, the level of fear of humans and the productivity of laying hens and meat chickens. Broiler chickens were most fearful of humans at farms in which stockpeople moved quickly, frequently tapped on objects in the facility and infrequently waved as they moved through the poultry facility (Hemsworth *et al.*, 1996; Cransberg *et al.*, 2000). It is possible that waving by the stockperson, which intuitively appears to be fear-provoking, may simply reflect a behaviour by the stockperson necessitated by non-fearful birds remaining closer to the stockperson as he/she moves slowly through the poultry facility, requiring the frequent use of this behaviour to move birds from under the stockperson's feet (Fig. 15.1).

Cransberg *et al.* (2000) found no evidence of a relationship between stockperson attitude and behaviour in the broiler industry, but the range of attitudinal variables

Fig. 15.1. The avoidance behaviour of broiler chickens to humans has been used to assess their levels of fear of humans. This photograph shows an experimenter filming the withdrawal responses of chickens as he moves through a broiler chicken unit in a standard manner.

targeted in the study may not have been sufficiently extensive. An improved understanding of the effects of routine behaviours used by stockpeople on the bird's fear of humans is necessary to reduce the effects of fear on bird welfare. This is particularly important because some of these interactions identified to date are not intuitively obvious. Furthermore, the attitudes underpinning these human interactions will also only be identified through rigorous research examining stockperson attitude–behaviour relationships.

Direct Effects of Human Factors on Animal Welfare

Technical skills and knowledge

One of the standard approaches to selecting individuals for jobs has been to identify the knowledge, skills and abilities (KSAs) needed for effective job performance and then to use these as a basis for identifying the appropriate selection tools. No detailed job analysis has been done for people working in the poultry industries. Hemsworth and Coleman (1998) attempted to identify a set of generic KSAs that would characterize a good stockperson. These included:

- A good general knowledge of the nutritional, climatic, social and health requirements of the farm animal.
- Practical experience in the care and maintenance of the animal.
- Ability to quickly identify any departures in the behaviour, health or performance of the animal and promptly provide or seek appropriate support to address these departures.
- Ability to work effectively independently and/or in teams, under general supervision, with daily responsibility for the care and maintenance of large numbers of animals (Hemsworth and Coleman, 1998).

While these KSAs appear to have some intuitive validity, there are no empirical data to indicate whether they are sufficient to identify good stockpeople in the poultry industry or how they relate to job performance in general or the care of birds in particular. In order to identify those stockpeople best suited to managing birds and to develop targeted on-the-job training programmes, the appropriate empirical work needs to be carried out. In one unpublished study of stockpeople in the pig industry, Coleman (2001) found a significant correlation ($r = 0.33$, $P < 0.05$) between technical knowledge and empathy and a moderate but non-significant correlation ($r = 0.23$) between technical knowledge and positive attitudes towards pigs. These limited data suggest that an investigation of the relationship between specific KSAs and stockperson performance may be a worthwhile exercise.

Job motivation and commitment

Job motivation and commitment in intensive animal industries refer to the extent to which a person applies his or her skills and knowledge to the management of the animals under his or her care (e.g. how reliable, thorough and conscientious a person is). Factors such as job satisfaction, meaningfulness of work and the utilization of skills will affect work motivation and commitment. High job performance in any industry relies on a combination of motivation, technical knowledge and skills, and the opportunity to perform the job. Clearly, low motivation will limit job performance regardless of the technical skills and knowledge of the individual.

Once again, empirical data from agricultural industries are sparse and there appear to be no data from the poultry industries. Nevertheless, O'Driscoll and Randall (1999), in a study of dairy workers from both Ireland and New Zealand, found strong positive correlations between job involvement (that is, the extent to which an individual identifies with his or her specific job) and the intrinsic rewards (those rewards that exist in the job itself, such as variety, challenge and autonomy) and extrinsic rewards (pay, fringe benefits and promotion opportunities) of the job. Significantly, they also found that job commitment was greater at higher management levels, although they did not report the specific data. This suggests that work motivation among stockpeople who are at the lower levels may be characterized by poor levels of job commitment. These results are consistent with a study of male part-time farmers by Coughenour (1995), who found that the intrinsic rewards of farm work were associated with their level of commitment. Furthermore, even the commitment of these part-time farmers to the non-farm jobs were affected by the intrinsic rewards derived from farming.

Job satisfaction

Job satisfaction is related to other job-related characteristics, such as job motivation and commitment, motivation to learn new skills and knowledge, and thus in turn technical skills and knowledge, etc. (Hemsworth and Coleman, 1998). Job satisfaction refers to the extent to which a person reacts favourably or unfavourably to his or her work and is considered to derive from the extent to which a person's needs or expectations are being met by the job.

As is the case with job commitment, it is generally considered that job satisfaction is influenced by rewards (personal and financial), job design and enrichment (e.g. involvement in decision-making process), work performance, animal comfort and health, and the working environment. Both O'Driscoll and Randall (1999) and Coughenour (1995) found that satisfaction was related to the intrinsic and extrinsic rewards of farm work. It remains to be determined whether these results from part-time workers on the one hand and dairy farmers on the other generalize to stockpeople working in the poultry industries. As has been a consistent theme so far, it appears that the basic research in the poultry industry has yet to be carried out. It is generally recognized that job performance in any industry is influenced by job satisfaction via its effects on job motivation and commitment, motivation to learn new skills and knowledge, etc. Several authors have suggested that a decline in job satisfaction is associated with staff turnover and absenteeism (e.g. Rusbult *et al.*, 1988).

Personality

Personality factors appear to be useful in matching people to some kinds of jobs (e.g. Barrick and Mount, 1991) and there is limited evidence of this in agriculture. For example, discipline and conformity may be important factors in some jobs in which routine tasks are performed by teams of people, while independence, introversion and self-motivation may be important in others in which the tasks are more problematic and where the individual may at times work alone.

Indeed, some limited studies in the Canadian pig industry suggest that the personality of the stockperson may be associated with piglet performance in the farrowing shed. The results suggest that the importance of personality variables may vary according to the working place, the relative importance of the traits depending on the type of farm. Ravel *et al.* (1996) found significant relationships between personality types of stockpeople, based on an evaluation using the Sixteen Personality Factor Questionnaire and productivity in farrowing units. While self-discipline was a trait that appeared to be important at all farms studied, high insecurity and low sensitivity were favourable traits in relation to piglet survival at independent owner-operated farms, while stockpeople who were highly reserved and bold, suspicious, tense and changeable were associated with higher piglet mortality at large integrated farms. These personality factors accounted for 17 and 26% of the variation between units in piglet survival at the independent and integrated farms, respectively.

In a study of some UK dairy herds, the highest-yielding herds were those where the stockpeople were introverted and confident (Seabrook, 1972). These results support the general view in the current literature that personality factors may influence stockperson performance in ways that are relevant to animal welfare.

Attitude to the animal

Evidence presented earlier in this chapter indicates that the attitude of the stockperson can affect animal welfare and productivity by influencing the stockperson's behaviour and in turn the animal's fear of humans. Empirical data from agricultural industries are sparse but there is limited evidence in the pig industry that stockperson attitudes might also influence animal welfare by affecting a number of job-related variables. Coleman *et al.* (1998) found that attitudes towards pigs and towards most aspects of working with pigs were correlated with a number of measures of work motivation of stockpeople. Attitudes showed similar relationships with job enjoyment and opinions about working conditions. Thus, the stockperson's attitudes may be related to aspects of work apart from handling of animals, and these influences may affect animal welfare.

The Influence of the Stockperson on the Effectiveness of a Welfare Audit Procedure

An effective welfare quality assurance programme should aim to systematically and regularly monitor and record the key welfare inputs and outputs to achieve a high level of compliance with defined welfare standards or targets. In addition, it should also facilitate prompt and effective intervention to appropriately adjust conditions when the desired welfare standards are not being achieved. Thus, compliance with the animal welfare standards in a quality assurance system will often involve verification of essential daily work tasks, procedures and record keeping. A combination of farm records and inspections of critical inputs (such as temperature, stocking density, inspection frequency, feed and water delivery and light regimes) and critical outputs (such as morbidity and mortality) can be used for such verification.

As reviewed in this chapter, a number of job-related characteristics of the stockperson will affect animal welfare. The attitude of the stockperson can affect animal welfare by influencing the stockperson's behaviour and in turn the animal's fear of humans. High and consistent standards of animal welfare will also rely on a combination of motivation, technical knowledge and skills; clearly, any deficiency in these job-related characteristics will limit animal welfare. While records and inspections of critical welfare inputs and outputs may identify suboptimal welfare standards, the underlying basis of the problem and the opportunity for correction may often rest with the stockperson. Furthermore, conventional records and inspections may not identify the cause of the welfare problem if it is a consequence of limitations in the work-related characteristics of the stockperson. For example, observing human behaviour in audit situations may not reflect routine behaviour, and records of inspection frequency or physical inputs may not highlight an attitudinal problem with the stockperson. Systematic and ongoing training that targets the range of job-related characteristics that have been highlighted in this chapter needs to be adopted as an essential element in any quality assurance programme that aims to improve animal welfare. For instance, sustained change in the behaviour of most stockpeople will only occur by specifically targeting the key attitudes and behaviour of stockpeople for improvement (Hemsworth and Coleman, 1998).

Assessing the key job-related characteristics of stockpeople may provide industry with the potential to monitor the potential impact of individual stock-people on animal welfare. Screening aids such as attitude and job motivation questionnaires may identify both weaknesses in individual stockpersons and targeted training for these individuals. The potential value of such screening is illustrated by a recent study in the Australian pig industry in which 144 inexperienced stockpeople were studied after they had completed a series of computerized job-related questionnaires (Coleman, 2001). Attitudes and a number of other job-related characteristics were found to be predictive of the subsequent job performance of the stockpeople.

Monitoring the behaviour of the bird in a standard manner to measure fear of humans may be useful in identifying deficiencies in the frequency and nature of stockperson behaviours and may thus complement the use of attitude questionnaires in predicting stockperson behaviour.

Conclusions

The sequential relationships between human and animal variables found in the livestock industries indicate that there is an excellent opportunity to target stockperson attitudes and behaviour in order to reduce limitations imposed by human–animal interactions on poultry welfare. Understanding stockperson behaviour appears to be the key to manipulating these human–animal interactions to improve poultry welfare. Empirical data from the agricultural industries, particularly the poultry industries, on the relationships between other job-related characteristics of stockpeople and animal welfare are sparse. Research on these topics is clearly required if the broiler industry is to minimize the limitations that human–animal interactions impose on poultry welfare. With such knowledge, stockperson selection, screening and training programmes addressing these job-related characteristics are likely to offer the poultry industries substantial potential to improve poultry welfare. While welfare audits incorporated into industry quality assurance programmes are likely to improve animal welfare, the potential impact of such programmes will only be realized by recognizing the limitations of stockpeople and providing systematic and ongoing training to target the key job-related characteristics of the stockperson.

References

Barnett, J.L., Hemsworth, P.H. and Newman, E.A. (1992) Fear of humans and its relationships with productivity in laying hens at commercial farms. *British Poultry Science* 33, 699–710.

Barrick, M.P. and Mount, M. (1991) The big five personality dimensions and job performance: a meta-analysis. *Personnel Psychology* 44, 1–26.

Breuer, K., Hemsworth, P.H., Barnett, J.L., Matthews, L.R. and Coleman, G.J. (2000) Behavioural response to humans and the productivity of commercial dairy cows. *Applied Animal Behaviour Science* 66, 273–288.

Coleman, G.J. (2001) Selection of stockpeople to improve productivity [abstract]. In: *4th Industrial and Organisational Psychology Conference, Sydney, NSW 21–24 June*, p. 30

Coleman, G.J., Hemsworth, P.H., Hay, M. and Cox, M. (1998) Predicting stockperson behaviour towards pigs from attitudinal and job-related variables and empathy. *Applied Animal Behaviour Science* 58, 63–75.

Coleman, G.J., Hemsworth, P.H., Hay, M. and Cox, M. (2000) Modifying stockperson attitudes and behaviour towards pigs at a large commercial farm. *Applied Animal Behaviour Science* 66, 11–20.

Coughenour, C.M. (1995) The social construction of commitment and satisfaction with farm and nonfarm work. *Social Science Research* 24, 367–389.

Cransberg, P.H., Hemsworth, P.H. and Coleman, G.J. (2000) Human factors affecting the behaviour and productivity of commercial broiler chickens. *British Poultry Science* 41, 272–279.

Gross, W.B. and Siegel, P.B. (1979) Adaptation of chickens to their handlers and experimental results. *Avian Diseases* 23, 708–714.

Gross, W.B. and Siegel, P.B. (1980) Effects of early environmental stresses on chicken body weight, antibody response to RBC antigens, feed efficiency and response to fasting. *Avian Diseases* 24, 549–579.

Gross, W.B. and Siegel, P.B. (1982) Influences of sequences of environmental factors on the responses of chickens to fasting and to *Staphylococcus aureus* infection. *American Journal of Veterinary Research* 43, 137–139.

Hemsworth, P.H. and Coleman, G.J. (1998) *Human–Livestock Interactions: the Stockperson and the Productivity and Welfare of Intensively-farmed Animals.* CAB International, Wallingford, UK.

Hemsworth, P.H., Barnett, J.L., Coleman, G.J. and Hansen, C. (1989) A study of the relationships between the attitudinal and behavioural profiles of stockpersons and the level of fear of humans and reproductive performance of commercial pigs. *Applied Animal Behaviour Science* 23, 301–314.

Hemsworth, P.H., Coleman, G.J. and Barnett, J.L. (1994a) Improving the attitude and behaviour of stockpersons towards pigs and the consequences on the behaviour and reproductive performance of commercial pigs. *Applied Animal Behaviour Science* 39, 349–362.

Hemsworth, P.H., Coleman, G.J., Barnett, J.L. and Jones, R.B. (1994b) Fear of humans and the productivity of commercial broiler chickens. *Applied Animal Behaviour Science* 41, 101–114.

Hemsworth, P.H., Coleman, G.C., Cransberg, P.H. and Barnett, J.L. (1996) Human factors and the productivity and welfare of commercial broiler chickens. Research report on Chicken Meat Research and Development Council Project, Attwood, Australia.

Hemsworth, P.H., Coleman, G.J., Barnett, J.L. and Borg, S. (2000) Relationships between human–animal interactions and productivity of commercial dairy cows. *Journal of Animal Science* 78, 2821–2831.

Hemsworth, P.H., Coleman, G.J., Barnett, J.L., Borg, S. and Dowling, S. (2002) The effects of cognitive behavioral intervention on the attitude and behavior of stockpersons and the behavior and productivity of commercial dairy cows. *Journal of Animal Science* 80, 68–78.

Jones, R.B. and Waddington, D. (1992) Modification of fear in domestic chicks, *Gallus gallus domesticus*, via regular handling and early environmental enrichment. *Animal Behaviour* 43, 1021–1033.

O'Driscoll, M.P. and Randall, D.M. (1999) Perceived organisational support, satisfaction with rewards and employee job involvement and organisational commitment. *Applied Psychology: an International Review* 48, 197–209.

Ravel, A., D'Allaire, S.D. and Bigras-Poulin, M.R. (1996) Associations between preweaning performance and management, housing, personality of the stockperson on independent and integrated swine farms in Quebec. *Proceedings of 14th Congress of the International Pig Veterinary Society, 7–10 July, Bologna, Italy.* Faculty of Veterinary Medicine, University of Bologna, Bologna, p. 505.

Rusbult, C.E., Farrell, D., Rogers, G. and Mainous, A.G. III (1988) Impact of exchange variables on exit, voice, loyalty and neglect: an integrative model of responses to declining job satisfaction. *Academy of Management Journal* 31, 599–627.

Seabrook, M.F. (1972) A study to determine the influence of the herdsman's personality on milk yield. *Journal of Agriculture Labour Science* 1, 45–59.

16 Auditing Systems: Are We Effective?

P. COOK

RI Consulting Ltd, The Field Station, Wytham, Oxford, UK

Types of Audit

I believe that there are three basic types of audits:

- *Assurance audits.* These audits are performed by an independent third party to assess compliance against a recognized set of standards. The standards might be those set by the farmer's customer or perhaps by an industry body.
- *Internal audits.* These audits are normally performed by the farmer himself, or by an internal quality team. The audit will be designed to highlight aspects of the farm systems that ensure that the farm remains compliant with all the necessary standards, and perhaps to specifically prepare for an imminent third-party inspection.
- *Production audits.* These are perhaps not recognized as classic audits, but the term 'production audit' is used in this context to describe the daily checks that a farmer performs every time he visits his site. Such audits are far less formal but will certainly involve key checks, such as the house temperature, the functioning of the equipment and the condition of the stock.

Is Auditing Necessary?

Auditing commonly plays a role in three key processes within an effectively run assurance scheme:

- Approving the members of the scheme.
- Maintaining the standards within the scheme.
- Ensuring that the scheme can demonstrate due diligence.

Let us consider each of these three processes in more detail and try to decide, in each case, whether auditing is a key part of the process.

Approving the members of a scheme

Different schemes have many different approval mechanisms for allowing individuals or companies to join. The mechanism that is selected can be defined by considering the risk that the new member poses to the club or scheme. There are many examples of this in our everyday lives.

For example, children can become part of many comic book clubs by simply sending off a registration form that details their name, age and address, along with a nominal registration fee. The membership is relatively uniform and, one would hope, poses little threat of bringing the reputation of that organization into disrepute. The mechanism is, therefore, appropriate.

A farm assurance scheme would lie towards the high end of the spectrum of risk. The membership is certainly not uniform in its production methods and exposure of poor practice is highly likely to bring the assurance standards into disrepute. Production bases are becoming increasingly international, which brings yet further variation of practice. It is also crucial to remember that the farmer is a key part of the food chain and is consequently in an extremely responsible role. The product that is sent from the farm can, in an extreme case, present a risk to the end consumer if not farmed responsibly.

It therefore becomes increasingly important that the assurance body collects sufficient information about the new applicant to enable an informed decision to be made about whether they are an appropriate member. This process might include the submission of an application form, a self-assessment questionnaire and other key data. However, it is only possible to fully comprehend the production systems and standards of the new farmer by visiting the site and understanding the approach that is being taken.

Therefore, for the process of selecting new scheme members, I would conclude that auditing is indeed necessary.

Maintaining scheme standards

Having identified the correct members for the production base, it is essential that the standards are effectively communicated to the farmers and checks are made that these standards are employed at all times.

It is a simple process to send the standards out to the farmer in the form of a text document. However, an audit of the site allows the assurance body to explain the ethos of the scheme in more detail and to interpret standards that are not entirely clear. This inspection visit will often be the only opportunity for the farmer to interact with the assessment body directly. Through this interaction there is a unique opportunity for the assessment body to gather feedback about the standards from those responsible for putting them into practice.

There are many ways of monitoring the maintenance of standards. These include the use of indirect measures, looking at the finished product, and routine self-assessment. However, as discussed earlier, the method of monitoring should be appropriate to the risk, and I would strongly support the need for production sites in the food industry to be visited by those responsible for maintaining the standards. Therefore, I again conclude that auditing is necessary.

Demonstrating due diligence

Poor practices on broiler farms continue to generate negative publicity for the farmer and for the assurance body. Litigation is also becoming increasingly commonplace. Therefore, if the need should arise, it is important that one can demonstrate that all reasonable steps are being taken to prevent poor practice within a scheme membership.

The use of a third party to conduct audits allows the scheme administrator to demonstrate that the standards are being policed independently and will ensure transparency in the eyes of those challenging poor practice.

This is a somewhat negative reason for auditing to be necessary, but a company must ensure that it is effectively protected. I again conclude, albeit slightly more reluctantly, that auditing is necessary to ensure due diligence.

Is Auditing Effective?

Looking at the initiatives that have been developed over the last decade, I feel that the broiler industry can confidently step back and point to many improvements. Key issues, such as stocking density, leg health and antibiotic usage, have been the subjects of a number of studies. Husbandry practices have evolved with the introduction of new lighting and feeding regimes, and technology has allowed greater monitoring and control within our poultry houses.

These advances have been driven not only by the desire of the industry to bring greater efficiencies to their businesses, but also by the codes of practice that have evolved. Retailer schemes have become competitive in their desire to be at least equivalent in the marketplace. National schemes, such as the UK's Assured Chicken Production scheme, have defined minimum production standards for the industry. Auditing has also helped to administer assurance schemes and has provided the poultry retailer with an effective due-diligence defence should the need arise. At these levels we would quickly move to the conclusion that auditing is effective.

However, incidences of poor practice continue to be exposed in the broiler industry, even on the farms of those producers who are operating within an apparently well run scheme. Examples of such incidences have included media exposure of birds with severe leg problems and incidences of poor practice during catching. It is for these reasons that I feel that we must conclude that auditing is not fully effective in eradicating these problems.

Effective Auditing

At this point we must remember that auditing is only one weapon in the armoury when we are fighting to eliminate bad practice. The producer obviously has a key role in this process and must be capable of training the staff and communicating the desired production standards.

Turning to the audit itself, I believe that there are four aspects of auditing that should be considered in our desire to increase our effectiveness:

- The standards
- The auditor
- The audit process
- The audit value

We will look at each of these in more detail.

The standards

When we write the standards we must ensure that each is technically correct. Some of the requirements in modern-day farming are extremely complex, particularly in areas such as understanding airflows in buildings and specifying the requirements for the enormous range of equipment on the market.

We must also ensure that each standard is necessary. If we cannot justify a sound technical, welfare, food safety or marketing reason for demanding a standard, then this standard should not be expected of the farmer.

Standards should be practical. There is nothing more frustrating for a farmer than having a non-compliance raised because information is perhaps recorded in the 'wrong' format or kept at the 'wrong' location. It seems reasonable that the system on a farm should work effectively for the farmer and the staff rather than being designed around ease of inspection for the auditor.

We should ensure that the standards can be applied universally in different climates and in areas that have different cost bases. This means that it is often better to define the ultimate aim of a standard rather than being totally prescriptive in the way in which the aim is achieved. Many systems can be equivalent in the standards of care that they achieve for the birds even though the degree of automation, for example, might vary considerably.

The standards should expect achievable best practice, although this does not stop the code recommending targets for the future. The farmer must have proven technology available in order that the goals that are set can be reached.

If all these steps are followed, the farmer will accept that those who really understand the complexities of poultry farming have written the standards, and that everything that is being asked for is indeed good practice that will improve performance and efficiency on the farm. In my experience I am confident that there is a direct correlation between the farms that achieve the highest standards of welfare, biosecurity and record-keeping and those that achieve the lowest cost of production. The best farmers understand this and fully accept the benefits that a well-run quality system can bring to their farm. Where there is real attention to the detail of the system paperwork, the birds will invariably also be being managed in a precise and effective manner. It is often hard work for the farmer to reach the standards demanded from an assurance scheme, but once there it is not so difficult to maintain them.

It is perhaps important that, for a scheme to be successful, the farmer must believe that it is technically and ethically correct to work to a set of standards rather than to do so because there is no choice and that the standards must be adhered to, even though the farmer may believe the standards are wrong. The standards must be 'live' and be rapidly adapted to new technology and changes in best practice.

The auditor

Should the auditor be proficient at auditing or poultry farming, or both? I feel that it is essential that the auditor understands the range of technical, practical and commercial issues associated with broiler farming as well as being able to audit. This knowledge will allow the auditor to gain the respect of the farmer and allow for more effective interaction.

Communication is key. It is often difficult, particularly with the modern international supply base, to achieve complete understanding immediately. Audits are often conducted through interpreters, and in these circumstances it is essential that the auditor has a sound knowledge of the production systems of the industry. Great patience is required by both parties when one is trying to fully understand the other's approach to technical issues.

I do believe that a technical approach should be taken to the audit. If the auditor can clearly explain or demonstrate to the farmer that a change to the operating system will bring benefit to the birds and an improvement in the efficiency of the farm, then changes will be accepted more readily. It might also be appropriate to implement a change in stages, so that the benefit of the improvement can be fully understood. For example, if the farmer applies the change to one house on the farm whilst leaving the others as a control, he might more easily monitor the true effect of implementing this 'better' practice.

The audit process

Many producers will own several farms and the auditor will need to be content with visiting a representative sample of the sites. These sites might be chosen entirely at random or by tracing back from a finished product supplied under the auspices of the assurance scheme. However, many audits require a considerable degree of logistic organization and sites are frequently selected because they are conveniently located, allowing more than one site to be visited in a day. When visiting for the first time it may be appropriate for the supplier to select a specific farm. Auditing a company's best site provides a clear insight into their ultimate ethos, whereas a poor site selected at random might not initially be as informative.

A second problem in selecting the site for the audit is that a single visit allows the inspector to see only a snapshot of the crop cycle on the farm. There are many processes that occur during the life of the birds, such as chick placement, brooding, the growing period, and depopulation. It is important to target the audits at the correct stage of the cycle. Many farmers seem to prefer an auditor to visit a site when the birds are in their second or third week of age, but this provides little information about the likely leg health of the birds and conditions in the house towards the end of the growing period. If an auditor is scheduled to visit more than one farm in an organization, farms can be selected so that the birds will be at a range of ages. One might be able to view older birds and catching at the same farm, at the same time. In any case, the auditor must be extremely clear about the purpose of the visit and the age of birds that need to be observed.

This brings us to another interesting aspect of the audit process. Should the visit be announced or unannounced? There is, perhaps, more likelihood of seeing the farm in its true state if the visit is unannounced. However, this may be difficult for the reasons discussed above, along with the specific requirements for personal bio-security that must always be observed. At the present time, I would suggest that most third-party audits that are conducted are announced. This means that auditors should also use other available techniques to make their assessments.

An example of this is the use of key performance indicators (KPIs). These KPIs comprise data that can be monitored between successive farm visits, such as stocking densities, mortalities, levels of hock burn and amounts of antibiotic used. These data will help the experienced auditor decide whether the standards that are being seen are truly representative of the company. I would suggest that, if we are reliant on such an approach, we will never avoid the incidence of poor practices. It is far more effective to try to ensure that our farmer fully subscribes to the standards that are being proposed and that the benefits of adopting best practice are recognized.

The value of the audit

The farmer and the assurance body may hold very different views about the value that is derived from the audit. We have reviewed the many reasons why the audit is essential for the audit body, but have not really considered the value of the audit to the farmer.

The daily production audits that we referred to above (see Types of Audit) are seen as essential by the farmer. Every time a poultry house is visited, certain checks will be made and records taken. This simple audit will take place every day and ensures the health, welfare and efficiency of the farm.

The farmer will receive many visits to the farm that he will consider to be of great value. For example, there will normally be a regional manager or 'fieldsperson' with responsibility for the site. This person will provide advice on the husbandry practices on the farm, discuss flock performance, and generally work to improve the efficiency of the site. There will also be a veterinary surgeon responsible for the birds. The vet will map out the ideal vaccination programme and perhaps provide a monitoring service to review the effectiveness of the house cleanout. Both of these visitors provide the farmer with an essential service and are effectively performing a form of internal audit, as discussed above (see Types of Audit).

However, when we come to the assurance auditor, the more cynical farmer may see this type of audit as a one-off test that must be passed to allow product from the farm to be supplied to the retailer of choice. The farmer will have to take much time and effort to prepare for the audit, there will then be a cost for the inspection, and there may then be a list of non-compliances that must be actioned before the farm can gain its accreditation. If the standards are not recognized by the farmer as being best practice, there will almost certainly be a feeling that there is little value in the exercise, just a large financial penalty.

On the other hand, if the standards are well conceived and the auditor has a robust understanding of broiler production systems and evolving technology, our cynical farmer should be able to derive real value from the visit. Effectively, the auditor works with the farmer, so that, in time, all aspects of the assurance audit become accepted as a part of either the daily production audits or the internal audits, both of which the farmer deems essential in the running of an efficient business. It would seem logical that a good system should expect a farmer only to make changes that are for the good and will improve the health, welfare and performance of the stock.

We have alluded to the fact that this would require the auditor to adopt a more consultancy-based inspection style and to work with the farmer in a true partnership to solve the issues that are identified. This approach, however, depends on the audit process allowing the auditor to work in this way.

It is recognized that some systems do not allow the auditor to go any further than raising non-compliances, as it is felt that this may erode the true third-party independence of the inspection and dilute any due-diligence defence that might be required. Unfortunately, this can lead to the farmer feeling that the farm has not only taken on the cost of the visit itself but has also been left with a further raft of work (or cost) to solve the problems that have been set.

If a consultancy style is adopted, the auditor must naturally be very conscious of the commercial sensitivity of the data to which privileged access has been given. However, there is a considerable amount of research and development that is in the public domain that the auditor can use to help drive systems forwards. There are also accepted practices used internationally and for other species that translate extremely well into local poultry production systems. The inspector might even be able to simply put the farmer or company in touch with the correct organization to help solve the issue.

Overall, I feel that it is a disappointment if we are not able to fully utilize the considerable knowledge and experience that lies within the audit teams for the good of the farmer.

Summary and Conclusions

Auditing is necessary in order to approve the membership of a scheme, to allow companies to demonstrate due diligence, and as an aid in maintaining standards.

Auditing is effective in certain respects, but does not stop the occasional breaches of the standards that lead to poor publicity for the industry.

Auditing might become more effective by ensuring that:

- Standards are technically correct and represent best practice.
- Auditors fully understand the processes and practicalities of poultry farming.
- The audit process inspects representative farms at the right times.
- The audit adds value to the farmer and is not perceived merely as a cost.

It is the last of these points that I would recommend as being the most important in our quest to achieve a real partnership between the auditor and the farmer.

17 Is UK Farm Assurance Delivering Good Welfare?

D. MAIN AND H.R. WHAY

School of Veterinary Science, University of Bristol, Langford, Bristol, UK

Introduction

There are numerous examples of farm assurance schemes and their likely impact on welfare is influenced by the objectives that the schemes were developed to meet. Hence, a scheme that has a stated aim of improving welfare is more likely to have a greater influence on welfare than a scheme designed to assure consumers about the fulfilment of a minimum welfare standard. In addition, a scheme that has goals other than welfare, such as food safety or environmental concerns, may encompass inherent conflicts with welfare. Obviously, examination of the scheme's goals is a superficial assessment of its likely impact on animal welfare. This chapter outlines the three key aspects that should be examined: the content of the standards, the implementation of the standards, and the evaluation of their impact on animal welfare.

All types of certification scheme, whether designed for the mass or the niche market, will need to follow similar principles. The key components are the technical standards and some form of assessment system that ensures all farmers within the scheme are compliant with these standards. Standards in this context refer to the written requirements that farmers must comply with in order to be members of a scheme. Farms that have demonstrated compliance with the standard will then normally receive appropriate certification. Most bodies that administer UK farm assurance schemes have been accredited to the European Standard EN45011 (equivalent to ISO Guide 65). Hence, the farms are certificated and the schemes are accredited. Accredited bodies will have demonstrated a certain level of competency and impartiality. The accreditation process does not set specific minimum thresholds for the scheme standards. However, the standard must be developed in consultation with relevant stakeholders, such as retailers and farmers. We will outline the key principles that are useful in assessing the ability of a scheme to provide a credible welfare assurance to consumers.

©CAB International 2004. *Measuring and Auditing Broiler Welfare*
(eds C. Weeks and A. Butterworth)

Standards for Current Quality Assurance Schemes

Farm assurance standards have been developed for most livestock sectors, including broiler chickens. The requirements set out in the standards of each scheme vary enormously in both content and scope. For example, organic schemes give quite specific detail on the content and origins of the diet while others have more general statements about diet suitability. Some standards deal only with the management of the animals on the farm, whereas others include transport and slaughter. For some countries there are mechanisms in place to ensure that minimum levels are set among certification schemes. For example, in the UK, Assured Food Standards moderate standards for several schemes.

The detailed content of the standard for each scheme depends, of course, upon agreement between interested parties (e.g. the farmer and the retailer). So for national schemes that are aimed at a large number of producers the standards would not be able to prohibit a husbandry practice of welfare concern that is intrinsic to a common husbandry system. For example, the largest current UK farm assurance scheme, Assured Chicken Production (Assured Chicken Production, 2002), does permit some level of space restriction as it is considered an intrinsic component of the commercial system:

> 5.23 As a base standard, this scheme supports the stocking density recommended by FAWC (Farm Animal Welfare Council) of 34 kg/m². However this recommendation is based on best practice over 30 years ago, and this code supports the practice of higher stocking regimes which reflect modern husbandry best practice, modern environmental control techniques and the modern genotype.

The ability of a scheme to deliver good welfare also depends on the extent to which this can be assured by the husbandry provisions defined in the standard. For example, a standard that defines the access to a certain number of drinkers and minimum water flow rates should ensure that animals are free from thirst. However, it would be more difficult to define standards of provision that would minimize more complicated multifactorial health and welfare problems, such as lameness in broilers.

The following are some key criteria that are useful in assessing the likely effectiveness of the standards.

- *Level of standards.* A scheme whose standards are designed to allow any farmer entry is not likely to impress the discerning customer. As a minimum, any accredited quality assurance scheme must include (and assess) all legislation that is relevant to the stated objectives of the scheme. However, it is also reasonable that the standards do not exceed those that are potentially achievable by those operating to best possible practice. An unachievable standard will alienate farmers.
- *Scope of the standards.* The standards should encompass all resources and husbandry practices on a farm that could affect the welfare of an individual of the species covered by the scheme. This should include all classes of stock and include all key activities, such as the management of cull animals. Most schemes base their standards on measures of provision; that is, the husbandry resources that should be provided to the animal rather than animal-based measures of outcome, such as disease incidence or the prevalence of cannibalism. Some

schemes include a requirement for a health plan, which is a type of management standard (see Box 17.1). A health plan can require farmers to plan their own husbandry procedures in order to minimize certain welfare concerns, and then monitor their effectiveness by recording and reviewing animal-based measures of performance over time. Thus, a particular health or welfare concern should be detected by the farmer and a specific action plan relevant to that farm instigated.

- *Formulation of the standard*. The standards should be clearly defined, understandable and unambiguous. They should be regularly and frequently updated. They should be auditable and enforceable, since a standard that cannot be verified on-farm is unhelpful and could lead to false claims by the scheme.

Implementation of Standards

A scheme cannot deliver assurance to the consumer unless there is a credible system for ensuring that the producer is complying with the requirements in the standard. There are usually five key stages to this process:

- *Farmer awareness*. Farmers will only implement standards if they are fully aware of the detailed requirements of the standard, so it is important that they receive updated copies of the standards. Some schemes use a self-assessment system, partly to draw the attention of the farmer to the standard requirements and partly to provide evidence that the farmer is complying with the standards, since an external auditor can verify the accuracy of the self-assessment.
- *Advisory input*. Advisers can assist the producer to comply with the standard. For example, the veterinary surgeon is well suited to perform this task, ideally in association with regular consultations on herd health and preventive medicine.

Box 17.1. Management standards.

Quality management systems such as ISO 9000 provide a model for management systems in any industry. The key principles of these management systems can be incorporated into livestock systems. For example a veterinary health plan (Main and Cartledge, 2000) might contain the following key elements:

- Plan the specific prevention and treatment procedures.
- Record the incidence of key health and welfare parameters.
- Review at regular intervals the levels of these parameters.
- Action plans to modify husbandry conditions in order to deal with specific problems.

Health plans may be specifically mentioned in farm assurance standards, e.g. Assured Chicken Production (2002).

2.2 Each site must have a written health and welfare programme tailor-made to the needs of the unit, and must contain a strategy for the prevention and control of common diseases. As a minimum the programme must be annually reviewed and updated. The programme must set out health and husbandry procedures covering the whole of the production cycle.

- *Assessment.* A representative of the scheme (assessor) will usually need to visit the farm to verify compliance with the standard. In general terms, the assessor gathers evidence (visual, verbal or written) that verifies compliance with the standard. The assessor gathers this evidence from observation of records, resources and management, structured dialogue with stockpersons and, of course, observation of the animals in their environment. The competency of the assessor in terms of animal welfare knowledge and assessment is, therefore, critically important. Also, the impartiality of the assessor is important for the credibility of the scheme. The assessor will also need inspection and audit skills (see Box 17.2). Inspection is used to assess the husbandry resources that are specified in the standard. Audit skills are needed to verify long-term compliance with the management aspects of the standard, such as the veterinary health plan.
- *Frequency and duration of visits.* Increasing the frequency and length of visits can increase the credibility of the assessment procedure, but this obviously has a direct impact on the cost of the scheme. Assessors with suitable auditing and inspection skills and access to records should be able to make some assessment of the management of the unit over a reasonable period, such as 1 year prior to the visit.
- *Non-compliance management.* The certification system will also need to ensure that any non-compliances are resolved. Usually, membership of such schemes is withheld until evidence is provided or another visit is arranged to demonstrate full compliance. Existing members will usually have a certain defined period in which to resolve non-compliances. If the issue is very serious or is not resolved, the farmer can be expelled from the scheme. Withholding membership from a scheme can have significant financial consequences, as membership of a farm assurance scheme is often a key customer requirement.

Evaluation of the Animal Welfare Impact of Schemes

There has been very limited investigative work conducted to see whether farm assurance schemes actually deliver these assurances. However, the Royal Society for

---

Box 17.2.   Audit versus inspection.

- *Inspection* is defined in international standard ISO 8402 as 'Activity such as measuring, examining, testing or gauging one or more characteristics of an entity and comparing the results with specified requirements in order to establish whether conformity is achieved for each characteristic'.
- *Auditing* in the same standard is defined as 'Systematic and independent examination to determine whether qualities, activities and related results comply with planned arrangements and whether these arrangements are implemented effectively and are suitable to achieve objectives'.

So for farm assurance standards that are designed to assure the consumer on welfare standards, the husbandry resources in the standard are *inspected* during a visit whereas the management requirements that are defined in the health plan (see Box 17.1) are *audited*.

the Prevention of Cruelty to Animals (RSPCA) has commissioned the University of Bristol to evaluate the impact on animal welfare of the RSPCA Freedom Food scheme. This is a UK-based farm assurance scheme designed to achieve high standards of animal welfare on farm during transit and at slaughter (RSPCA, 2001). The Freedom Food scheme requires members to adhere to welfare standards set by the RSPCA in association with species-specific technical working groups. These groups include producers, industry experts, veterinary surgeons and animal welfare scientists. The RSPCA has produced standards for dairy cattle, sheep, chickens (laying hens and broilers), turkeys, ducks, beef cattle and pigs. The scheme is operated by Freedom Food Ltd, a wholly owned subsidiary of the RSPCA, and farms are certificated by an independent body operating under the European Standard EN 45011. Farms applying for membership of the scheme receive a copy of the relevant standards and are inspected by Freedom Food assessors. Farms that comply with the standard are provided with a membership certificate and are monitored by assessors, usually on an annual basis. In addition, Farm Livestock Officers from the RSPCA farm animal department inspect a proportion (30%) of farms every year. Any aspects of the farm or its management identified during visits as constituting non-compliance with the standard are recorded. The Freedom Food certification panel then informs the farmer of the action required either to join or to remain a member of the scheme. The evaluation of the Freedom Food scheme has been completed for dairy cattle; it is summarized below as the process used is equally applicable to broiler chickens.

The first stage of the evaluation of the Freedom Food scheme involved the development of a credible welfare assessment protocol. Assessing welfare can be a contentious area of scientific investigation and there are various possible approaches to it at a group level (Johnsen *et al.*, 2001; Main *et al.*, 2001). Broadly, welfare assessment can be achieved by evaluating both the husbandry provisions (e.g. diet, housing and management) and/or evaluating the animal-based indicators of welfare outcomes (e.g. disease and behaviour). Farm assurance schemes (including the Freedom Food scheme) tend to examine the provisions as defined in their standards and interpret their assessment as a simple pass/fail result. Since the objective of this study was to assess the effect of the Freedom Food scheme on the welfare of dairy cattle, the assessment needed to be not only independent of the current standard and inspection system (Main *et al.*, 2001) but also an animal-based approach to the evaluation of welfare outcomes. The most important elements of the state of welfare of dairy cattle were first established through an iterative review of expert opinion (Whay *et al.*, 2003a), then incorporated into a systematic programme of observations and records-related welfare measures (e.g. production, reproduction, disease and behaviour) compiled during a single farm visit. Welfare assessment protocols have also been developed for broiler chickens (see Chapter 14).

Once the dairy cattle welfare assessment protocol had been established using expert opinion, the impact of the Freedom Food scheme on dairy cattle welfare was assessed after visits to 28 Freedom Food and 25 non-Freedom Food farms in England during the winter of 2000–2001. Expert opinion was also used to assess the significance of the welfare measures. The full results of this study have been published elsewhere (Main *et al.*, 2003; Whay *et al.*, 2003b). For dairy cattle, the data analysis showed that lameness and housing/lying area discomfort occurred at high levels in Freedom Food as well as non-Freedom Food farms, and should be of high priority for improvement by the scheme. These priorities were identified by evaluating the number of Freedom Food farms above intervention levels that had been derived from expert opinion. The number of farms above these intervention levels did not reflect the extent of compliance with the Freedom Food standard. The Freedom Food standards are largely resource-based (e.g. housing conditions) and the University of Bristol assessment was animal-based (i.e. behaviour, physical condition and health records). It is therefore possible that a farm performed poorly according to the University of Bristol intervention levels for one or more welfare measure but was still fully compliant with the Freedom Food standards.

An important conclusion to be drawn from this study is that welfare problems and priorities for action are specific to individual farms. It is the view of the authors that these will not be resolved through modification of the general standards of provisions (i.e. the assurance scheme standards). Resolution of specific welfare problems on individual farms needs to be addressed at a local level using a health plan format. For example, a reduction in lameness will not necessarily be achieved by imposing a standard foot-bathing routine on all farms. Each farm's lameness problem has specific causes and needs to be managed at a local level using the appropriate advice. It is very likely that individual broiler chicken units also have specific welfare problems that require specific solutions.

On receipt of the results of this study, the RSPCA and Freedom Food immediately set up a lameness initiative focusing attention on lameness and related discomfort issues on Freedom Food dairy farms. This initiative will consist of evening meetings to increase farmer awareness together with on-farm inspections that concentrate on relevant elements of provision and indices of welfare. The farmers' meetings will raise awareness of the welfare and financial implications of lameness and discomfort and discuss the role of health plans in reducing this problem. RSPCA Farm Livestock Officers will then visit Freedom Food dairy farms to assess and produce a report on the prevalence of lameness and other discomfort-related measures on the farms. This report will contribute to raising farmer awareness and be used as an advisory/management tool by advisers and farmers. It will not be used directly for certification purposes. Finally, the on-farm inspection system is being reviewed to ensure close attention is paid to compliance with lameness and discomfort-related standards, especially the relevant health plan requirements. Similar studies in broiler chickens are also likely to identify specific priorities for action, and the solutions at a scheme level could follow a similar format.

Conclusions

The influence of farm assurance schemes for animal welfare can be examined systematically. A review of the standards and the implementation procedures provides essential information. However, the ultimate assessment requires an evaluation of animal welfare on farm-assured farms. Assessment of welfare in this context can answer specific questions about any scheme's effectiveness in assuring or improving animal welfare. However, these relatively simple welfare assessments can and should be incorporated into the farm's own management system. Furthermore, it is the opinion of the authors that particular efforts must be made to promote encouragement initiatives that reward good performance in animal welfare. The existing system of permitting farmers to be members of a scheme is a particularly blunt instrument for improving welfare.

Acknowledgements

The dairy cattle study was supported and funded by the Royal Society for the Prevention of Cruelty to Animals.

References

Assured Chicken Production (2002) Poultry standards: chickens. Assured Chicken Production, Long Hanborough, UK.

Johnsen, P.F., Johannesson, T. and Sandøe, P. (2001) Assessment of farm animal welfare at herd level: many goals, many methods. *Acta Agriculturae Scandinavica Supplement* 30, 26–33.

Main, D.C.J. and Cartledge, V. (2000) The veterinarian's role in farm assurance schemes. *In Practice* 22, 335–339.

Main, D.C.J., Webster, A.J.F. and Green, L.E. (2001) Animal welfare assessment in farm assurance schemes. *Acta Agriculturae Scandinavica Supplement* 30, 108–113.

Main, D.C.J., Whay, H.R., Green, L.E. and Webster, A.J.F. (2003) Effect of the RSPCA Freedom Food scheme on dairy cattle welfare. *Veterinary Record* 153, 227–231.

RSPCA (2001) *RSPCA Welfare Standards for Dairy Cattle*. Royal Society for the Prevention of Cruelty to Animals, Horsham, UK

Whay, H.R., Main, D.C.J., Green, L.E. and Webster, A.J.F. (2003a) Animal based measures for the assessment of welfare state of dairy cattle, pigs and laying hens: consensus of expert opinion. *Animal Welfare* 12, 205–219.

Whay, H.R., Main, D.C.J., Green, L.E. and Webster, A.J.F. (2003b) Assessment of dairy cattle welfare using animal-based measurements. *Veterinary Record* 153, 197–202.

18 Using Welfare Outcomes to Control Intensification: the Swedish Model

C. BERG AND B. ALGERS

Department of Animal Environment and Health, Faculty of Veterinary Medicine, Swedish University of Agricultural Sciences, Skara, Sweden

Introduction

The animal welfare programme for broilers in Sweden was developed in the mid 1980s by the producers' organization, the Swedish Poultry Meat Association (SPMA), together with production advisers, veterinarians and representatives from the Swedish National Board of Agriculture. The programme aims to improve the general standard of broiler rearing, including housing facilities, equipment and management. It aims at higher levels than the basic standards set by the animal welfare legislation, which is controlled by the National Board of Agriculture and the local animal welfare inspectors (Algers and Berg, 2001). On the basis of a set protocol, an annual evaluation is made of the management and the standard of buildings and equipment on each broiler farm.

This programme has attracted attention both from broiler producers in other countries and from other types of livestock growers within Sweden.

Background

Approximately 80 million broilers are grown in Sweden annually. They are all kept in climate-controlled houses on litter and are usually slaughtered between 28 and 45 days of age, with a mean of 36 days. Litter is discarded and replaced and the house is thoroughly cleaned between batches of birds. Broiler production is relatively standardized, with little variation in rearing methods or marketing strategies. The production of 'alternatively' grown or organic broilers in Sweden has been on an extremely small scale so far. Two international hybrids, Ross and Cobb, are currently used as commercial broilers.

Broiler growers, hatcheries, abattoirs and other companies with broiler-related activities can become members of the producers' organization, SPMA. Currently 98% of Swedish broiler producers (approximately 130 growers) belong to this organization. Also most turkey producers and a considerable proportion of goose and duck producers are members (Svensk Fågel, 2003). There are three hatcheries for commercial broilers and seven broiler abattoirs, each slaughtering 1–35% of the total annual Swedish broiler production. There is a similar welfare programme running for fattening turkeys, but not for ducks or geese.

Organization of the Animal Welfare Programme

All broiler producers who are members of the SPMA have their houses evaluated at regular intervals, usually once a year. The SPMA National Standards Officer carries out the classification, together with a local official veterinarian appointed by the Board of Agriculture. The poultry manager at each slaughterhouse is usually also present, together with the producer. At these visits a set protocol is used to evaluate the management and the standard of buildings and equipment on each broiler farm (Ekstrand, 1993; Svensk Fågel, 2002). Each broiler house is evaluated separately. Because of the limited number of broiler farms in the country, a single person can carry out all the farm visits within the welfare programme. This decreases the risk of setting different standards in different regions or for different companies. However, it also makes the National Standards Officer a person whose judgement of a farm will have a huge impact on the conditions for production in the future. Therefore, it is important that his integrity cannot be questioned.

Avoiding theoretical discussions about different definitions of animal welfare but still being aware of the problems that arise when attempts are made to measure animal welfare in any way, the designers of the programme chose to mainly measure and evaluate the basic conditions on which good animal welfare, as perceived by the producers and consumers, can be based.

The protocol is divided into three main categories: bird area, surrounding areas and management practices. In total, 31 points are evaluated. Each point is scored from 0 (unsatisfactory) to 4 (very good), both in relation to the basic technical standard and to the actual management of the equipment. These scores are then weighted by multiplying them by a factor ranging from 1 to 11, depending on the relative importance of the item concerned. The total score for each broiler house is then presented as a percentage of the total possible score.

Any equipment or housing facility with which the birds have direct contact is evaluated under the heading 'bird area'. This includes heating, feeding and water equipment, the ventilation system and the type of lighting (Fig. 18.1). Furthermore, the type and quality of floor, walls and ceiling (including the height, i.e. the air volume) are evaluated, along with manure-handling strategies, security against wild birds and rats, and the facilities for rapidly and smoothly receiving or catching and loading the birds.

Considered under the category of 'surrounding areas' are the general hygienic conditions, the layout of the entrance and the biosecurity area, the personnel room, the auxiliary power supply system and the alarm system (required to detect power failure, too high or too low temperature and failure of the alarm itself; often available also for other features). Also, the equipment used for medication and the administration of vitamins, etc. is evaluated.

Fig. 18.1. Within the animal welfare programme all types of equipment possibly influencing bird health, such as feeding equipment, will be evaluated with respect to technical and maintenance standards.

Judged under the category of 'management' are, for example, environment-animal care, litter and air quality, outdoor conditions, record-keeping routines and slaughter quality (rejects and downgradings).

The protocol has been updated continuously since its inception in 1988 and new or stricter requirements have been added. In 1994, a new variable was added to the welfare programme: footpad dermatitis. However, this is not included in the on-farm protocol but is evaluated at the slaughterhouse and added to the scoring results separately (Ekstrand *et al.*, 1998).

Farmers outside the regulatory programme are, according to the national animal welfare legislation, not allowed to exceed a stocking density of 20 kg/m^2 in their houses (SJV, 2003). The maximum stocking density allowed for farms within the programme is 36 kg/m^2, and anything between 25 and 36 kg/m^2 is set in relation to the score achieved within the animal welfare programme (Fig. 18.2).

Bird-related Outcome Variables

As mentioned above, a number of outcomes that are possibly related to welfare have been included in the programme to ensure some type of follow-up from an animal welfare perspective.

The category 'environment-animal care' in the on-farm protocol encompasses both basic data, such as total mortality, first-week mortality and leg culls, and less concrete aspects, such as how the stockperson moves around among the broilers, how often he or she searches for and culls sick or dead birds, and a general evaluation of his or her eye for animals (Fig. 18.3). This evaluation is a delicate matter for the

Fig. 18.2. The highest stocking density allowed within the animal welfare programme is 36 kg/m^2 on the day of slaughter. For producers not participating in the programme the maximum stocking density is 20 kg/m^2, according to the national animal welfare legislation.

Fig. 18.3. The rapid removal and culling of sick birds, such as birds with severe leg problems, is an essential part of the animal welfare programme, although the main emphasis is on the prevention of disease.

National Standards Officer, as in a way it implies an evaluation of the character and personality of the stockperson.

Evaluations of litter quality and air quality are included in the programme because these parameters indicate the outcome of a number of aspects that are all highly relevant to bird welfare, such as the efficiency of the heating and ventilation system and the litter management strategy.

The direct link between record-keeping and bird welfare may appear less obvious. However, it is possible to argue that proper record-keeping, including anything from daily mortality to feed and water consumption, will give the producer early warnings about bird health problems. For example, an increase in water consumption may indicate an outbreak of diarrhoea among the birds before any clinical signs are seen. This makes it possible for the producer to take the necessary measures to identify the cause or treat the disease before it has affected too many birds. The same reasoning can be applied to the early identification of any type of increased morbidity or mortality.

The inclusion of production outcomes such as slaughter quality (rejects and downgradings) in the animal welfare programme can be explained by viewing these parameters as an evaluation of the health of the birds when still alive. Of course, post-mortem injuries acquired at the slaughterhouse must be distinguished from those arising during the growing period and those originating from catching and transport procedures. A total figure for rejects is of little value if it cannot be split up into different categories of lesions or diseases. It can be argued that the disease recording done by the meat inspectors at the abattoirs focuses mainly on meat hygiene aspects, and that signs of general unthriftiness or maltreatment in broiler flocks do not receive enough attention.

A monitoring programme for footpad dermatitis was linked to the welfare programme in the mid-1990s. This was done in order to incorporate a parameter that could give information on the rearing standard during the bird's entire life. Footpad dermatitis is a type of contact dermatitis that results in erosions or ulcers on the ventral footpads of poultry. The major risk factor for footpad dermatitis in broilers is wet litter, which is linked to a number of management and housing factors. Thus, the lesions can be used as an indicator of the standard of these factors. The monitoring programme for footpad dermatitis has been linked to an advisory programme in which risk factors have been identified on problem farms and relevant action has been taken (Berg, 1998; Ekstrand *et al.*, 1998).

The Effects of the Welfare Programme

The idea behind the Swedish animal welfare programme is that the improvements in facilities and management should clearly overrule any disadvantages from the increase in stocking density. By positively correlating the total score received in the programme to the maximum stocking density allowed at the time of slaughter in each broiler house, a type of incentive is created to encourage improvements in housing and management. The programme focuses mainly on evaluating the conditions on which good bird welfare relies. This means that items such as well-designed feeders and drinkers, well-functioning ventilation systems, efficient alarm systems, rapidly activated auxiliary power supplies and proper biosecurity routines earn a high score in the programme. On the other hand, the presence of all these facilities does not guarantee a good outcome with respect to bird welfare. Therefore, the results with respect to bird welfare must be followed up in different ways.

The introduction of a programme such as the Swedish broiler welfare programme will have large effects on the entire industry. When a programme is first introduced to a group of producers, it forces them to decide whether to invest the time and money needed to meet the criteria for applying higher stocking densities and to try to achieve the highest production level, or to do as little as possible while still being able to run the operation at a profit because of low costs, even at a low stocking density. In Sweden, a number of producers decided to leave the industry completely, as they did not welcome the necessary interference by outsiders in their enterprises. In most cases, these producers had broiler houses and equipment that was both old and poorly maintained, and they knew that they would not get competitive scores in the programme. In fact, a few producers were more or less forced to quit. However, the result was that the broiler industry, within a few years after starting the programme, reached a higher and more uniform standard of production operations and general bird management. Apart from the obvious benefits for bird welfare, this has also made the industry less vulnerable to negative attention from the media, which often focus on poor animal conditions on farms with the poorest management.

The programme has meant that virtually all new producers use the printout from the scoring programme, much as they might consult an advisory leaflet, to ensure that their new broiler houses will fulfil the technical requirements. In this way, welfare aspects have come to have a direct influence on decisions made when constructing the housing environment. As a result of this defined welfare programme, producers are able to ensure, during the planning phase, that they will be able to meet the high standards required. The National Standards Officer also contacts new producers at the beginning of the construction phase and is available for discussion during the first batches of birds or whenever needed. This probably results in fewer beginner's mistakes than would otherwise be the case. Currently, 72.3% of Swedish broiler sheds (corresponding to 79.6% of the total housing area) are classified for 36 kg/m^2 (Svensk Fågel, 2003).

The local official animal welfare inspectors are free to visit any farm at any time and as often as they wish to. In reality, however, their time and resources are limited. Most inspectors are aware of the existence of the animal welfare programme for broilers and will often feel confident that the regular visits of the National Standards Officer will prevent or at least reveal any serious welfare problems on the farms included in the programme. They are therefore likely to pay more attention to broiler farms outside the programme, although such farms are required by law to use a lower stocking density, which should (in theory) be easier to manage.

The bird welfare outcome variables included in the programme have probably made producers more focused on how their management influences the well-being of the birds. This awareness, in combination with knowledge of how to improve management routines, can be seen as a way to decrease the risk of major animal welfare problems. After the addition of foot health as a parameter, it was possible for results regarding footpad dermatitis to influence the stocking density of subsequent flocks. As a result, the incidence of severe footpad lesions was reduced from 11 to 5% in 3 years, which can be seen as a direct effect of the surveillance programme. Focusing on footpad dermatitis as an outcome parameter has also led to more attention being paid to the link between feed composition and bird health, as producers have noticed differences in litter quality – and consequently in foot health status – between batches of feed. This has led to more active and rapid communication between producers and feed suppliers, which can be expected to benefit the birds over time.

The problem with highlighting some welfare parameters is that lower priority might be given to other aspects of welfare which do not contribute directly to the financial outcome of production. For instance, focusing on and giving high priority to footpad dermatitis might distract attention from improving the situation regarding leg weakness. Therefore, there is also a need for other outcome-based parameters, reflecting the bird's welfare status during the entire rearing period. Efforts to find useful parameters should continue.

With stocking density as a way of rewarding 'good' broiler producers, it is the monetary return per square metre of housing area – and thereby the return per flock – that is being used to encourage farmers to improve welfare standards. It is questionable whether the link to stocking density is the optimal solution, or whether paying a premium to farmers who apply high welfare standards would be better from an animal welfare point of view. To a greater extent than today, payment could also be linked to the numbers of downgradings and rejections, which are related to the health of the birds. This would, however, require improved standardization of the rejection and downgrading criteria used at different abattoirs.

Welfare programmes in animal production have, over the years, gone through a generation shift. Originally, programmes were designed to assess welfare using parameters in the animals' environment. The second generation of welfare programmes, of which the Swedish broiler welfare programme is one of the first, have included health criteria measured by the effects on the animals themselves. However, good welfare also implies a positive state (e.g. Duncan, 1996), something that is currently not really evaluated within the programme. If, how and when such a third-generation welfare programme can be created and implemented is an open question.

References

Algers, B. and Berg, C. (2001) Monitoring animal welfare on commercial broiler farms in Sweden. *Acta Agriculturae Scandinavica Section A, Animal Science Supplementum* 30, 88–92.

Berg, C. (1998) Footpad dermatitis in broilers and turkeys – prevalence, risk factors and prevention. PhD thesis, Swedish University of Agricultural Sciences, Uppsala, Sweden. *Acta Universitatis Agriculturae Sueciae, Veterinaria* 36.

Duncan, I.J.H. (1996) Animal welfare defined in terms of feelings. *Acta Agriculturae Scandinavica Section A, Animal Science Supplementum* 27, 29–35.

Ekstrand, C. (1993) Effects of stocking density on the health, behaviour and productivity of broilers – a literature review. Report 32, Department of Animal Hygiene, Swedish University of Agricultural Sciences, Skara, Sweden.

Ekstrand, C., Carpenter, T.E., Andersson, I. and Algers, B. (1998) Prevalence and prevention of footpad dermatitis in broilers in Sweden. *British Poultry Science* 39, 318–324.

SJV (2003) SJVFS 2003:6 (L100). Statens Jordbruksverk, Jönköping, Sweden. Legal text.

Svensk Fågel (2002) Hjälplista vid bedömning av kycklingstallar. Svensk Fågel, Stockholm, Sweden.

Svensk Fågel (2003) Matfågelproduktion i Sverige. Årsrapport avseende 2002 avlämnad till Statens Jordbruksverk från branschorganisationen. Svensk Fågel, Stockholm, Sweden.

19 Developing and Implementing a Welfare Audit in the Australian Chicken Meat Industry

J.L. BARNETT[1] AND P.C. GLATZ[2]

[1]Animal Welfare Centre, Department of Primary Industries, Primary Industries Research Victoria, Werribee, Victoria, Australia; [2]South Australian Research and Development Institute, Pig and Poultry Production Institute, Roseworthy, South Australia, Australia

Introduction

In the Australian chicken meat industry ISO quality assurance programmes have been implemented by most of the large companies, particularly in the processing plants. Because of increasing consumer concern about the way meat chickens are farmed, the industry saw value in expanding their quality assurance programmes to cover welfare. By incorporating animal welfare checks in the quality assurance system, it places more emphasis on the importance of daily work tasks, procedures and record-keeping in maintaining good welfare for meat chickens. The Australian broiler welfare audit (Barnett *et al.*, 2001) has the main aim of assisting industry meet the community's expectations of the need to maximize the welfare of broiler chickens. The welfare of meat chickens was examined during all stages of production from the hatchery to processing. Many countries have codes of practice for welfare and there are audits associated with marketing of systems that are perceived to be 'welfare-friendly'; an example is Freedom Foods in the UK, for laying hens (RSPCA, 1999). This chapter describes the processes used to develop the documentation for the welfare audit for the Australian chicken meat industry and an evaluation of part of the audit at grower farms.

Developing the Audit Documentation

The documents were developed by a management group with representatives from commercial chicken meat companies, farmer groups, welfare groups, and teaching, research and legislative organizations. The management group had clear terms of reference, met frequently and debated controversial welfare issues after seeing first-hand such activities on farms and at processing plants. The aim of the audit project was to identify and encourage best practice and to highlight issues in the codes of practice that required further debate or research, rather than to provide a quick change in current industry practice or to resolve controversial issues. The codes of practice used in the development of the documentation were produced by the Standing Committee on Agriculture and Resource Management and the Bureau of Animal Welfare of the Department of Primary Industries, Australia (Standing Committee on Agriculture and Resource Management, 1987, 1998, 2001, 2002; Bureau of Animal Welfare, 1998, 2001).

An approach based on hazard analysis critical control processes (HACCP) was used to identify the key welfare issues. While HACCP validation is normally based on obtaining measurements (e.g. temperature), farmers need to make value judgements on bird welfare, and this was taken into account in defining the records to be kept by farmers and staff. The management group decided to include two types of questions in the audit. The first type consisted of critical questions which define a process or situation in which the birds' health and welfare are permanently damaged if something goes wrong. The second type – 'good practice' questions – was designed to identify current practices being used in the industry. There were two types of good practice questions: (i) those based on the code of practice, which can be answered by measurement or direct observation and refer to items that must be complied with to pass an audit; and (ii) questions that indicate that attention is required but refer to items that are difficult to verify and do not have to be complied with to pass an audit. In the latter category are questions such as 'Were chicks tipped gently from the tray at placement?' and 'Was the care shown by pick-up crews satisfactory?'.

The Audit Documentation

Five booklets were produced to cover all sectors of the broiler industry: (i) hatchery; (ii) broilers; (iii) breeder rearers; (iv) breeder layers; and (v) pick-up, transport and processing. Each booklet comprised audit questions, background information on the purpose of the questions and how they relate to welfare, and recording sheets and checklists for farmers/unit managers to complete. Most of the audit questions were developed from company documentation and thus farm staff were already largely following company policy. The major difference was the need to demonstrate compliance by keeping additional records. The areas covered by audit questions for each industry sector are shown in Table 19.1.

Wherever possible, targets were developed for specific issues, such as temperature requirements, stocking density, inspection frequency, and feed and light regimes. In most cases these targets reflect good industry practice and many are already being achieved by the industry. However, in some cases the targets in the code of practice were not applicable. For example, in some older, poorly insulated breeder sheds, when outdoor temperatures are low overnight the shed temperature may fall to around 14°C, well below the target recommendation of $20-22 \pm 2$°C. Similarly, in some areas using bore water or even mains water, it may not be possible to achieve the target for total dissolved solids in the water (500 p.p.m.; there was a note in the text of the welfare audit documentation (Barnett et al., 2001) that values of 1500 p.p.m. have been found in bore water with no adverse effects). In these situations the company veterinarians or technical advisers will need to specify revised

Table 19.1. Areas covered by audit questions for each sector of the broiler industry.

Hatchery	Rearing	Broiler growing	Breeding	Pick-up, transport and processing
Facilities and hygiene	Set-up prior to birds' arrival	Set-up prior to birds' arrival	Set-up prior to birds' arrival	Pick-up and loading
Egg management	Bird arrival and placement	Bird arrival and placement	Bird arrival and placement	Transport from farm to processing plant
Chick management	Routine husbandry procedures		Routine husbandry procedures	Holding facilities at the processing plant
Animal care staff	Equipment and housing	*Week 1*	Equipment and housing	Unloading crates from truck
Record-keeping	Animal care staff	Routine husbandry procedures	Egg management	Removal of birds from crates
checklist	Pick-up	Equipment and housing	Animal care staff	Catching and shackling
	Pullet transport	Animal care staff	Pick-up	Stunning procedure
	Record-keeping checklist	Record-keeping checklist	Record-keeping checklist	
				Killing and processing
		Week 2 to end of batch		Neck cutting
		Routine husbandry procedures		Skin condition, bruises and injuries
		Equipment and housing		Emergency procedures
		Animal care staff		
		Record-keeping checklist		
		Pick-up		
		Daily procedures for days 32 to 42+		

targets that do not compromise welfare. Targets were based on both the scientific literature and company recommendations.

Supporting information was also provided in the booklets on the purpose of the questions and some explanatory notes, for training purposes, on the impact of various issues on welfare (Table 19.2). Appendices in the documentation also provided the recording sheets, checklists, further reading, information on welfare training and background information on the size, structure, value and employment levels in the Australian chicken meat industry.

Evaluation of the Broiler Grower Farm Audit Documentation

Twenty-four broiler farms contracted to one company were used to evaluate the welfare audit. They were all located in a region about 100 km south-west of Melbourne (south-eastern Australia). The farms varied in size from 46,500 to 200,000 birds and had from two to six sheds with an average of 29,600 birds per shed. Thirteen farms had solely fan-assisted ventilation in their sheds, three had solely tunnel ventilation, one had solely natural ventilation, one had both natural and tunnel-ventilated sheds, and six farms had both fan-assisted and tunnel-ventilated sheds. The company provided production data and the farms were ranked according to their performance. Farmers were individually approached to participate in the study and, on the basis of the company performance data, were ranked from 1 (best) to 24 (worst). Farms with similar production performance for the previous three batches of birds were paired and allocated to either the treatment group (12 farmers who received the audit document and filled in recording sheets) or the control group (12 farmers who did not receive a copy of the audit document). The farmers in the control group were asked to continue recording their normal batch card details, such as mortalities, culls, feed supplied and body weight.

The participants in the treatment farm group were asked to complete the recording sheets for three batches. At the end of the third batch, the audit was conducted at all farms for birds for the period from 2 to 5 weeks of age; this period was chosen to avoid variation due to pick-up schedules. The audit involved asking the farmers a subset of 31 audit questions from the documentation. Twenty-four questions related to general bird production and everyday animal husbandry factors and were focused mainly on routine husbandry procedures, equipment and housing, while seven questions related to staff issues.

There were a number of differences between the treatment and control farms that were related to improved record-keeping by the treatment farms. There were six procedures where the proportion of treatment broiler grower farms could provide evidence of compliance that was greater ($P < 0.001$) than that on the control farms. These were 'Were the birds checked at least daily?', 'Were unthrifty birds culled?', 'Were drinkers regularly adjusted to be at the correct height?,' 'Was water pressure gauge/height checked daily?', 'Was availability of water to all drinkers checked daily?' and 'Were feeders regularly adjusted to be at the correct height?'. In all cases the difference arose because the control farms were unable to provide evidence of their actions, although there was visual evidence that the tasks were conducted. At the control farms there were no diary entries or logs of how often the farmers

Table 19.2. Background information provided for each sector of the broiler industry.

Hatchery	Rearing	Broiler growing	Breeding	Pick-up, transport and processing
Sanitizing facilities and equipment	Water and drinkers	Water and drinkers	Water and drinkers	*Pick-up and loading*
Maintaining clean building surrounds	Feeders, feed, body weight and single-sex rearing	Feeders and feed	Feeders, feed, body weight and separate-sex feeding	Biosecurity and disinfectants
Pest control and biosecurity	Sanitizing sheds and equipment, maintaining clean shed surrounds	Sanitizing sheds and equipment, maintaining clean shed surrounds	Sanitizing sheds and equipment, maintaining clean shed surrounds	Shed preparation for pick-up
Building control and alarm systems	Litter	Shed control and alarm systems	Litter	Light
Ventilation, temperature and humidity	Shed control and alarm systems	Pest control and biosecurity	Shed control and alarm systems	Responsibilities during pick-up
Egg management	Pest control and biosecurity	Bird placement and behaviour, and appearance during the growing phase	Pest control and biosecurity	Number of birds carried by hand and automated pick-up
Chick management, health concerns, seeking technical advice and vaccinations	Light levels and light regime	Health concerns and seeking technical advice	Light levels and light regimes	Deaths, injuries and meat quality
Culling surplus and unthrifty chicks	Bird placement and behaviour, and appearance during brooding and rearing	Frequency of inspection	Mating and mating management	Temperature control and problems of high and low temperatures
Transport of day-old chicks	Health concerns, seeking technical advice, vaccinations and parasite control	Culling unthrifty birds and mortalities	Bird placement and behaviour and appearance	Space allowance in containers
Record keeping	Frequency of inspection	Temperature requirements	Health concerns, seeking technical advice, vaccinations and parasite control	Stress during pick-up
Staff training	Beak trimming and feather pecking, removal of the claw and spur	Ventilation rates, fan adjustments and cooling systems	Frequency of inspection	Partial pick-up
	Culling unthrifty birds and mortalities	Odours (ammonia) and gases	Culling unthrifty birds, culling excess males, and mortalities	Cervical dislocation
	Stocking density	Light levels and light regimes	Stocking density	Training for pick-up crews
	Temperature requirements	Body weight, evenness and stocking density	Temperature requirements	*Transport*
		Lameness and mobility	Odours (ammonia) and gases	Biosecurity and disinfectants
		Pick-up		Responsibilities of the driver
				Planning and contingency plans
				Overheating of birds and truck specifications
				Temperature control and problems of high and low temperatures
				In-transit inspections
				Shelter during transport
				Deaths, injuries and meat quality
				Driver training

continued

Table 19.2. *Continued.*

Hatchery	Rearing	Broiler growing	Breeding	Pick-up, transport and processing
	Odours (ammonia) and gases	Record-keeping	Ventilation rates, fan adjustments and cooling systems	*Processing plant*
	Ventilation rates, fan adjustments and cooling systems	Staff training	Egg management	Extreme ambient temperatures and time in crates
	Pick-up		Pick-up	Unloading crates from trucks and birds from crates
	Transport of pullets		Record-keeping	Stress and novelty
	Record-keeping		Staff training	Cervical dislocation
	Staff training			Shackling
				Removal of crates/containers requiring repairs
				Reasons for stunning
				Water bath stunning
				Gas stunning
				Religious slaughter
				Efficacy of stunning
				Humane slaughter
				Bleeding out
				Meat quality
				Industrial disputes
				Planning
				Management responsibilities

checked the birds. Similarly, culls were not recorded, despite the company batch card requiring that culls and mortalities be recorded separately; instead the farmers recorded them all under the category of total mortalities. When asked if they checked water pressure, all control farmers said that it was the first thing that they looked at on entry to the sheds; however, again there was no written evidence that this was the case. Similarly, there was no written evidence that the control farms were regularly adjusting the height of the feeders and drinkers to the correct height or checking the availability of water at the drinkers each day, although this was visually evident on the day of the audit.

Mortality in the first week of life was lower ($P < 0.001$) at the treatment than at the control farms (Table 19.3). The closer attention to detail required by the audit procedure may have been in part responsible for the lower mortality in the first 7 days.

Comments from farmers about the concept of a welfare audit were generally favourable, most recognizing the industry benefits. While the document used for the audit was considered to be large, the participants at the treatment farms soon found that the tasks were not onerous, particularly because the questions (and hence the records) were oriented to routine tasks. The feedback received from farmers has resulted in simplification of the recording sheets included in the booklets. Nevertheless, it is expected that the recording sheets will only be used as examples and it is likely that the broiler companies will modify their own batch cards and perhaps include additional recording sheets for farmers to demonstrate compliance.

Discussion

The processes outlined in this chapter are considered a good model for the development of animal welfare audits for any animal industry. In developing the documentation, we have taken a prescriptive approach that is pragmatically based around the tasks of stockpeople. A similar approach has been used for some sectors in other industries, such as identifying critical control points for stunning cattle and pigs at abattoirs (Grandin, 1999, 2000). An alternative approach is to use a scoring system

Table 19.3. Production data for the 12 treatment and 12 control broiler grower farms.

	Farms		
Variable	Treatment	Control	LSD$_{(P = 0.05)}$
7-day mortality (%)	1.37^x	1.74^y	0.176
Total mortality (%)	5.05	5.35	0.475
Food conversion ratio	1.88	1.87	0.021
Growth rate index*	41.2	42.5	2.13
PIF**	251.4	253.2	4.19

^{xy}Different letters denote a significant difference at $P < 0.001$; *number of days to reach 2.1 kg; **performance indicator factor for the farm (calculated by combining the feed conversion ratio (kg feed/kg live weight), mortality and growth rate for each batch into a single value). PIF = [(weight(kg)/age (days)/feed conversion ratio) × live%] × 100, where live% is the number of birds alive at the end of grow out/number of birds housed × 100. A higher score indicates better overall performance.

developed by a team of experts, such as that for sow housing systems (Bracke *et al.*, 1999). We prefer the prescriptive approach as it does not involve attributing a weighting to any aspect. The documentation was designed to fulfil several aims, most importantly to enable the industry to demonstrate high standards of animal welfare. However, in achieving this objective there were also a number of subsidiary aims, including improved awareness by farmers and industry personnel of the interactions between production and welfare; that is, awareness that bird welfare is an important component of farm and animal management and not a separate issue. Other aims were to establish processes to continually improve welfare and for industry to have spokespersons for their industry on both welfare and quality assurance issues. Additional potential benefits include improvements in product quality, the benefits of a training aid focusing on practical welfare issues, and improvements in work-related characteristics in stockpeople, such as job satisfaction. Improvements in job satisfaction may arise from increased knowledge and/or improved productivity and fewer crises. Furthermore, providing a regular forum to help in the achievement of defined goals enables interactions between industry and welfare groups helping to improve relationships between groups, and a better understanding of each other's position. The evaluation study also showed some production and welfare benefits.

Another benefit of the audit is its relationship to the code of practice. While industries generally comply with the relevant codes, they are seen as being based on minimum standards and any breaches of the code are used to discredit entire industries. An audit system that is based on good practice rather than minimum standards should provide greater public confidence. There should only be occasional breaches of good practice, which should provide a learning experience that helps to minimize the risk of a recurrence.

Some issues were not resolved in the welfare audit. For example, there is evidence that mortalities are reduced if stocking density is lowered (Ekstrand, 1993; Hall, 2001). Maximum stocking density in the Australian code of practice (CSIRO, 2002) is higher than that in many other countries (maximum of 40 kg/m^2; range of 28–40 kg/m^2 depending on the environmental conditions), and clearly this aspect requires further research under Australian conditions to determine optimum stocking densities under different environmental conditions. Another issue is lameness. Research data suggests the incidence of lameness is high (Kestin *et al.*, 1992) and that pain is involved (Danbury *et al.*, 2000; Lunam and Gentle, 2001). In contrast, anecdotal evidence from the present project suggests that a culling rate greater than 1% of birds per week due to lameness around the time of pick-up would be of concern. This suggests that the incidence of lameness may be relatively low because of improvements in genetics and management and the use of an appropriate on-farm culling policy. Obviously, data on the percentage of lameness in local flocks are required in order to resolve this issue. The code of practice requires rewriting to be more age-specific in its recommendations on temperature limits, rather than using the current figures of 19–33°C. In addition, the minimum requirement of 8 h of light per day is at variance with intermittent lighting programmes (e.g. 15 min per hour over a period of 16 hours per day, giving a daily total of 4 hours) designed to reduce feed intake in adult hens, or extended light periods (e.g. 23 hours) intended to maximize growth rate (Manser, 1996). The welfare implications, if any, of such modified light regimes are unknown.

Following release of the documentation in February 2001, 40% or more of broilers are covered nationally by the audit. One reason why uptake has been slower than hoped for is that some companies believe that implementation should only occur if there is a price benefit for the product. However, there is no evidence from around the world that this will occur except for niche markets, and, as indicated above, welfare needs to be seen as a product quality issue. One company that has incorporated the welfare audit nationally, across all industry sectors, reported recently that it was not difficult to incorporate the welfare audit into their quality assurance programme. There were no reported difficulties in meeting targets and none were changed. Staff using the revised quality assurance recording sheets reported that, as they were already monitoring animal welfare regularly, the programme simply affirmed their practices. They reported that the only adjustment they had to make was the increased amount of recording, but this was straightforward given the modified company recording sheets.

The reaction to the programme as a whole was one of general acceptance within the company. Implementation generally focused awareness and was not seen as revolutionizing farm practices. It was reported that there were instances of altered mentality of staff, who were now considering birds more as individual animals. Staff commented that they were more informed about animal welfare and that they had access to better information (i.e. the audit background information on issues) on animal welfare.

Acknowledgements

This project was funded by the Australian Rural Industry Research and Development Corporation and the collaborating research organizations. We are indebted to research colleagues Andrew Almond, Paul Hemsworth, Ellen Jongman and Elizabeth Selby, to members of the management group, and to the farmers who participated in the evaluation experiment. The feedback provided by the participating farmers on the structure and content of the booklets has been incorporated into the final documentation. We are also indebted to Bartter Pty. Ltd for providing details and records of the contract farmers who participated in the study.

References

Barnett, J.L., Glatz, P.C., Almond, A., Hemsworth, P.H., Cransberg, P.H., Parkinson, G.B. and Jongman, E.C. (2001) *A Welfare Audit for the Chicken Meat Industry*. Rural Industries Research and Development Corporation, Canberra, Australia. (Available at: www.animal-welfare.org.au – click on 'What's new?')

Bracke, M.M.M., Metz, J.H.M., Spruijt, B.M. and Dijkhuizen, A.A. (1999) Overall welfare assessment of pregnant sow housing systems based on interviews with experts. *Netherlands Journal of Agricultural Science* 47, 93–104.

Bureau of Animal Welfare (1998) *Code of Accepted Farming Practice for the Welfare of Poultry*. Revision Number 1, AG007. Bureau of Animal Welfare, Department of Primary Industries, East Melbourne, Australia.

Bureau of Animal Welfare (2001) *Code of Practice for the Land Transport of Poultry (Victoria)*,

AG0888. Bureau of Animal Welfare, Department of Primary Industries, East Melbourne, Australia.

Danbury, T.C., Weeks, C.A., Chambers, J.P., Waterman-Pearson, A.E. and Kestin, S.C. (2000) Self-selection of the analgesic drug carpofen by lame broiler chickens. *Veterinary Record* 146, 307–311.

Ekstrand, C. (1993) *Effects of Stocking Density on the Health, Behaviour and Productivity of Broilers: a Literature Review*. Department of Animal Hygiene, Faculty of Veterinary Medicine, Swedish University of Agricultural Sciences, Report No. 32, Skara, Sweden.

Grandin, T. (1999) Critical control points of animal handling and stunning to improve meat quality and welfare. In: *Proceedings of the 45th International Congress on Meat Science and Technology, section 2-P4, Yokohama, Japan*, pp. 66–67.

Grandin, T. (2000) Effect of animal welfare audits of slaughter plants by a major fast food company on cattle handling and stunning practices. *Journal of the American Veterinary Medical Association* 216, 848–851.

Hall, A.L. (2001) The effect of stocking density on the welfare and behaviour of broiler chickens reared commercially. *Animal Welfare* 10, 23–40.

Kestin, S.C., Knowles, T.G., Tinch, A.E. and Gregory, N.G. (1992) Prevalence of leg weakness in broiler chickens and its relationship with genotype. *Veterinary Record* 130, 190–194.

Lunam, C.A. and Gentle, M.J. (2001) Arthritis and chicken ankle joint. In: *Proceedings of the Australian Poultry Science Symposium 13*, 234.

Manser, C.E. (1996) Effects of lighting on the welfare of domestic poultry: a review. *Animal Welfare* 5, 341–360.

RSPCA (1999) *Welfare Standards for Laying Hens and Pullets*. Royal Society for the Prevention of Cruelty to Animals, Horsham, UK.

Standing Committee on Agriculture and Resource Management (1987) *Australian Code of Practice for Poultry Processing*. Standing Committee on Agriculture, CSIRO, East Melbourne, Australia.

Standing Committee on Agriculture and Resource Management (1998) *Model Code of Practice for the Welfare of Animals. Land Transport of Poultry*. Standing Committee on Agriculture and Resource Management, Animal Health Committee, CSIRO, East Melbourne, Australia.

Standing Committee on Agriculture and Resource Management (2001) *Model Code of Practice for the Welfare of Animals. Livestock at Slaughtering Establishments*. Standing Committee on Agriculture and Resource Management, Animal Health Committee, CSIRO, East Melbourne, Australia.

Standing Committee on Agriculture and Resource Management (2002) *Model Code of Practice for the Welfare of Animals. Domestic Poultry*, 4th edn. Standing Committee on Agriculture and Resource Management, Animal Health Committee, CSIRO, East Melbourne, Australia.

20 Does Broiler Welfare Matter, and to Whom?

D.B. MORTON

Centre for Biomedical Ethics, Division of Primary Care, Public and Occupational Health, University of Birmingham, Birmingham, UK

Introduction

In this chapter I try to show the link between chicken welfare and our moral values and ethical perspectives. I will discuss the importance of moral agency in determining who should take action and the philosophical bases on which appropriate actions can be suggested (see also Morton, 1998). Finally, I will briefly look at why chickens seem to have a lower moral value than other sentient animals.

Important Links

Animal welfare science, the scientific basis of the recognition and assessment of animal well-being, is an emerging scientific discipline. Animal welfare has long been of concern to the public, and evidence elsewhere in this book indicates that the welfare of many broiler chickens is compromised during their short lives. It is encouraging that some solutions have been suggested so that poor animal well-being may be mitigated or prevented altogether, and that the industry is trying to promote improved welfare through quality assurance schemes. In the developed world, but increasingly worldwide, once an animal welfare problem is recognized something is normally done about it, although this may take some time. In the present context, if broiler chickens are thought to suffer excessively during their production, then government (or sometimes a supranational organization such as the EU) may issue guidance, codes of practice (Farm Animal Welfare Council, 1992), or even make legal requirements for those who have a duty of care for the animals. This illustrates the clear link between science and law via our moral values. It is therefore important that we identify and analyse these moral values to see if they are concordant with the way we treat animals, including ourselves, and, if they differ, to be able to justify the differences.

©CAB International 2004. *Measuring and Auditing Broiler Welfare*
(eds C. Weeks and A. Butterworth)

Do Broiler Chickens Matter?

This question is subsumed by a more general question: Do animals matter and, if so, which animals do matter and which do not? As a subsidiary question we may ask in what ways they matter, as all animals may matter in some regard (e.g. they are all living creatures and some can suffer) but they may not matter in other ways (e.g. it is acceptable to kill and eat them). Opinion polls, letters to politicians and societal norms (as evidenced, for example, in laws) show that cruelty to animals is generally unacceptable; that is, it is unacceptable to cause animals to suffer more than is necessary for an acceptable reason (e.g. confinement in a battery cage) or to cause animals to suffer for an unacceptable reason (e.g. cock-fighting). If this is to hold true for chickens, then we have to show that chickens are sentient and are able to experience emotions such as pain, distress, frustration and boredom. On the basis that any suffering has to be balanced by, and in proportion to, any (human) gain, we need to clarify what are the benefits that outweigh that suffering. For example, in veterinary research, in which a few experimental chickens may be deliberately infected in order to test for safety a vaccine that will be used to protect millions of other chickens against that disease in the future, it may be argued that the suffering of the few is offset by the prevention of suffering of the many. In other cases, such as in meat production, the position is less clear because we live in a culture in which meat-eating is the norm and some degree of animal suffering is accepted – but how much is accepted is not clear. We may look to alternative systems of production, but all seem to have advantages and disadvantages and it is difficult to determine the best. Consequently, if we wish to continue to eat chicken meat cheaply (this is often not the same as eating it affordably) a system with disadvantages for the chicken will have to be tolerated.

Moral Agency, Autonomy, Responsibility and Accountability

All (human) societies have moral values that are reflected in the law, and members of these societies are held responsible for their actions. We may have to account for why we did not do something as much as why we failed to do something, which may be as culpable as acting illegally. What we decide to do is the very core of our self-determination, and it has become an important cornerstone in ethics that we should respect the autonomy of others; for example, doctors should gain consent before treatment and maintain confidentiality. However, to be truly autonomous a human being has to have the 'Three Fs' – freedom of thought, freedom of will and freedom of action – only then can he or she be free to determine alternative courses of action, to be able to choose one of them, and then to try to enact it. We also have to be competent to make such decisions, but this is another matter; suffice it to say that children and mentally handicapped persons may not always be able to make reasoned or reasonable decisions.

Are Animals Moral Agents?

Autonomy and moral agency are fundamental concepts in explaining why humans differ from animals in their moral accountability. We are guided by our moral values and are accountable for our actions, but animals seem not to have such basic concepts, or to have them in a very limited way. Consequently, it is nonsense to talk about animals having vices, or being cruel, cunning, wicked, etc. The fox that kills a chicken or evades the hunt is no more morally culpable than a child who knows no better than to pull the wings off a butterfly, or cries when it is hungry. At the end of the day, humans are judged by their actions, which may be moral, immoral or even amoral, whereas animals are non-moral agents.

How Does One Choose the Right Action?

Because humans are moral agents and accountable for their actions, they would normally wish to act in accordance with the law and choose to do the right thing. In determining what to do, they might discard some thoughts and entertain others, depending on the motivation of the person (overall goals) and their intention (what is achieved by their choice of particular action). An ethical analysis might help in determining what actions are acceptable, or what is the best one, or at least what is the most appropriate in a difficult situation. The resulting action may not always be to promote good, but an analysis may help to determine what is the least harmful of the alternative actions. Motivation is an important determinant of our actions, and knowledge of a person's motivation can alter one's opinion of whether their action was right or wrong. For example, a person may kill another, but the reason they do so can lead to a variety of judgements: murder, self-defence, defence of others, euthanasia, a just war, and so on.

So what are the ethical principles that should motivate us when dealing with animals? Presumably we do not want to deliberately cause harm to animals, but rather to promote good in some way, and by so doing promote the best interests of the animals in our care and to which we may have a duty of care (e.g. as farmer, veterinarian). Is there also not an obligation to treat all animals of a similar type equally or, if not equally, then at least according to their needs? I now want to look at three of the many philosophical theories that could help one to choose the right action: deontological ethics, utilitarian ethics and virtue ethics.

Deontological ethics determines whether an action is inherently right or wrong through a set of rules like the ten commandments: it is wrong to kill, wrong to tell lies, steal, and so on. Such a system determined the early rules for an orderly human community. However, there is another side to this philosophy, which leads to rights-based arguments. If it is wrong to kill, then surely those that it is wrong to kill have a right to a life? It may also be argued that it is wrong to cause vulnerable humans who are unable to fend for themselves to suffer, and so there is a duty on those who are moral agents not to cause such humans to suffer unnecessarily. So we may ask if this applies to animals as well as to humans, because animals, too, are vulnerable in many ways. One line of argument is that we do have such a duty unless we can point to a significant morally relevant difference between animals and ourselves. And if we are morally serious, then we ought to treat vulnerable humans and animals as beings worthy of our concern and not objects purely for our utility. We have a duty of care to those animals for which we are responsible, to treat them humanely and according to their needs. If we do not have this duty, then we have to find a substantive reason that allows us to treat humans and animals differently. If we cannot identify such as reason, then we are just being prejudiced, as has happened in the past against women (sexism) and against other ethnic groups (racism, as in slavery and tribal wars). (This prejudice by humans towards other species has been termed 'speciesism', but goes beyond that inasmuch as we favour some animals above others; for example, dogs and cats over mice and rats.) Speciesism demands, therefore, that we do not do things to animals that we would not be prepared to do to other humans, such as kill them and eat them, or experiment on them. The rub arises not from the generality of the argument but from the necessity of treating each being as an individual, just as each one of us is equally important in our own right; one counts the same as another, and so deserves equal consideration.

Thus, we should give the same consideration to a non-moral chimpanzee, rat or broiler chicken as to a non-moral baby or mentally retarded child; it is because we are moral agents that we ought to guard their best interests, taking their biological attributes into account. Producing books in Braille is pointless for chickens but not for the blind, but if both chickens and blind people can suffer pain and have an interest in avoiding pain, then moral agents should not cause pain to either. This is the position of those who believe that animals have rights. They argue that it is wrong to cause animals to suffer and wrong to take their lives as this infringes their natural (i.e. not legal, religious or cultural) rights; and so we, as humans, have a duty not to do certain things that are not in an animal's best interests. Believers in animal rights attribute at least two basic rights to animals – to a life, and not to suffer – and so it is always wrong to kill them and to make them suffer in any way; for example, through confinement on a farm or in a household as a pet.

The second theory, *utilitarianism*, does not look at whether an action itself is right or wrong (in fact in its strictest sense it ignores the morality of any action completely) but decrees that the right action is the one that is likely to give the best outcome, normally for humans. Put another way, the right action is the one that leads to the greatest benefit for the greatest number, with the least amount of harm. When animals are involved, a utilitarian often assumes the ascendancy of humans (so called speciesism, but see above) but this does not mean that the life of an animal has no value, or that causing animals to suffer is of no consequence. These factors have to enter the calculation and be offset against the overall benefits; thus, causing animals to suffer is not an acceptable action unless there are some compelling justifying reasons. Moreover, it is not ethically acceptable to cause avoidable suffering; that is, to cause more suffering than is strictly necessary to achieve a specified goal; for example, through poor or less than good practices. It is this minimum degree of suffering required to achieve the goal that has to be justified and balanced against the benefits.

Before I introduce the third theory, let us look at the strengths and weaknesses of the two approaches described above. Do they accord with how we behave today (not that this would make it right)? Are they credible? Do they seem to accord with our intuition? Deontological reasoning seems to be too rigid. For example, if someone comes to your door and threatens to kill your partner and asks where he or she is, deontology would require you to tell the truth, whereas a utilitarian would probably tell a lie (at least you hope they would!). On the other hand, if your partner required a transplant and someone killed you (not an immoral action for strict utilitarians), not only would your partner benefit but so also would many others; your organs and tissues could be used to save their lives and improve their quality of life – definitely producing greater happiness for a greater number of persons. In practice we seem to use both theories, with a bias towards utilitarianism but incorporating some prima facie rules, such as 'killing humans is off-limits unless there is some major justification' (note here the recent political and religious arguments for a just war).

So is there another way that combines both these approaches as this seems to be the way that most people intuitively reason. The third theory, *virtue ethics*, may do this. It is based on Aristotelian philosophy and is summed up by asking questions such as 'What sort of person should I be?' and 'How does one lead a good and flourishing life?' Aristotle thought the answers involved notions of wisdom, justice, temperance and courage, but today they might include beneficence and non-maleficence as well as qualities such as altruism, empathy and caring for the vulnerable. Even this approach still leaves unanswered the question of what is the best action to achieve these goals, but it does move us away from strict rules and the seductively unpredictable consequences of utilitarianism.

Ethical Frameworks

While these philosophical theories offer us some guidance in our journey to find the right action, there are also various ethical frameworks that can help us ask important questions about ethical dilemmas. There are many such frameworks for many situations in human relationships, such as the doctor–patient relationship, business and journalism. There are also other frameworks that look at our relationship with animals in research, farming wildlife, etc. The most relevant in the present context of broiler welfare are the 'Five Freedoms' (Webster, 1995) and the 'Three Rs' (Russell and Burch, 1959). A combination of these philosophical approaches and ethical frameworks needs to be used in practice, and it is important to ask to whom these considerations apply (e.g. the various stakeholders) and who judges what actions are ultimately good, bad, acceptable or unacceptable. Let us now look at the broiler industry to see what some of the ethical concerns are and how they may be met.

The Industry

Some 44 billion broilers are reared each year worldwide (800 million chickens in the UK, 4 billion in Europe, 8 billion in the USA) and this figure is likely to rise in the coming years (SCAHAW Report, 2000). Estimates of lameness in broilers, a potentially painful condition, have varied from 2 to 90%; conversely, estimates of soundness in these birds have varied from 98 to 10% (Yogaratnam, 1995; Sørensen *et al.*, 2000; Weeks *et al.*, 2000; Dall'Aqua and Pfeiffer, 2001; Farm Animal Welfare Council, 2002). Other estimates have put lameness at 26% (in some 200 million birds in the UK) for the heavy strains for the last 2 weeks of their lives (Kestin *et al.*, 1992,

1999; McNamee *et al.*, 1998; Sanotra *et al.*, 2001). Not all lame animals may be in pain but simply have a mechanical problem in walking. Moreover, there is some evidence that the incidence of lameness is decreasing year on year. However, lameness aside, each year up to 6% or 48 million die during rearing; 4% or 32 million have chronic arthritis; 3% (24 million) sustain a bone fracture while they are conscious and a further 30% show other injuries; 1% have cardiorespiratory problems; and 2 million die during transport (Gregory and Wilkins, 1990; Gregory and Austin, 1992; Warriss *et al.*, 1992; Thorp and Maxwell, 1993). Catching and killing of these animals is also far from ideal and could be made more humane (Bayliss and Hinton, 1990). Even as a percentage, many of these losses in the broiler industry at equivalent times are greater than the losses in any red meat production system. However, it also has to be recognized that broiler chickens in other ways are better off than other farm livestock in that they are freer to move around (albeit through a mass of other birds) and to forage, and are not mutilated by castration, tail docking or beak trimming. Nevertheless, it is most unlikely that such death rates and disease levels would be tolerated in any other area of farming, or indeed by the owner of a pet dog or cat, or a racehorse. So why do we allow it for broiler chickens? Several questions arise from these practices. First, does the public have a right to buy cheap and wholesome food? Should the public temper its desire for this type of meat with an obligation to promote and ensure humanitarian practices? Can it be done at a lower cost to the animals in terms of the harm done to them? If so, are we not morally obliged to reduce the suffering to the minimum level whilst still rearing them to eat, even though it may mean paying more?

Pain and Suffering

Is our current legal tolerance of the situation to do with ignorance of what is happening on farms or something to do with the nature of the animal concerned, such as its size, or the industry and the margin of profit, or has it to do with the moral standing of the broiler chicken? One starter question would be to ask: Do birds feel pain and can they suffer? If they do not, then perhaps the perceived harms would not be as great (but may still be aesthetically offensive?). There is no evidence that chickens are not able to experience pain, and a considerable amount of physiological and behavioural evidence that they do (e.g. Duncan, 1996; Gentle, 2001), and they are also able to experience other unpleasant mental states, such as fear, anxiety, boredom, hunger, thirst, discomfort and distress. Evidence exists to support the view that some species of bird display signs of intelligence, and animals with highly developed nervous systems may be more likely to suffer simply as a consequence of their advanced mental abilities. For example, they may be able to predict what may happen to them in the light of their earlier experiences, or they may have their desires frustrated in some way, and such anticipation or mental frustration may make their suffering worse than for animals that lack these abilities.

Take, for example, the work of Irene Pepperberg with Alex the African Grey parrot (discussed in Dawkins, 1993). Alex has demonstrated that he has the ability to count, to be able to identify shapes, colours and objects, and even to use words to get what he wants. He is able to identify objects that are presented in different shapes (e.g. paper), link novel shapes and colours and even identify shades of colour correctly. In this way the bird shows he is able to make new, previously unlearned connections, confirming that he is not simply learning by rote. More remarkably, he apparently uses words like 'no' to express feelings of annoyance, displeasure and non-cooperation, rather than his native language of a squawk or a screech. Experiments on other species of bird have shown that pigeons are also capable of limited thought. For example, when given a set of related items: A is greater than B, B is greater than C, and C is greater than D, and asked about the relationship of A and D, pigeons were able to order them correctly, thought processes that would challenge some humans.

So if birds are able to experience pain, fear, distress, etc. and have limited intelligence, are we justified in treating them differently from other animals, such as sheep and cattle (which, to date, have not shown such advanced development, although this is not to say they could not have such attributes)? If chickens are able to suffer, ought we to be farming them in this way? Surely the production of broilers constitutes a disregard for our obligation to avoid causing suffering to intelligent animals whenever possible: should we be trying to reduce it?

Desensitization and Lack of Social Connection

Perhaps it is the vast scale on which the birds are reared that desensitizes producers, veterinarians and all those connected with, or who know about, the trade. The care of one animal is different from the care of tens or hundreds of thousands kept in this way. The same desensitization may have occurred in the treatment of some humans (for example, in prisoner-of-war camps, bombing raids, or in schools or universities) or for busy physicians or politicians. Conversely, it is possible that the care of a few animals can be too protective – but can a lack of care be justified or rationalized in any way? A duty of care should extend to all as much as to one, but this is plainly impossible with broiler chickens in view of the way they are reared today, and so a compromise has to be sought. Apart from ensuring best practices, such as humane handling and effective culling, if animals are predictably going to suffer (e.g. die of dehydration due to lameness), greater care should be taken of them during the critical periods.

Perhaps the lack of concern is to do with the size of the animal: the bigger the animal the more we take notice of it. This may be because large animals show more obvious signs of pain; for example, they make louder noises (pigs) or cause more damage to the surroundings when trying to escape the pain (horses with colic) than small animals. Humans may find it easier to relate to and recognize signs of pain in the larger animals, especially if these animals live in close proximity, such as dogs. Alternatively, we may consider that small things do not feel pain because we forget how limited our understanding of their response to painful situations may be. They may struggle but be relatively powerless in our restraining grasp; they may cry out in ultrasound frequencies that we are unable to hear; they may remain immobile as a response, which we perhaps wrongly interpret as showing that they do not feel any fear or pain; or we may simply not recognize when animals are afraid, distressed or in some form of pain. Finally, small animals may find some things painful that we cannot conceive of as being so, such as high ultrasound frequencies, odours, and low or high temperatures.

Does this lack of concern reflect the intrinsic moral worth of chickens or is it more about some extrinsic value, such as commercial value (compared with a cow, sheep, pig or racehorse), companionship value (pet dog or cat) or replacement value (endangered species such as chimpanzees)? Is it because they appear so different from ourselves and other mammals that we cannot easily identify (empathize) with their suffering? In some countries there are laws that give primates, dogs and cats special protection over and above that afforded to other mammals, whereas other animals are not deemed worthy of consideration at all. Are any of these arguments that reduce our concern strong enough for us to turn a blind eye to the suffering undergone by these birds? If not, what can be done about it? Here are some suggestions.

Practical Questions and Alternatives

First, we could stop eating chicken – at once. Undoubtedly, in a short time this would decrease animal suffering. Not eating chicken raises the issue of whether people should become vegetarians or simply eat more animal-welfare-friendly meat products. We do not need to eat meat to survive – we do it because of tradition and because it adds to our pleasure in life. Moreover, chicken has become a necessary and cheap protein for many nations. But is the human benefit outweighed by the cost to the animals? Believers in animal rights could accept only vegetarianism as a way forward because they believe that animals have a right to a life and should not be caused to suffer in any way. Animal welfarists, on the other hand, tend to be utilitarians who acknowledge that humans have a duty not to cause animals avoidable harm; thus, they wish to avoid causing suffering whenever possible but are prepared to use animals for human benefit. They accept that some minimal level of suffering may be inherently necessary to produce their food, and do not see animal life as sacrosanct. They may, therefore, choose free-range chickens as opposed to broiler chickens as the suffering and benefits are more proportionate.

Secondly, codes of practice have been introduced in some countries for broiler birds to help ensure good welfare farming standards, facilities and practices. However, as far as I am aware there is no required welfare benchmarking or auditing for levels of mortality, lameness, fractures or ineffective killing such that breeders, farmers, catchers or processors would be penalized if agreed limits of suffering were found to be exceeded. Self-auditing could be required, so that producers are encouraged to confront and assess their own performance on the basis of score ratings for relevant criteria. Independent farm assurance schemes would help ensure that certain criteria were being met; for example, those embodied in the Five Freedoms. But these welfare criteria have to be rigorously monitored to really reflect good welfare. A simple statement of compliance would not be adequate.

On a practical basis it would be possible to feed the birds less. The weight gain would then not be so rapid and the disparity between body weight and skeletal growth not so great. This may reduce leg problems by as much as 50%, but the profit per bird would be less and the birds may be chronically hungry. Perches could be provided for the birds, which would increase exercise and strengthen their legs for as long as they were able (motivated?) to get on and off the perches, but this might also increase breast damage from perching. In addition, periods of darkness could be provided, which has been shown to improve skeletal strength. Perhaps, most importantly, breeders should select for birds that have good leg health and should incorporate more welfare-related criteria into their selection programmes rather than just production traits; I understand that this is now starting to happen.

Thirdly, the public could pay more for chicken so that broilers could be reared more humanely and profitably. The public could be educated about production methods and the suffering and losses, and chicken products could be labelled to indicate the system of production, thus giving consumers a choice. If consumers were not prepared to pay more, or not that much more (perhaps not much more is needed?), then other avenues would have to be explored. But which group has the greatest financial flexibility to invest in the improvement of broiler welfare? Is it the breeders, the producers, the processors or the retailers? No matter, something should be done to shoulder this responsibility: it must not be simply a wringing of hands and a passing of the buck to others.

The retailers would have to be prepared to work together to promote welfare. Animal welfare is currently becoming a marketable commodity, and more retailers are advertising products as 'animal-friendly' in one way or another (exemplified in phrases such as 'beauty without cruelty', 'not tested on animals', 'dolphin-friendly tuna' and 'free-range' eggs and chicken).

Fourthly, in Europe, under the Treaty of Amsterdam all farmed animals are now classified as sentient beings and so the ability of animals to feel pain and to suffer has been recognized in law and has differentiated animals from other traded goods. In the long run this change should help strengthen and increase welfare legislation so that chickens can be better protected and, therefore, the harm inflicted on them reduced. The engagement of more states in the EU trading community will ensure a level playing field on which all will have to play by the same rules. Commissioner David Byrne, in an address to Eurogroup on 30 November 2001, gave a considerable boost to animal welfare: 'Animal welfare is part of the Community agenda and it is here to stay. The protocol on animal welfare in the Treaty of Amsterdam has ensured there will be no turning back in this process.' He noted that the World Trade Organization drives trade and has excluded animal welfare to date, but the OIE (Office International des Epizooties) has proposed that welfare be included in their remit, and so things may change. Byrne also observed, quite correctly, that the single most effective, immediate and practical measure to promote animal welfare would be the strict implementation of existing legislation.

In conclusion, there is no good scientific or ethical reason why the broiler chicken should be treated differently from any other farmed species. What is so shocking is the absolute number of animals involved, even if the percentage of animals injured or diseased may be similar to those for other methods of rearing animals for their flesh. It is not easy to see how the situation can be improved if we are going to feel we have to provide cheap food for humans regardless of animal welfare, and so perhaps it is the attitude of humans that will make the real difference. This can probably only be changed by bringing such problems to their attention, through empowering their choice by labelling, and encouraging the development of more humane conditions for these creatures from birth to death.

References

Bayliss, P.A. and Hinton, M.H. (1990) Transportation of broilers with specific reference to mortality rates. *Applied Animal Behaviour Science* 28, 93–118.

Dall'Aqua, F. and Pfeiffer, A. (2001) An analysis of a national broiler chicken leg weakness survey in the United Kingdom. Royal Veterinary College, University of London, London.

Dawkins, M.S. (1993) *Through Our Eyes Only? The Search for Animal Consciousness*. W.H. Freeman, Oxford, pp. 119–127.

Duncan, I.J.H. (1996) Animal welfare defined in terms of feelings. *Acta Agriculturae Scandinavica, Section A, Animal Science, Supplement* 27, 29–35.

Farm Animal Welfare Council (1992) *Report on the Welfare of Broiler Chickens*. BP 0910. MAFF, London.

Farm Animal Welfare Council (2002) *Report on the Welfare of Broiler Chickens*. MAFF, London.

Gentle, M.J. (2001) Attentional shifts alter pain perception in the chicken. *Animal Welfare* 10, S187–S194.

Gregory, N.G. and Austin, S.D. (1992) Causes of trauma in broilers arriving dead at poultry processing plants. *Veterinary Record* 131, 501–503.

Gregory, N.G. and Wilkins, L.J. (1990) Broken bones in chickens: effect of stunning and processing in broilers. *British Poultry Science* 31, 53–58.

Gregory, N.G., Wilkins, L.J., Alvey, D.M. and Tucker, S.A. (1993) *Veterinary Record* 132, 127.

Kestin, S.C., Knowles, T.G., Tinch, A.E., Greifinger, R.B. (1992) Prevalence of leg weakness in broiler chickens and its relationship with genotype. *Veterinary Record* 131, 190–194.

Kestin, S.C., Su, G. and Sorensen, P. (1999) Different commercial broiler crosses have different susceptibilities to leg weakness. *Poultry Science* 78, 1085–1090.

McNamee, P.T., McCullagh, J.J., Thorp, B.H., Ball, H.J., Graham, D., McCullough, S.J., McConaghy, D. and Smyth, J.A. (1998) Study of leg weakness in two commercial broiler flocks. *Veterinary Record* 143, 131–135.

Morton, D.B. (1998) Fowl deeds. In: Orlans, F.B., Beauchamp, T., Dresser, R. and Morton, D.B. (eds) *The Human Use of Animals. Case Studies in Ethical Choice*. Oxford University Press, New York, pp. 255–269.

Russell, W.M.S. and Burch, R.L. (1959) *The Principles of Humane Experimental Technique*. Universities Federation for Animal Welfare, Horsham, UK.

Sanotra, G.S., Lund, J.D., Ersboll, A.K., Petersen, J.S. and Vestergaard, K.S. (2001) Monitoring leg problems in broilers: a survey of commercial broiler production in Denmark. *World's Poultry Science Journal* 57, 55–69.

SCAHAW (Scientific Committee on Animal Health and Animal Welfare) (2000) The welfare of chickens kept for meat production (broilers). SANCO.B.3/AH/R15/2000. European Commission, Brussels.

Sørensen, P., Su, G. and Kestin, S.C. (2000) Effects of age and stocking density on leg weakness in broiler chickens. *Poultry Science* 79, 864–870.

Thorp, B.H. and Maxwell, M.H. (1993) Health problems in broiler production. In: Savory, C.J. and Hughes, B.O. (eds) *Fourth European Symposium on Poultry Welfare*. University Federation for Animal Welfare, Horsham, UK, pp. 208–218.

Warriss, P.D., Bevis, E.A., Brown, S.N. and Edwards, J.E. (1992) Longer journeys to processing plants are associated with higher mortality in broiler chickens. *British Poultry Science* 33, 201–206.

Webster, J. (1995) *Animal Welfare. A Cool Eye Towards Eden*. Blackwell Science, Oxford, p. 163.

Weeks, C.A., Danbury, T.D., Davies, H.C., Hunt, P. and Kestin, S. (2000) The behaviour of broiler chickens and its modification by lameness. *Applied Animal Behaviour Science* 67, 111–125.

Yogaratnam, V. (1995) Analysis of the causes of high rates of carcase rejection at a poultry processing plant. *Veterinary Record* 137, 215–217.

21 Public Attitudes and Expectations

M.C. APPLEBY

The Humane Society of the United States, Washington, DC, USA

Introduction

A common tendency in developed countries over the last 50 years has been the drive for efficiency in agriculture, for cutting the cost of producing each egg or kilogram of meat or litre of milk. This was initiated by public policies before, during and after World War II in favour of more abundant, cheaper food (Williams, 1960). In the post-war period it was typical for people to spend between a quarter and a third of their income on food but now about 10% is usual. In practical terms the success of the techniques used to improve efficiency has been spectacular, and broiler production probably provides the strongest example. In the post-war period a meat bird took more than 13 weeks to grow to 2 kg and cost the equivalent of what is now about US$50. Nowadays, because of genetic selection and changes in management, it takes less than 6 weeks and costs under US$3. However, in recent years many different concerns have been expressed over the impact of such increases in the efficiency of food production on animal welfare, the environment, food safety and quality, food security, small-scale producers, farm workers, rural communities and developing countries (Appleby *et al.*, 2003). This chapter will address the question of public attitudes and expectations with regard to these issues in relation to the production of food from animals in general and broiler production in particular.

Cheap Food Production

While pressure for cheap food production was initiated by public policy, it subsequently became market-driven, with competition between producers and between retailers to sell food as cheaply as possible, and thereby it acquired its own momentum. This pressure is sometimes described (by the animal production industry as well as others) as consumer demand for cheap food, but this is an oversimplification, implying that people want cheapness at the expense of all other considerations and that cutting prices is an end that justifies all possible means. It is not surprising – indeed, it is reasonable – that, offered two otherwise similar products, most shoppers will buy the cheaper. However, some people are willing to seek out and pay more for food produced by alternative methods, such as free-range, that are perceived to be better for welfare or for other areas of concern, such as food quality or the environment. This is the basis of the growth of the free-range broiler market in France and other European countries. Furthermore, the proportion of people who say they want the welfare of farm animals to be improved, even if this increases food prices, is larger than the proportion who actually buy higher-priced, welfare-friendly products (Bennett, 1997). The disparity between what people say they want and how they spend their money is sometimes portrayed as hypocrisy, but it seems more reasonable to conclude that they are behaving as citizens when they answer the questionnaire but as consumers juggling varied priorities when they do their shopping. The only case in which people have actually been asked to vote on legislation to improve animal welfare, with associated higher costs, was in Switzerland, and they did approve that legislation: battery cages for laying hens were banned in Switzerland as the result of a referendum.

The increasing numbers of people who are concerned about farm animal welfare do not merely want improvements in the welfare of the animals that supply them personally with food, but improvements for all farm animals. Consequently, the fact that a significant proportion of people – albeit still a minority – are willing to seek out opportunities to pay to support their principles is particularly important. In Europe, Canada and Australasia this has been taken as grounds for politicians to introduce more widespread improvements to the welfare of farm animals. As of 2003, the European Commission is planning new legislation on broilers, following the report from their Scientific Committee on Animal Health and Animal Welfare (2000).

Responsibility

The argument is sometimes made that if society wants animal welfare – or other matters of concern, such as the environment – to be safeguarded, then a higher proportion should buy products such as those produced by free-range methods. To some extent this is reasonable: it would be good if everyone could put their money where their mouth is. On the other hand, it is not reasonable to expect shoppers to take day-to-day responsibility for animal welfare at the point of sale any more than they are expected to do so for other issues that are of concern to society, such as pollution. It is increasingly recognized that people who do not look after farm animals themselves expect those who do to take responsibility for doing so well, either voluntarily or involuntarily. For example, they may be expected to ensure that the food on sale is safe to eat and produced under humane conditions, and that the producers earn a reasonable living. In the USA this is being recognized by the retail sector. A senior executive of one of the major fast food chains has commented that their customers expect them – the restaurant company – to ensure that the animals providing them with food are properly looked after (England, 2002). Similar comments have been made by executives of the Food Marketing Institute, which represents the major US supermarket chains (Associated Press, 2003). Also in the USA, the National Council of Chain Restaurants and the Food Marketing Institute have developed a collaborative programme, coordinating husbandry guidelines for their suppliers of animal products in 2002. These do not go as far as European legislation, but they are important in acknowledging the importance of animal welfare and in forming a basis for the possible future raising of welfare standards.

As part of that programme, KFC (formerly Kentucky Fried Chicken) and its parent company Yum! Brands announced poultry welfare guidelines in 2003 (KFC, 2003). Some parts of these guidelines are significant; for example, the use of antibiotics to promote the growth of healthy chickens is prohibited where such antibiotics are significant for human health. Other parts of the guidelines, however, seem more concerned with publicity than with actually improving welfare. Thus, the guidelines state that KFC prohibits its suppliers from trimming the beaks of any poultry that will be sold in their restaurants. Most broilers are not beak-trimmed anyway. Nothing is said about avoiding beak trimming in breeding stock, where it is commonly practised.

In fact broiler breeders are rarely considered in discussions of poultry welfare, despite major issues such as feed restriction and hunger (Savory et al., 1993). The public seems to be unaware of these issues, and even of the existence of these birds. To some extent this may be explained by the idea that people devolve responsibility for taking care of their food production. An alternative or overlapping interpretation, however, is that members of the public avoid such responsibility: they 'don't want to know the gruesome details' (Associated Press, 2003).

Producers

The matter of whether the food production system provides a reasonable living to food producers is apposite. As well as a decline in the proportion of their income that people spend on food, there has also been a decline in the proportion of that spending that reaches the producers; an increasing percentage goes to marketing, and there is therefore an even steeper total decline in farm income. Furthermore, in conventional agriculture the income is unevenly distributed, with increasing success of large producers at the expense of small ones. Thus, many small producers either leave the business or go into contract work for large companies. These developments are also contrary to public expectation, as demonstrated by the fact that another attraction of specialist markets such as free-range foods, to both producers and customers, is that remuneration often goes directly to the producer rather than to intervening retailers.

In this regard, the Fairtrade Foundation, hitherto solely concerned with ensuring that producers in developing countries obtain a fair price for their products, announced in 2003 that they would consider marketing food from UK farmers (Lamb, 2003).

The quid pro quo for this, however, is again that producers may be expected to take a responsible attitude to welfare and related issues. One recent development provides a counter-example that is unlikely to be acceptable to the public: the announcement that a scientist had produced a featherless broiler that would lose heat more readily in hot conditions, and therefore perhaps grow faster and cost producers less (Young, 2002). The scientist also implied that the development might be advantageous for welfare, as the bird might suffer less heat stress. However, many people are likely to agree with the spokesperson for Compassion in World Farming, who described this project as 'disgusting' (Young, 2002). Part of the reason for this divergence of opinion is that different people have different concepts of welfare (Duncan and Fraser, 1997). Producers, and many scientists working to increase agricultural efficiency, tend to emphasize the physical aspects of welfare, such as health and growth. The general public, by contrast, tends to emphasize both mental aspects, such as suffering, and aspects such as naturalness and animal integrity. While integrity is difficult to define, it is clearly not upheld in the breeding of a featherless bird (Hitt, 2002). In any event, there is ever-increasing evidence that the welfare of broilers is severely compromised by the normal breeding practices of the industry, even if only the physical aspects are considered: fast growth is associated with poor rather than good health (Corr *et al.*, 2003a,b).

Responding to Public Expectations

Finding mechanisms for responding to the many public expectations discussed here is difficult in an industry largely driven by competition, especially as that competition is intensifying with the burgeoning international trade in agricultural produce. Ironically, however, the fact that only a small proportion of expenditure on food reaches farmers may make finding such mechanisms more likely: the shift towards the sale of preprocessed food in developed countries offers hope for the improvement of farm animal welfare in general and broiler welfare in particular. If a meal containing animal products is bought in a supermarket or restaurant, those products account for only about 5% of the price. So an increase in cost of animal production by, say, 10% would only increase the cost of such meals by 0.5%. Most customers would not notice such a change and would approve it if asked, to benefit animal welfare or the environment.

McInerney (1998) has analysed the financial impact of banning certain livestock systems. He estimates that banning the intensive rearing of broilers and keeping them in smaller groups, larger areas and more varied conditions would increase meat production costs by 30%. However, such a ban would increase retail prices by only 13%, because these prices include transport, packing, marketing and so on: changes in production costs are diluted by the further costs of bringing products to market, plus the mark-up added by retailers. As broiler meat is not a major component of household food expenditure, this would add only a few pence to weekly food bills. Meanwhile, it should be possible for the farmers to maintain their profits, offsetting increased costs with increased selling prices.

We need to know considerably more about the sociology of public attitudes to animal welfare and related issues in order to take expectations properly into account. However, one thing is clear: decisions about the structure of agriculture will in future have to take greater account of public opinion than hitherto. This should not be a burden on farmers, who are stewards of our animals and our environment on behalf of society. On the contrary, it should ensure farmers a more valued place in society and a more reliable income. The importance of public opinion is emphasized in numerous discussions of agriculture, especially in view of what is widely regarded as a crisis in agriculture in the early 21st century. Thus, one of the main recommendations of the UK's Policy Commission on the Future of Farming and Food (2002) is greater integration and communication between all stages of the food chain, from producer to consumer. And in the USA the National Research Council (2002), reviewing the research programme of the US Department of Agriculture, recommends increased public accountability; for example, by holding a public discussion forum every 2 years. It also recommends that government-funded research should in future be devoted less to productivity and more to public goods such as environmental stewardship – and we can add animal welfare as another example of such public goods.

References

Appleby, M.C., Cutler, N., Gazzard, J., Goddard, P., Milne, J.A., Morgan, C. and Redfern, A. (2003) What price cheap food? *Journal of Agricultural and Environmental Ethics* 16, 395–408.

Associated Press (2003) Producers address issues related to treatment of chickens. *Jefferson City News Tribune* 27 January.

Bennett, R.M. (1997) Economics. In: Appleby, M.C. and Hughes, B.O. (eds) *Animal Welfare*. CAB International, Wallingford, UK, pp. 235–248.

Corr, S.A., Gentle, M.J., McCorquodale, C.C. and Bennett, D. (2003a) The effect of morphology on the musculoskeletal system of the modern broiler. *Animal Welfare* 12, 145–157.

Corr, S.A., Gentle, M.J., McCorquodale, C.C. and Bennett, D. (2003b) The effect of morphology on walking ability in the modern broiler: a gait analysis study. *Animal Welfare* 12, 159–171.

Duncan, I.J.H. and Fraser, D. (1997) Understanding animal welfare. In: Appleby, M.C. and Hughes, B.O. (eds) *Animal Welfare*. CAB International, Wallingford, UK, pp. 19–31.

England, C. (2002) Burger King and animal welfare: why did this company get involved? In: *Proceedings, Canadian Association for Laboratory Animal Science and Alberta Farm Animal Care Conference*. Canadian Association for Laboratory Animal Science, Edmonton, Canada, p. 13.

Hitt, J. (2002) The featherless chicken. *New York Times*, 15 December.

KFC (2003) *KFC Poultry Welfare Guidelines (An Overview)*. http://www.yum.com/community/animalwelfare_guidelines.htm, accessed May 2003.

Lamb, H. (2003) Fair trade for all. *The Guardian*, 12 March.

McInerney, J.P. (1998) The economics of welfare. In: Michell, A.R. and Ewbank, R. (eds) *Ethics, Welfare, Law and Market Forces: The Veterinary Interface*. Universities Federation for Animal Welfare, Wheathampstead, UK, pp. 115–132.

National Research Council (2002) *Frontiers in Agricultural Research: Food, Health, Environment and Communities*. National Academies Press, Washington, DC.

Policy Commission on the Future of Farming and Food (2002) Farming and food: a sustainable future. Cabinet Office, London.

Savory, C.J., Maros, K. and Rutter, S.M. (1993) Assessment of hunger in growing broiler breeders in relation to a commercial restricted feeding programme. *Animal Welfare* 2, 131–152.

Scientific Committee on Animal Health and Animal Welfare (2000) The welfare of chickens kept for meat production (broilers). European Commission Health and Consumer Protection Directorate-General, Brussels, Belgium.

Williams, H.T. (1960) *Principles for British Agricultural Policy*. Oxford University Press, Oxford.

Young, E. (2002) Featherless chicken creates a flap. NewScientist.com news service, 21 May.

22 A Global Perspective on Broiler Welfare Standards

P. LYMBERY

World Society for the Protection of Animals, London, UK

Introduction

Today, we live in an era of globalization. Commercial companies, brands of goods and trading rules are all becoming increasingly globalized. There are also global animal welfare organizations tackling animal welfare problems worldwide. The World Society for the Protection of Animals (WSPA) is the world's largest animal welfare federation, with over 440 member societies in more than 100 countries. WSPA is seeking a world in which the principles of animal welfare are recognized, respected and protected worldwide. One area that raises serious animal welfare concerns worldwide is that of commercial broiler chicken production for meat. A large and increasing proportion of the world's broiler chickens are reared intensively. Key factors affecting the welfare of these birds include genetics, housing, and feeding regimes, all of which are becoming increasingly standardized throughout the world.

Using case studies from major global players in the chicken production industry, this chapter examines the effects on both the people and the birds involved in the poultry population explosion worldwide.

Global poultry production is going through a period of spectacular expansion. There is scarcely a corner of the earth that remains unaffected. Intensive chicken rearing began in the Western world during the middle of the last century. In many developed countries, chicken has been transformed from an occasional treat eaten on special occasions to a cheap, everyday commodity. Chicken meat is often seen as little more than a culinary base for the day's choice of sauce or topping. Output, in terms of numbers of birds produced per year, continues to expand. Over the last 30 years, chicken production has doubled in developed countries. The surge in chicken production has been particularly marked in developing countries, as they too have adopted the industrial model of poultry rearing, all too often with serious consequences for rural livelihoods and food security. In the least developed countries over the past 30 years, chicken production has tripled, and in developing countries it has increased by six times. The explosion in poultry output has been particularly marked in South America, where production in the last 5 years alone has increased by 30%, and Asia, which has seen a 22% increase.

Commercial chicken production is dominated globally by just two or three companies, which supply an estimated 80% of the chicks. The white, fast-growing broiler chicken has become a standardized global 'product'. It follows that many of the serious health and welfare problems facing chickens of this type will be similar globally.

Fast-growing broiler chickens can suffer high rates of lameness, heart disease and skin lesions. Although broilers now put on weight very quickly, their musculo-skeletal capacity can sometimes be compromised. A doubling in the rate of weight gain has led to a 42-day-old bird's skeleton carrying the weight of an 84-day-old. In its report on the welfare of broiler chickens for the year 2000, the European Commission's scientific advisory committee on animal welfare (SCAHAW, 2000) concluded that:

- 'Leg disorders are a major cause of poor welfare in broilers.'
- 'Contact dermatitis [see Chapter 3] is a relatively widespread problem in the European broiler industry.'
- 'Ascites [see Chapter 4] has a serious negative effect on broiler welfare. The problem has increased in recent years.'
- 'The greatest threat to broiler welfare due to behavioural restriction would appear to be likely constraints on locomotor and litter directed activities caused by crowding, and consequences for leg weakness, poor litter quality and contact dermatitis.'
- Findings are 'indicative of poorer welfare at higher stocking densities'.

Commercial stocking densities for broiler chickens vary throughout the world and are often expressed in terms of the number of kilograms of bird reared per square metre of floor space. WSPA has found these to vary from about 20 kg/m^2 in the Philippines through 34–38 kg/m^2 in the UK to 50 kg/m^2 in Taiwan.

Free-range and organic systems, especially where more traditional, slow-growing strains of bird are used, may provide real alternatives to intensive broiler chicken production. In these systems, the enriched environment and greater space encourage the chickens to exercise and move around rather than spending a large proportion of their day squatting on the litter floor. This can reduce leg problems, hock burn, and pododermatitis resulting from wet litter, thereby contributing to a better quality of life for the birds.

Small-scale Poultry Rearing

Whilst industrial-style farming is behind the phenomenal surge in poultry production worldwide, the raising of small flocks of chickens still plays a key role in the survival of many farmers in the developing world. Some 80% of farmers in Asia and Africa raise small flocks of chickens (Garces, 2002). Small, community-level farms are often important in maintaining rural livelihoods and local food security.

As industrial chicken-rearing methods are adopted around the world, the animal welfare concerns are typically replicated. Fast-growing genetic strains of bird, with their attendant propensity to poor leg health and cardiovascular problems, are usually reared in large numbers at high stocking densities. These technically sophisticated, low-labour techniques can also threaten rural livelihoods and local food security. In East and South-east Asia, for example, industrial agriculture has been increasing, with greater use of machinery, chemical fertilizer and financial services, such as foreign loans. There has been a consequent shift from small-scale rearing of ruminant animals to the industrial production of pigs and poultry. As developing countries adopt mechanized livestock rearing, there is a parallel shift away from self-sufficiency towards a dependency on imports. Local food security can be threatened as a result. As Garces (2002) puts it,

> Grains, tractors, oil to fuel the tractors, fertilizers and special animal units and processors are all needed for intensive livestock rearing, none of which a developing country starts out by making itself. Asia now imports large amounts of grain to feed its factory-farmed animals.

It is tempting to believe that intensive farming needs less land to produce food for humans and animals. Yet intensive chicken production requires plentiful supplies of grain to feed to the birds. The result is that crop farming often becomes intensive, involving the use of large amounts of chemicals, such as fertilizers, herbicides, insecticides and fungicides. This has been linked to the loss of soil fertility and farmland wildlife. In the Philippines, for example, a serious consequence of the rise in intensive animal production has been the diversion of imported grain for human consumption to feed farm animals rather than people. A recent report on livestock development published by the World Bank comments that the 'shift to more grain-based production could seriously affect global and national food security' (de Haan *et al.*, 2001).

Case Studies from Around the World

The Philippines

Ranked as the world's 16th largest producer of broiler chickens, the Philippines is a developing nation where the rise of industrial chicken farming has had profound effects on rural livelihoods and food security.

Intensive chicken production has become increasingly dominant in the Philippines. The vast majority of the 540 million chickens reared annually for meat are now produced this way, with only 10% being reared in small, village-level enterprises or backyards. The rapid expansion of intensive poultry production in the Philippines has caused serious animal welfare problems. It has also seen farmers relegated to the status of contract growers to large farming companies, and has affected traditional village livelihoods. The average Filipino farming family has not benefited directly from the recent boom in the chicken business. Forty years ago, the nation's entire population was fed on native eggs and chickens produced by the Filipino family farmer. This traditional livelihood, using native strains of bird, is now threatened by an array of viral diseases due to the influx of intensive rearing of commercial chicken breeds (Teresa Farms, 2001) (Box 22.1).

There are a number of important differences between a typical Filipino industrial chicken shed and the intensive units of Europe. The birds are typically given significantly more space in the Philippines (0.09 m^2 per bird as opposed to 0.06 m^2 or less in the UK) and have the benefit of natural light and ventilation. Their cousins in Europe will be stocked more densely in dimly lit, windowless sheds with computer-controlled temperature, lighting and ventilation. However, there is pressure on the Filipino industry for the wider adoption of these even more highly intensive, fully enclosed units.

The Philippines is not the only country at risk from the spread of factory farms. Argentina, Brazil, Canada, China, India, Mexico, Pakistan, South Africa, Taiwan and Thailand are all seeing growth in industrial animal production.

Box 22.1.   Reviving village-level poultry rearing.

An edited extract from *Factory Farming in the Developing World* by Danielle Nierenberg, Worldwatch Institute, USA

In the Philippines, there is renewed interest in alternative, village-level methods of poultry rearing. On the farm of Bobby Inocencio in the hills of Rizal province in the Philippines, hundreds of chickens (a cross between native Filipino chickens and a French breed) roam around freely in large, fenced pens, pecking at various indigenous plants, eating insects, and fertilizing the soil.

Bobby's farm is anything but simple. What he has recreated is a complex and successful system of raising chickens that benefits small producers, the environment, and even the chickens. Once an industrial farmer, Inocencio used to raise white chickens for one of the biggest companies in the Philippines. Thousands of birds were housed in long, enclosed metal sheds that covered his property. Along with the breed stock and feeds he had to import, Bobby also found himself dealing with a lot of imported diseases and was forced into buying expensive veterinary antibiotics. Bobby also used growth promoters to decrease the time it took for his chickens to mature. At the same time Bobby noticed that fewer and fewer of his neighbours were raising chickens, which threatened the community's food security by reducing the locally available supply of chickens and eggs.

As the community dissolved and farms disappeared, Bobby became convinced that there had to be a different way to raise chickens and still compete in a rapidly globalizing marketplace. In the last two decades the Filipino poultry production system has changed from mainly backyard farms to a huge industry. In the 1980s the country produced about 200 million birds annually. Today that figure is more than 500 million. The large poultry producers have benefited from this population explosion, but unfortunately average farmers have not. So Bobby decided to revive village-level poultry enterprises that support traditional family farms and rural communities.

Since 1997, Bobby's Teresa Farms has been raising free-range chickens and teaching other farmers how to do the same. Bobby believes that the way he used to raise chickens, by concentrating so many of them in a small space, is dangerous. Diseases such as avian flu, leukosis J (avian leukaemia), and Newcastle disease are spread from white chickens to the Filipino native chicken populations. Now Teresa Farms chickens are no longer kept in long, enclosed sheds, but roam freely in large tree covered areas of his farm.

Bobby's chickens also don't do drugs. Bobby found the answer to preventing diseases in chickens literally in his own back yard. His chickens eat spices and native plants that

---

Box 22.1. *Continued.*

---

have antibacterial and other medicinal properties. Chilli, for instance, is mixed in grain to treat respiratory problems, to stimulate appetite during heat stress, as a wormer, and as an effective treatment for Newcastle disease. Native plants growing on the farm, including *ipil-ipil* and *damong maria*, are also used to treat disease and provide a low-cost alternative to antibiotics and other drugs.

(From *World Watch* magazine, May/June 2003, published by the Worldwatch Institute, Washington, USA)

Brazil

Brazil, the world's third largest broiler chicken producer, provides a particularly pertinent case study of how poultry production in a predominantly small-scale farming nation has changed. Compassion in World Farming, a WSPA member society, has described how, between 1970 and 1991, Brazil's poultry industry grew from small backyard farming to a multinational mechanized industry, becoming almost entirely vertically integrated. Originally, small family farmers in Brazil were provided with day-old chicks by major companies and paid to raise them. One family-owned company, for example, employed 14,000 small-scale farmers who raised chickens on a mixed farm basis. However, such family-owned companies have been taken over by financial interest groups and foreign companies. The small-scale mixed farms have given way to large production units, with a consequent loss of rural livelihoods.

Industrial poultry production methods in Brazil are similar to those used in Europe. Bird stocking densities tend to be slightly lower because of the heat, and the open-sided units provide natural lighting. Birds are slaughtered at 45 days of age. There is an increasing demand in Brazil for free-range and organic poultry, mainly because of the excessive use of antibiotics in intensive units and the consequent human health concerns (Cox and Varpama, 2000).

Taiwan

Taiwan ranks among the world's 30 largest broiler producers. Most of the 320 million broiler chickens slaughtered for meat each year in Taiwan (Department of Agriculture and Forestry, 1997) are produced intensively. Big integrator companies largely control production. A typical Taiwanese broiler chicken unit has natural lighting and ventilation and a deep-litter floor of rice husk. Each holds thousands of birds at stocking densities of up to 50 kilograms of live birds per square metre of floor space. This compares to a planned 34–38 kg/m^2 in the UK. At these densities, the birds carpet the floor when 4–5 weeks old.

The standard commercial white broiler chicken, seemingly ubiquitous the world over, is common in Taiwan. In the country's main chicken-producing province, Miao Li, WSPA observed birds as young as 11 days looking unsteady on their legs. These fast-growing genetic strains also appear to be susceptible to heat stress. At midday in July, the young birds were panting in the heat. Equally prevalent in

Taiwan is the buffy-brown 'colourful' broiler, which is also reared intensively, with up to 8000 birds per building.

Traditional poultry rearing continues to survive in Taiwan. Local breeds of chicken are reared, with up to 1 million birds being sold under the 'Taiwanese traditional breed approved' label. These are reared in open-sided, deep-litter houses at 10–15 birds per square metre, have access to the outdoors, and grow at a more natural rate. They are slaughtered on the farm at 5 months of age.

International Promotion of the Industrial Model

In the European Union (EU), public disquiet at the factory farming of animals has driven some reforms. It is therefore ironic to see companies from Europe and North America busily promoting the factory farm model around the world. For example, the major Asian event for pig and poultry producers, VIV Asia, held in Thailand in 2003, had a strong presence from Western European and North American companies selling intensive farming equipment and genetics. Battery cages and sow stalls, two systems now being phased out by legislation in the EU, were prominently on sale on the stands of Western companies. Although not yet subject to regulation on welfare, Western companies were also heavily promoting the fast-growing breeds of broiler chicken with their genetic propensity to serious leg and cardiovascular problems.

Europe

About 14% of the world's broiler chickens are housed in the EU. The vast majority of production is in large units housing thousands of birds of fast-growing strains at high stocking densities with artificial lighting and ventilation.

There is currently no EU-wide legislation to specifically protect the welfare of broiler chickens. In the EU, only Sweden and Denmark have legal limits on stocking densities. Germany and the UK have government guidelines (codes of practice), but most other countries rely on advice from breeding companies for their practices (SCAHAW, 2000).

Outside the EU, in Switzerland the law fixes an upper limit of 30 kg/m^2 for stocking density and requires that birds be given at least 8 hours of darkness when kept in windowless sheds (SCAHAW, 2000).

As in many parts of the world, increasing public and political importance is attached to animal welfare, partly as a result of greater awareness of how animals are raised on industrial or factory farms. In the UK, humane alternatives, such as the Freedom Food scheme of the Royal Society for the Prevention of Cruelty to Animals and the free-range and organic options, are now widely available. Nine out of ten of the major supermarket companies in the UK, for example, sell free-range or organic chickens. Nearly a third of the chickens sold by two major supermarket companies (Marks & Spencer and Waitrose) come from these high-welfare systems (Lymbery, 2002a).

However, the vast majority of supermarket chickens in the UK are produced intensively. The welfare standards of most of the UK top ten supermarkets allow chickens to be stocked at densities that exceed government guidelines. UK government guidelines specify a maximum of 34 kg of birds per square metre. Only Marks & Spencer stipulates this maximum, which in itself is too high to promote good welfare. Most other supermarkets will accept chickens kept at stocking densities up to 38 kg/m^2, thereby exceeding government guidelines (Lymbery, 2002a).

Farm Assurance in the UK

Farm assurance schemes were introduced by the UK food industry in the 1980s to boost consumer confidence in livestock products. Increasing concern about animal welfare was a major driving force behind the development of these schemes. The ten biggest supermarkets in the UK account for 60% of all grocery sales. The vast majority of fresh meat, milk and eggs sold through these supermarkets is produced under national farm assurance schemes. Farm assurance schemes therefore provide the baseline animal welfare standards for a large part of the fresh livestock products consumed in the UK.

A recent study found that the standards set by the main livestock farm assurance schemes in the UK fail to assure the use of high-welfare systems of breeding and rearing. The conventional farm assurance schemes studied failed to ensure the incorporation of the majority of key welfare determinants – the building blocks of animal-friendly rearing methods – into their schemes' farming systems. The national chicken scheme, Assured Chicken Production, was found to allow highly intensive methods of rearing broiler chickens in which birds could be crammed even more tightly than recommended by Government guidelines. The globally ubiquitous fast-growing strains of bird with a genetic propensity to serious leg and cardiovascular problems can also be used under this scheme (Lymbery, 2002b).

There is much talk within the European industry of the need for a so-called level playing field. In an increasingly global market place, the industry has a clear opportunity to differentiate its product on the basis of quality factors such as food safety and animal welfare in a bid to maintain market share. This is especially the case in the increasingly discerning domestic market in Europe. However, to achieve credible and lasting product differentiation on the basis of animal welfare, there must be a clear differential between the higher-welfare product and the current norm.

The industrial model of poultry production used so heavily in Europe, with its attendant animal welfare problems, is found throughout the world. The animal welfare problems caused by fast-growing genetic stock and intensive conditions are similar globally. It would therefore be disingenuous to suggest that standard British intensively reared broiler chickens, for example, are produced to a significantly higher welfare standard than those in the Philippines, Thailand or Brazil.

Is Mass Production Inevitable?

The long-term triumph of industrial farming worldwide is not inevitable. In 2001, the World Bank released a new livestock strategy. In an astounding reversal of its previous commitment to the funding of large-scale livestock projects in developing nations, the Bank said that as the livestock sector grows 'there is a significant danger that the poor are being crowded out, the environment eroded, and global food safety and security threatened'. It promised to use a 'people-centred approach' to livestock development projects that will reduce poverty, protect environmental sustainability, ensure food security, and promote animal welfare (de Haan *et al.*, 2001). This turn-around was not the result of pressure from environmental or animal welfare activists but came about because the large-scale intensive animal production methods the Bank once advocated are simply too costly. Past policies drove out smallholders because the economies of scale for large units do not internalize the environmental costs of producing meat. The Bank's new strategy includes integrating livestock–environment interactions into environmental impact assessments, correcting regulatory distortions that favour large producers, and promoting and developing markets for organic products. These measures are a step in the right direction, but more needs to be done by lending agencies, governments, non-governmental organizations and individual consumers.

Animal welfare is increasingly recognized as an integral part of the development of society. The overintensification of livestock farming, whereby large numbers of animals may be forced to grow super-fast in overcrowded conditions, not only causes scientifically proven suffering to the animals but also often has serious impacts on the environment, public health and rural livelihoods. WSPA believes that, in view of the serious animal welfare (including health) problems caused by industrial chicken rear-ing and the social and environmental impacts of such production methods globally, action is needed by governments, industry and other stakeholders to encourage an urgent move towards humane and sustainable poultry farming worldwide.

Acknowledgements

Grateful thanks are due to Danielle Nierenberg of the Worldwatch Institute in Washington, USA, for help and input into this paper.

References

Cox, J. and Varpama, S. (2000) *The 'Livestock Revolution': Development or Destruction?* A report into factory farming in 'developing' countries for Compassion In World Farming: Brazil report. Compassion in World Farming, Petersfield, UK.

de Haan, C., van Veen, T.S., Brandenburg, B., Gauthier, J., Le Gall, F., Mearns, R. and Simeon, M. (2001) Livestock development: implications for rural poverty, the environ-ment, and global food security. World Bank, Washington, DC.

Department of Agriculture and Forestry, Taiwan (1997) *Taiwan Agricultural Yearbook.* Taiwan Provincial Government.

FAO (2003) FAOSTAT Database. Agricul-tural data website, accessed 12 April 2003.

Garces, L. (2002) *The Detrimental Impact of Industrial Animal Agriculture.* Com-passion in World Farming, Petersfield, UK.

Kestin, S.C., Knowles, T.G., Tinch, A.E. and Gregory, N.G. (1992) Prevalence of leg weakness in broiler chickens and its relationship with genotype. *Veterinary Record* 131, 190–194.

Lymbery, P. (2002a) *Supermarkets and Animal Welfare: Raising the Standard.* Compassion in World Farming, Petersfield, UK.

Lymbery, P. (2002b) *Farm Assurance Schemes and Animal Welfare: can we Trust Them?* Compassion in World Farming, Petersfield, UK.

Qureshi, A.A. (2001) Disparities in world poultry production in 1999. *World Poultry* 17 (2), 22–23.

SCAHAW (2000) *The Welfare of Chickens Kept for Meat Production (broilers).* Report of the Scientific Committee on Animal Health and Animal Welfare. European Commission.

Teresa Farms (2001) Reviving village-level poultry enterprise: the Philippine experience. In: *Proceedings of the 3rd International Animal Feeds and Veterinary Drugs Congress, Metro Manila, Philippines, May/June 2001.* A.P. Inocencio Farms/Teresa Farms.

23 Measuring Broiler Chicken Behaviour and Welfare: Prospects for Automation

L.P.J.J. NOLDUS AND R.G. JANSEN

Noldus Information Technology, Wageningen, The Netherlands

Introduction

Other chapters in this book evaluate the extent to which the welfare of both broilers and their parent stock is compromised as a result of animal production methods. To evaluate how welfare is affected by current housing and breeding circumstances and the extent to which it is improved by changing these conditions, one must be able to objectively measure the welfare of broilers. Furthermore, welfare needs to be audited, or monitored, in the animals' actual habitat; that is, on the farm. Therefore, simple and reliable indicators for broiler welfare and tools to measure these are needed.

Welfare indicators for broilers

Numerous welfare indicators for poultry have been suggested in the literature:

- *Physical indicators*, such as bone strength, feather condition and cover, and claw length (Barnett *et al.*, 1997a,b; Rodenburg *et al.*, 2002).
- *Physiological indicators*, such as corticosterone level, heterophil:lymphocyte blood cell ratio (Barnett *et al.*, 1997a,b), body temperature and temperature of comb and wattles (B.M. Spruijt, personal communication).
- *Vocal indicators*, such as the incidence of gakel calls (Zimmerman and Koene, 1998; Koene *et al.*, 2001).
- *Behavioural indicators*, such as feather pecking, activity level (idle or active, type of activity; Barnett *et al.*, 1997a,b; Bizeray *et al.*, 2002b), the incidence of displacement preening and stereotyped pacing (Duncan and Wood-Gush, 1972), the tendency to range (Dawkins *et al.*, 2003), and walking style (Kestin *et al.*, 1992; Corr *et al.*, 2003; Savory *et al.*, 2003).

In order for a welfare indicator to be suitable for routine on-farm monitoring, several conditions must be met: it must be possible to measure it objectively and reliably, the measurement should be minimally invasive, and the measurement should be cost-effective. For the last requirement tools are needed that automate welfare measurements as much as possible.

If we limit ourselves initially to the criteria of reliability and non-invasiveness, we can say immediately that not all indicators listed above pass the test. The measurement of most physical and physiological welfare indicators requires a considerable amount of manipulation, which makes them unsuitable for large-scale on-farm application. Significant progress has been made in the *in vivo* monitoring of stress-related metabolites (Savenije *et al.*, 2003), but this technique is not yet ready for large-scale practical use. Body temperature can be measured remotely by infrared thermography, but the feasibility of this method in practice remains to be established. Obviously, vocalizations can be recorded without manipulating the birds. The same applies to overt behaviours, which can be observed without the need to touch or otherwise disturb the birds. Behavioural and vocal indicators of welfare seem to be most promising as the basis of tools for on-farm measurement of welfare. In the following sections, we will review the techniques and tools used for the computer-aided measurement of behaviour, and how these have been used in studies on chickens. We include computer tools that support the systematic observation of behaviour, and fully automated movement tracking and behaviour recognition systems. We will also discuss the cost-effectiveness of the tools available, the limitations of current techniques, prospects for further development, and the potential of other techniques for the automated measurement of welfare.

Computer-aided Behavioural Observation

Systematic observation of behaviour

The ideal behaviour monitoring system is completely automated and produces objective and reproducible measurements with a minimum of human effort. However, for many types of behaviour automated measurement techniques are not yet available and direct observation by human experts remains a necessity. Computers can improve the quality of such measurements and automate significant parts of the data collection and analysis process. Direct observation is also the starting point for the development and validation of automated behavioural measurement systems, which requires a solid knowledge of the temporal structure of the behaviour under study. In practice it means that a human expert needs to watch the animals and record the occurrence of behavioural events and interactions at the moment these are observed. Traditional pen-and-paper recording methods have gradually been replaced by computer event-recording programs, such as THE OBSERVER (Noldus, 1991; Noldus *et al.*, 2000). Such tools facilitate the data entry process and increase the accuracy of collected data by validating entries against a predefined coding scheme (Fig. 23.1) and by automating the event timing, using either the internal clock of the computer or time information extracted from video sources. The last method

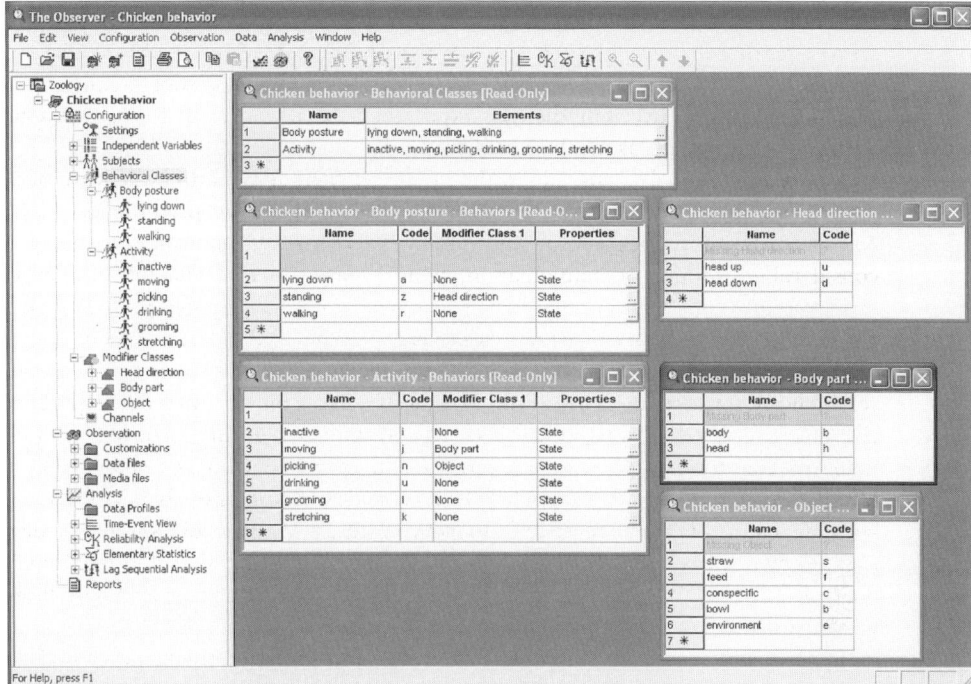

Fig. 23.1. Screen of THE OBSERVER showing the project 'Chicken behavior'. In the left-hand window, the folders named 'Configuration', 'Observation' and 'Analysis' represent the three stages in an observational study. The windows on the right show details of the coding scheme. There are two classes of behavioural elements ('Body posture' and 'Activity') and three classes of modifiers ('Head direction', 'Body part' and 'Object').

synchronizes observational data with video recordings, which allows post-test review and editing of behavioural records.

THE OBSERVER

With THE OBSERVER, detailed observations can be made from video recordings at any playback speed without loss of time information (Fig. 23.2). In poultry research, this technique is indispensable for accurate measurement of precise behaviours, such as pecking (Picard *et al.*, 1999). Another advantage is that it facilitates the use of continuous focal sampling (Altmann, 1974), which, compared with interval sampling, has the advantage that it registers all behaviours, including behaviours that occur rarely or momentarily. Furthermore, every aspect of behaviour can be studied this way, including vocalizations.

THE OBSERVER software runs on desktop or notebook PCs with Microsoft Windows 98 or higher. Special versions are available for handheld computers. These computers are available with waterproof, dustproof and shockproof housing and can be used for data collection on the farm. For detailed measurements, however, scoring

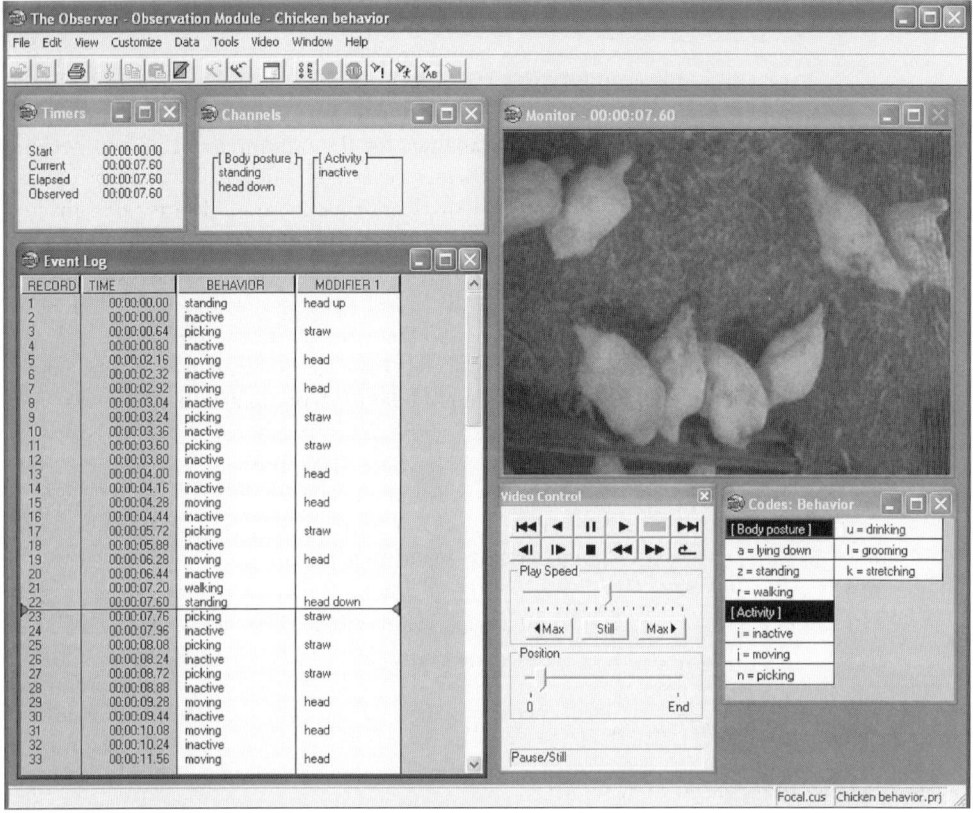

Fig. 23.2. Screen of THE OBSERVER during an observation session. The Event Log window contains the chronological list of observed events. The horizontal cursor indicates the time that corresponds with the current position in the video file, displayed in the Monitor window. (Video material by courtesy of M. Picard.)

behaviour from video is the preferred option. Recent advances in digital video technology have been extremely helpful in increasing the quality and feasibility of behavioural observations, and costs have decreased. With THE OBSERVER 5.0 (the latest release at the time of writing), behavioural data can be collected from CD, DVD and any video device connected to a computer via a FireWire interface (e.g. a digital camcorder hooked up directly to a notebook PC).

Once observational data have been collected, software tools can be used for analysis. For instance, THE OBSERVER includes functions for the creation of time–event tables and plots, calculation of the frequency and duration of behaviours, lag sequential analysis and reliability analysis, and the data can be exported to other behavioural data analysis programs, such as MATMAN (de Vries *et al.*, 1993), GSEQ (Bakeman and Quera, 1995) and THEME (Magnusson, 2000). THEME has proved particularly valuable for the analysis of the temporal structure of feed pecking (Martaresche *et al.*, 2000).

Examples of the use of THE OBSERVER in research on chicken behaviour

THE OBSERVER has been used for a wide variety of behavioural studies on chickens. A good example is the research on feeding behaviour by the group of Michel Picard, who used computer-supported video analysis for the measurement of feed pecking (Yo *et al.*, 1997a,b) and feed selection (Bessonneau *et al.*, 2001). The same group also used THE OBSERVER in studies of the activity of broilers around feeding troughs in a flock of 19,500 birds on a commercial farm (Richard *et al.*, 1997), food identification as a function of feeding situation (Vilariño *et al.*, 1998), behavioural responses to changes in diet quality (Haskell *et al.*, 2001), and the effect of sequential feeding on activity and time budgets (Bizeray *et al.*, 2002c). Activity and exercise has been the subject of several studies, as it is known to be related to the incidence of leg problems in broilers. This includes research on the detailed organization of general and locomotor activity (Bizeray *et al.*, 2002b), how this varies between genetic stocks of broilers with different growth rates (Bizeray *et al.*, 2000; Bokkers and Koene, 2003), and how it can be enhanced by environmental enrichment (Cornetto and Estevez, 2001; Bizeray *et al.*, 2002a). Other aspects of chicken behaviour and welfare that have been studied with the aid of THE OBSERVER include reproductive behaviour (Bilcik *et al.*, 2002; McGary *et al.*, 2003), feather pecking (Rodenburg *et al.*, 2002; Rodenburg and Koene, 2003) and aggressive interactions (Cornetto *et al.*, 2002).

Automated Observation of Behaviour

Techniques for automated observation

Direct observation of behaviour and manual event recording offer maximum flexibility to the researcher, because any behaviour that is observed can be recorded. However, this flexibility is also an inherent weakness of the approach, because scoring can suffer from subjectivity. Furthermore, prolonged direct observation is highly time-consuming (i.e. costly) and therefore not suitable for the routine monitoring of animal welfare. Automated observation systems provide very significant advantages and the time-saving aspect is the most important of these. Furthermore, an automated system records behaviour more reliably because it always works in the same way and does not suffer from observer fatigue or drift, so observations can continue almost infinitely (Spruijt *et al.*, 1998; Noldus *et al.*, 2001). None the less, manual, non-automated systems remain a logical precursor to automated systems, because one needs to know what should be measured before one can create a system that automates the recording and analysis of behaviour, and reference data are needed to validate the automated system against scoring done by human experts.

Technology for the automated detection and recording of animal behaviour and movement has evolved dramatically in the past two decades, from hard-wired electronics able to track a single animal in highly artificial environments, through activity meters based on infrared photobeams and ultrasound or Doppler radar, to versatile video tracking and behaviour recognition software (for a review, see Noldus *et al.*, 2001). Miniaturization and reductions in the price of sensors and transmitters continue to create new opportunities for animal monitoring and tracking.

Inexpensive passive infrared detectors (commonly used in burglar alarm systems) can be used to detect gross activity levels of animals, but these do not provide information about body posture, behavioural patterns or movement. Global positioning system (GPS) receivers have become small enough to be carried by birds for studies of ranging behaviour (Steiner *et al.*, 2000). Of all these techniques, video-based systems offer the greatest potential for automated on-farm behavioural observation and we will therefore elaborate on these in the next section.

Video tracking

Commercial video tracking systems were introduced in the early 1990s and offered clear advantages in flexibility, spatial resolution and temporal precision for many applications. An early method of automatic video tracking, still used in some commercially available systems, is to feed the analogue video signal to a dedicated tracking unit, which detects peaks in the voltage of the video signal (indicating a region of high contrast between the tracked animal and the background). These analogue systems had the disadvantage of being relatively inflexible (dedicated to particular experimental set-ups) and could track only one animal, in rather restricted lighting and background conditions. Greater flexibility is achieved by the use of a video digitizer (frame grabber), which enables real-time conversion of the entire video image to a high-resolution grid of pixels. This allows pattern analysis to be carried out on video images and so yields quantitative measurements of the observed animals' behaviour. The functionality of such a system is limited mostly by the computer's processing speed and the sophistication of the video tracking software.

The need for automated detection and recording of animal behaviour and movement has not been confined to the poultry industry. Welfare problems are also prevalent in pigs. Moreover, there is a growing need to be able to automate the recording of the behaviour and movement of laboratory rodents, preferably round the clock, for the purpose of welfare monitoring and behavioural phenotyping. These requirements are driving new developments in video tracking. Several digitizer-based video tracking systems are now commercially available. The authors are closely involved in the development of one system, ETHOVISION, which will be introduced below.

The ETHOVISION system

Development history

The ETHOVISION system has been in continuous development since the early 1990s, in collaboration with universities and industrial research laboratories. Its development was a response to the limitations of analogue systems that were predominant at that time, and the objective was (and still is) to provide a powerful general-purpose tool for video tracking, movement analysis and behaviour recognition. Combining the latest digital video hardware and feature-rich software, it is a versatile image processing system designed to automate behavioural observation and movement tracking of multiple animals against a variety of complex backgrounds. The first DOS-based implementation was released in 1993; since then the system has evolved continuously, driven by technological advances and customer demands, and this has led to the current implementation for Windows 98, 2000 and XP. The description in this chapter is based on ETHOVISION 3.0, the latest version at the time of writing.

Principle of operation

A video camera records the area occupied by the birds (an experimental enclosure, a pen or the floor of the poultry house). The video signal is fed into the computer, digitized by the frame grabber and passed on to the computer's memory. During data acquisition, ETHOVISION grabs video images at a user-defined rate (up to 30 Hz). The software analyses each frame in order to distinguish the object(s) to be tracked from the background, on the basis of either their grey scale (brightness) or hue and saturation (colour) values. The next step is object identification (when tracking more than one chicken per arena) on the basis of either size or colour differences. Having detected the objects, the software extracts relevant image features, including the coordinates of the geometric centre, the surface area, and the number of pixels that have changed position since the previous frame. The tracking process is displayed on the screen (Fig. 23.3). (For more technical details, see Noldus *et al.*, 2001.) After a data acquisition run, calculations are carried out on the image features in order to produce quantified measurements of the birds' behaviour. For instance, if the position of a chicken is known for each video frame and the whole series of frames is analysed, the average speed of locomotion of a chicken or the proportion of time spent moving during an experiment can be calculated. If multiple birds are present in one arena, the distances between several individually identified chickens can be computed for each frame. In addition, if certain regions are identified as being of interest (the centre and edges of a circular arena, for example), the proportion of time spent by the chickens in these regions can be determined.

ETHOVISION was designed as a generic tool that can be used in a wide variety of different set-ups and applications. Although most studies using ETHOVISION are carried out using laboratory rodents (for a review, see Spink *et al.*, 2001), the system is also used in research on insects, fish, birds, primates and other mammals. The ETHOVISION software has been described in detail by Noldus *et al.* (2001). The following paragraphs focus on recent innovations and those aspects of the software that are of special relevance for the study of poultry behaviour and welfare.

Zone definition

An essential aspect of all modern video tracking systems is the possibility of defining regions of interest (in ETHOVISION these are called *zones*). These can be used in the analysis (e.g. to compute the time spent in different concentric zones around a feeding nipple) and for the automatic control of experiments (see below). Zones can be combined to make a *cumulative zone* (e.g. if a test apparatus has more than one target area) or defined as *hidden zones* (to allow the system to deal with instances when the chicken is obscured by something between it and the camera, such as a laying box). Zones can be defined and altered either before or after data acquisition, allowing iterative exploratory data analysis of the effects of changing zone positions and

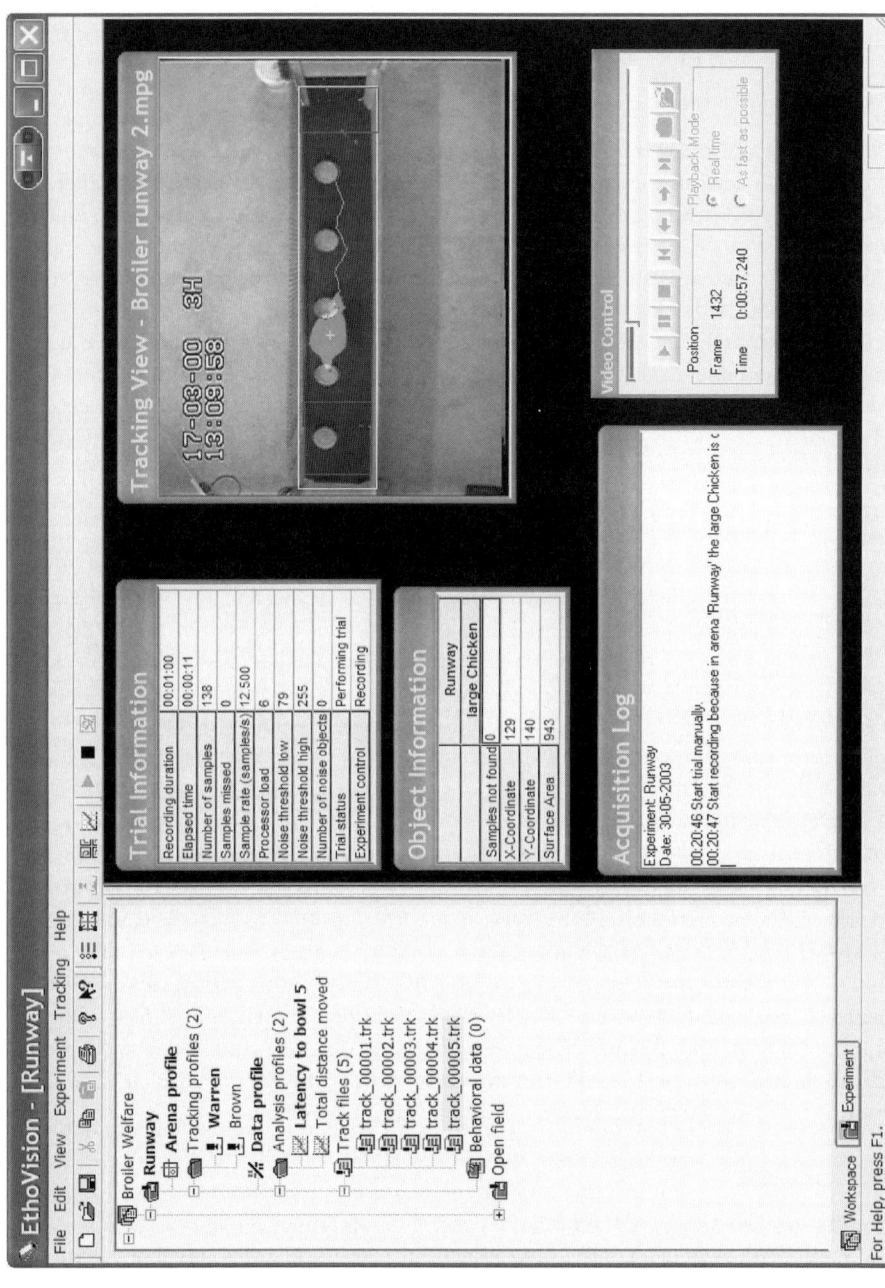

Fig. 23.3. Screen of EthoVision showing the experiment 'Runway'. During data acquisition, EthoVision displays the live video image, tracking statistics (elapsed time, number of samples, processor load, etc.), measurement values of the object (x and y coordinates and surface area) and characteristics of the trial (independent variables). (Video material by courtesy of E. Bokkers.)

shapes. Points of interest can also be defined (e.g. the locations of feeding bowls and drinking nipples).

Image resolution

The frame grabber used by ETHOVISION has a resolution of 768×576 pixels. Using the rule of thumb that an object in a digitized image should occupy at least three adjacent pixels in order to be reliably distinguished from noise, this implies that the system can track an animal in an arena up to about 200 times its size. Thus, if a broiler is 0.1 m wide, the maximum size of the area overseen by a single camera is 20×20 m. To obtain the maximum arena size one should take the smallest dimension of the bird's body (i.e. the width of a bird, not its length). Furthermore, the number 200 assumes optimal illumination and contrast; in practice the maximum is usually lower.

Digital video files

An alternative to the use of a frame grabber fed by a live video source or recording on tape is tracking on the basis of digital video files (e.g. MPEG-1, MPEG-2), a feature available in ETHOVISION 3.0. Storage of video images in digital files on disk has many advantages over the use of videotape, including duplication without loss of quality and fast random access. The latter is especially useful if video files are to be inspected visually with an observational tool such as THE OBSERVER VIDEO-PRO (Noldus *et al.*, 2000), thus synchronizing movement tracking with behavioural event recording. Digital video files also allow more advanced data management and project administration, and the possibility of combining the animation of a movement track with playback of the corresponding video clip.

Experiment control

ETHOVISION has a facility for automatic experiment control. This can be used to start or stop data acquisition depending on the location of the animal, or to trigger external events (by sending a TTL pulse to a computer port). For instance, in a test of motor ability or spatial orientation, the system can be set to start a trial when a chick enters an arm of a plus maze and stop the trial after it has been in a specific target location for more than a user-defined length of time (i.e. a spatial and a temporal criterion combined). One can also let the system perform trials in series, with the number of trials and the inter-trial interval (e.g. 72 consecutive trials, each with 5 min duration and separated by 55 min intervals) defined by the user. Such series are useful in, for instance, studies of rhythms of diurnal activity.

Data analysis

Tracks can be plotted on the computer screen, with or without the arena and zones or the background image of the experimental set-up. ETHOVISION can replay the tracks at a user-defined speed, allowing a detailed and interactive visual analysis of the data. For numerical analysis, one can group tracks by the values of independent variables; for example, the mean velocity can be calculated for all chickens that were reared on a particular diet. Data can be smoothed by activating a 'minimal distance moved' filter (to eliminate slight apparent movements) and down-sampling steps (to eliminate redundant data if a lower sample rate describes the path equally well or better). This is especially important when an accurate quantification of the path shape is required (Bell, 1991). A wide range of quantitative measures of behaviour is available in ETHOVISION, including parameters for location and time, path shape, individual behavioural states and social interactions, and these can be described with a full range of descriptive statistics. Parameters, statistics and raw data can be exported in a variety of formats for making graphs and carrying out further analysis in third-party programs, such as THE OBSERVER (Noldus *et al.*, 2000), Microsoft EXCEL, SAS and SPSS. For a detailed list of ETHOVISION's analysis parameters, see Noldus *et al.* (2001).

Examples of the use of ETHOVISION in research on broiler behaviour and welfare

Behavioural and movement parameters for quantification of hunger

De Jong *et al.* (2003) tried to establish objective parameters for the quantification of hunger in broiler breeders to evaluate new management or feeding systems that may alleviate hunger and thus improve the welfare of broiler breeders. Two strains of individually housed female broiler breeders (white Hybro G and brown JA57) were subjected to *ad libitum* feeding and feeding at 90, 70, 50, 35 and 25% of *ad libitum*. At 6–7 weeks of age, home pen behaviour, behaviour in the open field and baseline plasma concentrations of corticosterone and glucose were determined. Thereafter, birds were subjected to the feed intake motivation test (FIM test), which measures compensatory feed intake. Behaviour (walking, sitting, standing, peck feeder, pick nipple, foraging, comfort, peck object, and other) was studied in the home pen at 7 weeks (when the birds were still subjected to the feeding restrictions), and at 8 weeks (FIM test, i.e. after feeding restrictions had been removed) using scan sampling with THE OBSERVER. The open field consisted of a square of 2 × 2 m and the floor was covered with sawdust. The walls were 1.5 m high and covered with brown paper. ETHOVISION was used to measure the distance walked. The bird was placed in a corner of the open field, and the trial was started as soon as the researcher who placed the bird in the open field was no longer in the camera's view. At 1 sample per second the trial lasted 5 minutes. Grey-scaling was used to distinguish the birds from the background (i.e. sawdust). The program was set to track objects brighter than the background in the case of the white Hybro G birds, whereas for the brown JA57 birds, the setting 'object darker than background' was used. Lighting was direct, using luminescent tubes providing about 200 lux. The behaviour of the birds was recorded on videotape and one observer sitting behind the wall of the open field scored the number of vocalizations. The following behaviours were scored from the videotapes using THE OBSERVER software: sitting, standing, walking, preening, foraging, pecking at the wall and defecating. A significant linear relationship was found between the level of restriction and the time inactive (i.e. sitting or standing without doing anything else) measured in the open field. The researchers have planned a future study of the locomotion of groups of 1- to 3-day-old broiler chickens.

Effects of sex and type of feed on motivation and ability to walk for a food reward

Bokkers and Koene (2002) conducted a runway experiment using ETHOVISION to investigate the effects of sex (4–7 weeks of age) and two types of feed (conventional versus free-range, i.e. less energy and protein) on motivation and the ability to walk for a food reward (mealworm). The runway measured 2.4 m long × 0.4 m wide × 0.6 m high and five black bowls were placed on it, 0.4 m apart (Fig. 23.4). The runway was painted black and had no roof. A bird entered the runway through a black wooden start box with a sliding door at one end of the runway.

The session started as soon as the door was opened. Images were processed at a rate of 16.6 samples per second using 'subtraction absolute' as the detection method (i.e. objects are detected if they are darker or lighter than the background) and birds could not be detected by the system until they came out of the box. White broilers (Ross) were individually tested in three sessions: (i) a control session, in which each bowl contained one mealworm; (ii) a frustration session (to measure motivation), in which the first four bowls were empty and the last bowl contained five mealworms; and (iii) an obstacle session (to measure walking ability), in which each bowl contained one mealworm and black-painted obstacles 0.1 m high were placed between the bowls, which the broilers needed to cross in order to obtain their reward. For data analysis, four zones were defined in ETHOVISION

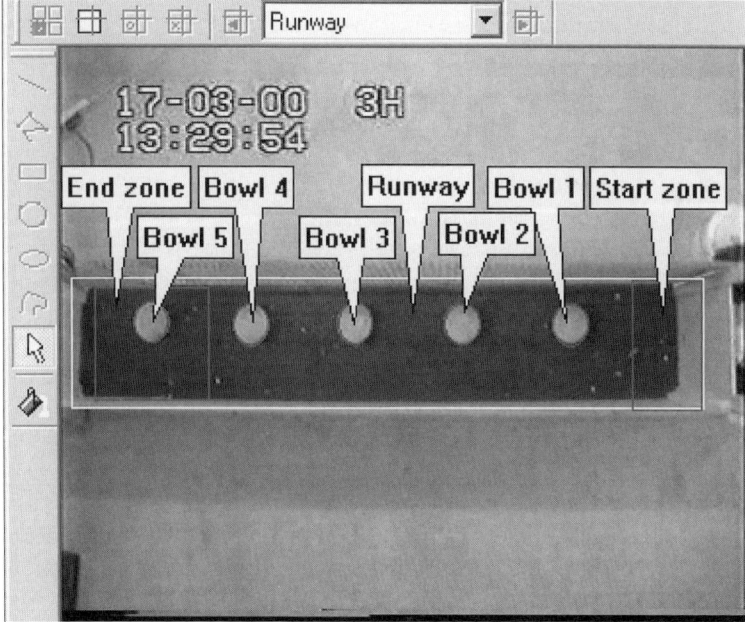

Fig. 23.4. ETHOVISION arena definition. The video image of the set-up is used to draw and define the arena and zones of interest. The rectangular arena contains a start zone that is close to a guillotine door through which the chicken enters the runway, five zones representing the feeding bowls, and an end zone around the furthest bowl. (Video material by courtesy of E. Bokkers.)

around the first four bowls. These were drawn runway-wide from 0.15 m in front of to 0.15 m behind the centre of each bowl. Latency to leave the start box, latency to reach the last bowl, walking speed, and walking speed in each zone were measured with ETHOVISION. Furthermore, sitting and preening behaviour and the number of vocalizations were scored manually. Male broilers were found to walk faster to a food reward and to vocalize more than females. Compared with free-range-fed birds, conventionally-fed birds vocalized less, sat more and had a greater latency to leave the start box and to reach the last bowl, but no differences in walking speed were recorded. No sex or feed differences were found for motivation and the ability effect. The authors concluded that frustration and obstacles had similar effects on both sexes and both types of feed, and that sex differences in walking speed were probably a result of physical differences between males and females.

Analysis of behaviour and production traits for quantitative trait loci

Schütz *et al.* (2002) used ETHOVISION to monitor a foraging–social maze to investigate phenotypic correlations in an F_2 intercross between red junglefowl and White Leghorns with respect to behaviour, i.e. foraging strategies, social behaviours and production traits. A further aim was to scan the genome for possible quantitative trait loci associated with the behavioural traits. They used four identical specially constructed mazes to analyse the behaviour of four birds simultaneously. Each arena consisted of four solid-sided arms, each measuring 0.8 m long × 0.5 m wide × 0.5 m high, and one central box measuring 0.5 × 0.5 × 0.5 m, arranged as in a plus maze (Fig. 23.5). Two arms contained novel food, to obtain which the birds had to spend energy, whereas the other two arms contained commercial laying hen food. Also, at the end of two arms, one with each type of food, a mirror was placed at an angle of about 20° outside the arms to simulate a social stimulus when the bird was deep into the arm. Wire mesh was used as a roof, both for the central box and the arms, and as a barrier between the mirror and the arms. The wire mesh was attached to a frame that was just outside the arena. The birds were able to see a companion bird through a window in the central box to prevent them from feeling socially isolated.

The light intensity was 70–75 lux during tests. The animals were introduced in complete darkness. The tests were started (four arenas at once) by turning on the light. ETHOVISION recorded the path of movement and (percentage of) time spent in each arm. Tracks lasted 20 minutes at a sample rate of 3.57 samples per second. The object was detected using background subtraction. The wire mesh roof was simply included in the background image. The researchers tried to track two birds at the same time, but this was very difficult because of the highly varying colour and size of the F_2 intercross. To eliminate disturbance from reflections (i.e. to prevent reflecting surfaces being confused with the birds), they placed the rough side of hardboard panels on the inside of the maze, covered reflecting surfaces with grey tape, and adjusted the 'minimal size of object' setting when needed. In all, Schütz *et al.* (2002) tested 773 birds, half of them male and half female, and principal components analysis revealed four behavioural patterns, which were labelled tentatively as 'social tendency', 'exploration', 'fear' and 'contrafreeloading' (the

Fig. 23.5. Foraging–social maze used by Schütz *et al.* (2002). (Picture courtesy of K. Schütz.)

tendency to search for food even if it can be obtained easily without searching). Some of these patterns were found to be clearly linked to production traits. For instance, exploration was found to decrease with increased egg weight. Furthermore, males with a fast growth rate tended to perform less contrafreeloading, supporting the hypothesis that contrafreeloading is likely to decrease as a result of selection for increased production.

Evaluation of ETHOVISION as a tool for broiler behaviour and welfare measurement

Suitability of video tracking for automated observation

Automated observation using video tracking is particularly suitable for measuring three types of behaviour: locomotor behaviour, expressed as spatial measurements (distance, speed, turning, etc.) that the human observer is unable to estimate accurately; behaviours that occur briefly and are then interspersed with long periods of inaction; and behaviours that occur over many hours, such as diurnal variation in behaviour (for references, see Noldus *et al.*, 2001). The resolution of modern image sensors, usually better than 500×500 pixels, allows movement tracking at a spatial resolution more than 10–15 times that of a typical grid of photobeams (e.g. one with 32×32 beams). This allows chickens to be tracked in arenas 100–200 times their size. On the other hand, this also sets an upper limit to the observation area. The temporal resolution, up to the video frame rate set by TV standards (PAL, 25 frames per second; NTSC, 30 frames per second) is much higher than that of other tracking techniques, such as radio tracking, radar tracking and GPS tracking. This makes video tracking ideally suited for detailed studies of movement patterns and interactions between chickens. On the other hand, video tracking using CCD cameras has the inherent drawback of requiring a minimum amount of light (visible, infrared or UV-irradiating fluorescent pigments) and being sensitive to occlusion of the object being tracked (caused, for example, by rooster bars or drinking nipples). One of the strengths of ETHOVISION is that it is flexible enough to be used in a wide variety of experimental set-ups and applications. Other video tracking software has been specifically designed for a particular experiment, such as the Morris water maze for rodents (e.g. Spooner *et al.*, 1994; Mukhina *et al.*, 2001), whereas ETHOVISION can work with any shape of arena and with a large variety of backgrounds and lighting conditions. Although there are, of course, practical limitations, in principle ETHOVISION can be used to track chickens in any situation where the bird is within sight of a camera and in a delimited and constant area that is no more than 200 times its size. Welfare-related measurements that can be automated with ETHOVISION include activity level (distance moved), activity in specific zones (around drinking nipples, food troughs): *viz.*, the number of visits, time in the zone, and abnormal movement patterns (path shape).

The importance of illumination

Proper illumination is an essential requirement for any video tracking set-up. Video tracking works best with homogeneous illumination across the scene and a high level of contrast between animal and background. In a practical farm setting, the distribution of light may vary because of shadows cast by sunlight falling into the pen. When using a video tracking system, this must be dealt with by choosing a simple and robust method for detecting objects (e.g. one based only on grey scales). For monitoring of poultry in the dark, an infrared light source in combination with an infrared-sensitive camera works well. In fact, most modern monochrome CCD cameras are sufficiently sensitive in the near-infrared (up to 1000 nm). If tracking needs to proceed around the clock, in daylight as well as in darkness, the camera should be fitted with a band-pass filter adapted to the infrared light source used. This way, the camera will not detect the daylight and – with the IR light left on continuously – ETHOVISION will track undisturbed for 24 hours a day.

Measuring interactions and body orientation

The ability of ETHOVISION to track multiple objects in an arena means that interactions between animals can be identified (Spruijt *et al.*, 1992; Sams-Dodd, 1995; Rousseau *et al.*, 1996; Sgoifo *et al.*, 1998). Poultry researchers can use this technique in studies on, for example, allo-pecking, social hierarchy and sexual behaviour. Marking and tracking separate parts of animals independently, as was done by Šustr *et al.* (2001) in their research on pigs, is also possible with chickens, provided that they are sufficiently large for dots of paint to be applied well apart. This method could be used to study the orientation responses of chickens to a variety of stimuli.

Other Computer Techniques

Gait analysis

Because of the susceptibility of fast-growing broilers to pathologies of the leg bones and joints, leading to lameness, there is a need for objective methods to assess walking ability in commercial flocks. A widely used method is the gait scoring system proposed by Kestin *et al.* (1992). With this method, walking ability is rated by a human observer by the use of six categories, ranging from completely normal to immobile. However, in recent years the Bristol gait score (Kestin *et al.*, 1992) has been criticized for being unreliable and producing inconsistent results (Garner *et al.*, 2002; Savory *et al.*, 2003). For this reason, improvements are being proposed. Garner *et al.* (2002) developed a modified gait scoring system with improved within-observer and between-observer reliability. Other groups are developing automated scoring systems. Gait patterns can be characterized with a pedobarograph, a device which allows pressure patterns to be established for various regions of the foot (Corr *et al.*, 1998), or with a force plate, which measures the ground reaction force in one or more directions (vertical, craniocaudal or mediolateral). According to Corr *et al.* (2003), a standard force plate can be used to objectively measure the ground reaction force in walking adult hens, but the large amount of variation in the measurements obtained renders the technique of limited practical use. Savory *et al.* (2003) are more optimistic. Experiments with their force plate system, a 1.6 m long runway equipped with four force transducers, showed that, out of 24 quantitative descriptors of walking style, two (step length and the standard deviation of vertical force) could correctly classify birds as either lame or healthy. However, several improvements of the technique are needed before it can be used in commercial practice. An alternative for a force plate may be the CatWalk, developed for gait analysis in laboratory rodents (Hamers *et al.*, 2001), which enables easy visualization of contact between the foot and the floor and detailed and quantitative analysis of many gait parameters. To our knowledge, this technique has not yet been applied to chickens.

Sound analysis

The study of vocalizations as an indicator of broiler welfare has received less attention than the overt behavioural responses discussed above. Most publications on chicken vocalizations concern the domestic laying hen (for references, see Zimmerman, 1999; Koene *et al.*, 2001). However, Siegel (1989) mentioned that poultry farmers routinely use sound to judge the well-being of a flock of birds, indicating that broiler vocalizations can be of commercial interest as a monitoring tool. More recently, advances in digital sound analysis technology have made it possible to unravel the acoustical structure of the chicken's vocal repertoire and to study the relationship between physiological state and vocal behaviour. With modern PC-based sound analysis tools, such as SIGNAL (Engineering Design, www.engdes. com) or AVISOFT-SASLAB (Avisoft Bioacoustics, www.avisoft.de), vocalizations can be quantified by means of, for example, call intensity (loudness), pitch (sound frequency), number of calls, mean length of calls and number of notes per call.

Vocalizations can be visualized as a 'sonagram', a graph in which sound frequency, intensity and temporal structure are integrated (Fig. 23.6), and spectral analysis can be used to classify specific call patterns. A chicken call that has received special attention is the gakel call, which typically consists of a whining, elongated note, followed by a variable number of short notes (Zimmerman and Koene, 1998; Zimmerman *et al.*, 2000b). In the laying hen, it has been shown that gakel calls are a sign of frustration: deprivation of food, water and dustbathing elicits gakel calls (Zimmerman *et al.*, 2000a) and the number of gakel calls is positively correlated with the degree of hunger (Zimmerman *et al.*, 2000b). Similar results have been obtained with broilers, in which the number of vocalizations increases with food deprivation (Koene *et al.*, 1999), which means that vocalizations are potential welfare indicators for group-housed broilers. Digital sound analysis offers fascinating prospects for the automated determination of the state of farm animals, which is being investigated for chickens, pigs and cows (e.g. Jahns *et al.*, 1997). The automated detection and quantification of calls in large broiler flocks awaits commercial application, but we

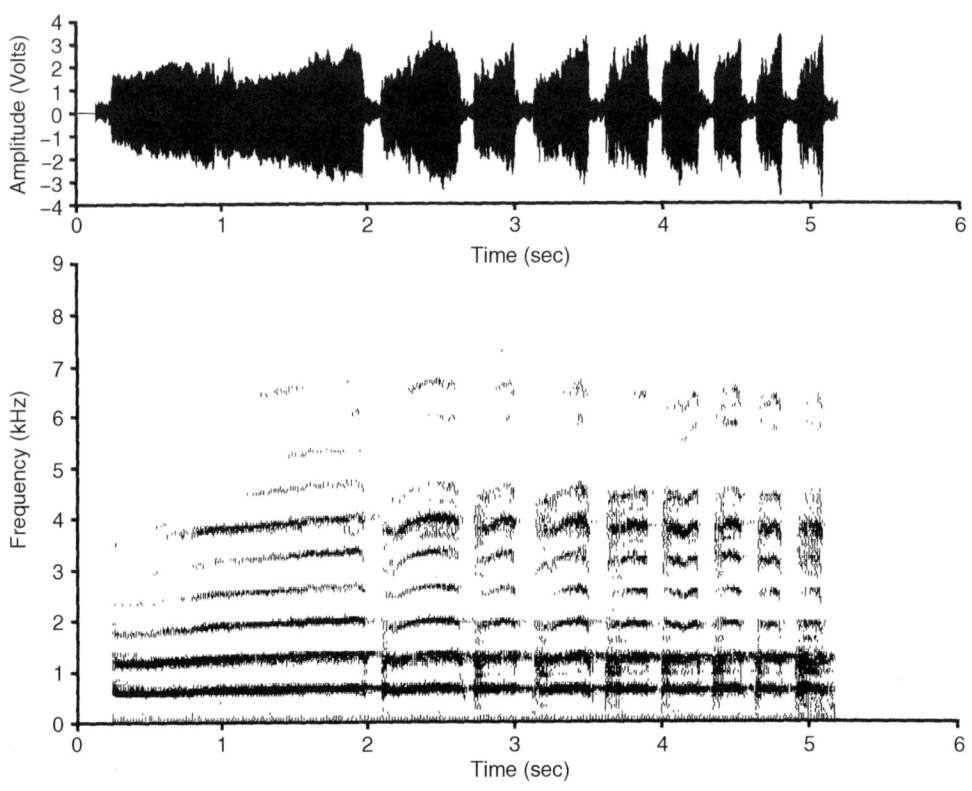

DF: 39 Hz   DT: 25.6 ms   T-inc: 10.4 ms   FFT: 512   Wind:HANN
Hi-Filt: OFF   Lo:–40 db   Hi: –6 db

Fig. 23.6. Visualization of a gakel call. On the *x*-axis is the time in seconds. Top: sound waveform, with the intensity (amplitude) on the *y*-axis. Bottom: sonagram, with the pitch (frequency) on the *y*-axis. A darker shade of grey indicates a higher intensity of a certain frequency. (Figure courtesy of P. Koene.)

agree with Koene *et al.* (2001) that vocalization-monitoring systems have a future in the broiler industry.

Infrared thermography

Advances in computer vision have contributed to the automatic classification of behavioural patterns. However, certain behavioural states of animals cannot be distinguished visually. It is known that animals exposed to stressors show various physiological responses, including changes in body temperature. Monitoring the body temperature of an animal may thus yield additional information about the state of the animal. Infrared thermography offers a non-invasive alternative to existing temperature measurement methods (e.g. implanted transmitters). A thermographic camera converts the thermal radiance emitted by the body into a video image. Thermographic cameras have evolved significantly in recent years. Earlier models required cooling with liquid nitrogen, which was very cumbersome and severely limited the positioning and handling of the device (e.g. they could not be pointed downwards at a scene). Modern thermal imagers are non-cooled, can be calibrated in degrees Celsius (they are accurate to 0.1°C) and produce a standard video signal (e.g. PAL at 25 frames per second), which allows computerized image processing. This is necessary to extract the temperature values pertaining to the objects of interest, i.e. the animal's whole body or a specific body part. Combining such a camera with a customized version of ETHOVISION produces a unique integrated and non-invasive system that can track movement, behaviour and body temperature simultaneously. This technique has been used successfully in studies on rats (Houx *et al.*, 2000; B.B. Houx and B.M. Spruijt, submitted for publication). In the case of chickens, comb temperature is known to vary with stress (Cabanac and Aizawa, 2000), so thermal imaging may be used to measure temperature changes in a non-intrusive manner. Broilers have relatively small combs, and the feasibility of this measurement technique remains to be established.

The price of the hardware remains an issue. Although thermographic cameras have come down in price from more than €100,000 to around €15,000 for a low-end model, this price tag is still prohibitive for most non-research applications. However, the cost–benefit balance may well swing to the positive side within the next 5 years or so.

Discussion

From laboratory to farm

In the previous sections we have reviewed a variety of techniques and tools, each of which can be used to measure specific welfare indicators for broiler chickens. We should realize, however, that each of these techniques has been developed for research purposes, for use on small numbers of animals in controlled environments. These conditions are quite different from those on commercial farms with many thousands of birds. When taking the tools discussed above from the laboratory to the farm, it is clear that not all can be used for the continuous monitoring of a flock of birds. Tools requiring the presence of an operator, such as THE OBSERVER (where the human observer scores the data) or a gait analysis system (where an operator is needed to place birds on a walkway), can only be used for welfare audits on small samples of birds, taken out of the flock for a specific test. Fully automated methods, on the other hand, have potential for round-the-clock monitoring. This includes video tracking and automated behavioural observation (e.g. using ETHOVISION), sound analysis and infrared thermography. Because video tracking is the most versatile technique for automated behavioural observation, we will discuss its potential and limitations at the farm level in more detail.

On-farm video tracking

The on-farm use of a video tracking system entails several problems that do not arise in the laboratory. For instance, low ceilings may limit the size of the arena, reflections and a poor distribution of light may disturb object detection, high humidity or a large amount of dust may affect electronic components, and rooster bars that are above ground level may cause additional tracking problems. Furthermore, despite the fact that video tracking technology has proved its usefulness in the study of poultry behaviour and movement, it is still not possible to measure welfare indicators in a large flock of birds. Each of these issues needs to be dealt with before automated on-farm behavioural measurement becomes a reality. We will now discuss some aspects that might be developed in the future, thus providing some idea of how long we expect it will take to accomplish this, taking account of bottlenecks and alternatives.

Size of the observation area

In theory, the image resolution of standard video equipment limits the size of the arena observed by a single camera to about 200 times the size of the chicken, which corresponds to a limit of 20×20 m for a chicken 10 cm wide. This is larger than any type of experimental maze. However, in practice the arena size will more often be limited by the height of the farm ceiling, since it may not be possible to place the camera high enough to view the entire arena, even when a wide-angle lens is used. So in reality the maximum observation area of a single camera may be as small as 5×5 m. This means that only a tiny fraction of the farm area can be monitored. A possible solution is to connect an array of cameras to a multiplexer, each observing a specific region of the pen. In this way surveillance over a much larger area becomes possible. Meyhöfer (2001) used this technique in field studies of insects, with 16 cameras and a time-lapse VCR. Meyhöfer analysed the videotapes manually (using THE OBSERVER VIDEO-PRO; see Noldus *et al.*, 2000), but it may be possible for a video tracking system to perform this task automatically, provided that the background, light conditions, contrast, etc., are adequate for video tracking.

Tracking large numbers of chickens

The study of free-ranging chickens is hampered by the number of birds that can be tracked at the same time. ETHOVISION can track up to 16 identified animals in a single enclosure. However, this can be done only if each bird is marked with a different colour. Obviously, even if the farmer allows the birds to be marked, this method is not suited for tracking individuals in a commercial flock, where the birds are crowded close together and the pattern of clustering is continually varying. Furthermore, 16 birds is a very small sample compared with the thousands of individuals in a commercial flock. Tracking large numbers of unmarked animals requires a different approach. Some progress has been made using track segmentation and reconstruction (Buma *et al.*, 1998) and the active modelling and prediction of the animals' shapes (Sergeant *et al.*, 1998; Bulpitt *et al.*, 2000) (Fig. 23.7). Using these approaches, it is no longer possible to recognize each bird individually. Once it becomes feasible to track large numbers of unidentified chickens, a video tracking system can be used to monitor chicken behaviour around feeding bowls, drinking nipples, perches and openings. The development of new ways to track large numbers of unidentified chickens may offer ways to automate the measurement of additional welfare indicators, such as the overall activity level, the degree of clustering, the tendency to range, and spacing.

Automatic detection of behaviours and movement patterns

In its current implementation, ETHOVISION is able to detect whether a chicken is moving or whether two individuals are approaching each other, but (in common with other commercial systems) it cannot automatically detect other behaviours or movement patterns (unless, of course, it is possible to define them in terms of the existing parameters). Researchers have used model-based pattern recognition, statistical classification and neural networks to detect body postures and behavioural

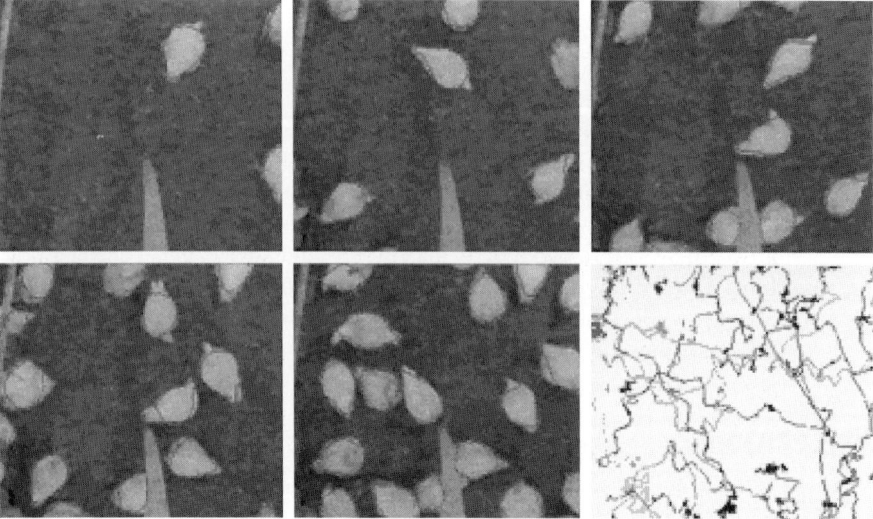

Fig. 23.7. Video tracking of multiple broilers as described by Bulpitt and colleagues. The figure shows five processed video frames and calculated trajectories. The system was able to track 30 birds simultaneously out of 93 individuals that entered the scene during a 5-minute period. Over 90% of the birds were tracked successfully. (From Bulpitt *et al.*, 2000.)

sequences in rats automatically (van Lochem *et al.*, 1998; Heeren and Cools, 2000; Rousseau *et al.*, 2000; Twining *et al.*, 2001) and pigs (Šustr *et al.*, 2001). Some of these techniques may also be applied to poultry. A complementary approach to automated measurement of behaviour focuses primarily on the detection of movements of the body or extremities (molecular behaviours), such as vertical head movement, lateral head movement, pivoting and head retraction, followed by the identification of patterns, within the sequence of movements, that represent behaviours with an identifiable goal, meaning or motivation (molar behaviours), such as allo-pecking, preening and stereotyped pacing. The THEME program (Magnusson, 2000) could possibly aid in identifying such behavioural structures. Further development and refinement of these algorithms is necessary before they can be incorporated into a standard video tracking system.

Multimodal data acquisition

It is hard to verify a particular welfare state based on a single indicator. For instance, running, a behaviour that is fairly rare in fast-growing broilers, could be associated with very different emotional/motivational states, such as anxiety and aggression. Of course, determining emotional states can be made more reliable by taking into account the context of the behaviour; that is, by looking at which behaviours (of the same bird or other birds) precede and follow a particular behaviour. Still, this might not always be sufficient. Emotional/motivational states can be identified much more reliably in the presence of other sources of information. Combining acoustic, visual and physiological information (e.g. body temperature) could result in far more reliable determination of emotional states. Also, by combining the level of sound in the yard with the presence of a particular pattern of clustering of birds, it may be possible to identify disturbances automatically.

Acknowledgements

Many colleagues at Noldus Information Technology have contributed to the development of THE OBSERVER and ETHOVISION over the years; their hard work and dedication is acknowledged here. We are also grateful to those users who provided us with research data, illustrations and advice, in particular Eddie Bokkers, Inma Estevez, Ingrid de Jong, Paul Koene, Michel Picard, Bas Rodenburg, Karin Schütz and Berry Spruijt. Thanks are due to Claire Weeks for valuable comments on earlier versions of the manuscript.

References

Altmann, J. (1974) Observational study of behavior: sampling methods. *Behaviour* 49, 227–267.

Bakeman, R. and Quera, V. (1995) *Analyzing Interaction: Sequential Analysis with SDIS and GSEQ.* Cambridge University Press, Cambridge.

Barnett, J.L., Glatz, P.C., Newman, E.A. and Cronin, G.M. (1997a) Effects of modifying layer cages with solid sides on stress physiology, plumage, pecking and bone strength of hens. *Australian Journal of Experimental Agriculture* 37, 11–18.

Barnett, J.L., Glatz, P.C., Newman, E.A. and Cronin, G.M. (1997b) Effects of modifying layer cages with perches on stress physiology, plumage, pecking and bone strength of hens. *Australian Journal of Experimental Agriculture* 37, 523–529.

Bell, W.J. (1991) *Searching Behaviour: the Behavioural Ecology of Finding Resources.* Chapman and Hall, London.

Bessonneau, D., Chagneau, A.M., Le Fur, C., Bouchot, C., Lessire, M. and Picard, M. (2001) SRABOX, un nouveau test de choix alimentaire chez le poulet: application aux aliments contenant ou non des matières premières d'origine animale. In: *Quatrièmes Journées de la Recherche Avicole, Nantes, 27–29 Mars 2001*, pp. 169–172.

Bilcik, B., Estévez, I. and Luque, M.R. (2002) Impact of male–male competition and morphological traits on reproductive behavior in broiler breeders. *Poultry Science* 80 (Supplement 1), 5.

Bizeray, D., Leterrier, C., Constantin, P., Picard, M. and Faure, J.M. (2000) Early locomotor behaviour in genetic stocks of chickens with different growth rates. *Applied Animal Behaviour Science* 68, 231–242.

Bizeray, D., Estevez, I., Leterrier, C. and Faure, J.M. (2002a) Effects of increasing environmental complexity on the physical activity of broiler chickens. *Applied Animal Behaviour Science* 79, 27–41.

Bizeray, D., Leterrier, C., Constantin, P., Le Pape, G. and Faure, J.M. (2002b) Typology of activity bouts and effect of fearfulness on behaviour in meat-type chickens. *Behavioural Processes* 58, 43–55.

Bizeray, D., Leterrier, C., Constantin, P., Picard, M. and Faure, J.M. (2002c) Sequential feeding can increase activity and improve gait score in meat-type chickens. *Poultry Science* 81, 1798–1806.

Bokkers, E.A.M. and Koene, P. (2002) Sex and type of feed effects on motivation and ability to walk for a food reward in fast growing broilers. *Applied Animal Behaviour Science* 79, 247–261.

Bokkers, E.A.M. and Koene, P. (2003) Behaviour of fast- and slow-growing broilers to 12 weeks of age and the physical consequences. *Applied Animal Behaviour Science* 81, 59–72.

Bulpitt, A.J., Boyle, R.D. and Forbes, J.M. (2000) Monitoring behavior of individuals in crowded scenes. In: *Proceedings of Measuring Behavior 2000, 3rd International Conference on Methods and Techniques in Behavioral Research, Nijmegen, 15–18 August 2000*, pp. 28–30.

Buma, M.O.S., Moskal, J.R. and Liang, D. (1998) EthoVision Multi-Pro: improved animal identification during automatic multi-object tracking. In: *Proceedings of Measuring Behavior '98, 2nd International Conference on Methods and Techniques in Behavioral Research, Groningen, 18–21 August 1998*, pp. 103–104.

Cabanac, M. and Aizawa, S. (2000) Fever and tachycardia in a bird (*Gallus domesticus*) after simple handling. *Physiology and Behavior* 69, 541–545.

Cornetto, T. and Estevez, I. (2001) Behavior of the domestic fowl in the presence of vertical panels. *Poultry Science* 80, 1455–1462.

Cornetto, T., Estevez, I. and Douglas, L.W. (2002) Using artificial cover to reduce aggression and disturbances in domestic fowl. *Applied Animal Behaviour Science* 75, 325–336.

Corr, S.A., McCorquodale, C.C. and Gentle, M.J. (1998) Gait analysis of poultry. *Research in Veterinary Science* 65, 233–238.

Corr, S.A., McCorquodale, C.C., McGovern, R.E., Gentle, M.J. and Bennett, D. (2003) Evaluation of ground reaction forces produced by chickens walking on a force plate. *American Journal of Veterinary Research* 64, 76–82.

Dawkins, M.S., Cook, P.A., Whittingham, M.J., Mansell, K.A. and Harper, A.E. (2003) What makes free-range broiler chickens range? In situ measurement of habitat preference. *Animal Behaviour* 66, 151–160.

de Jong, I.C., van Voorst, A.S. and Blokhuis, H.J. (2003) Parameters for quantification of hunger in broiler breeders. *Physiology and Behavior* 78, 773–783.

Duncan, I.J.H. and Wood-Gush, D.G.M. (1972) Thwarting of feeding behaviour in the domestic fowl. *Animal Behaviour* 20, 444–445.

Garner, J.P., Falcone, C., Wakenell, P., Martin, M. and Mench, J.A. (2002) Reliability and validity of a modified gait scoring system and its use in assessing tibial dyschondroplasia in broilers. *British Poultry Science* 43, 355–363.

Hamers, F.P.T., Lankhorst, A.J., van Laar, T.J., Veldhuis, W.B. and Gispen, W.H. (2001) Automated quantitative gait analysis during overground locomotion in the rat: its application to spinal cord contusion and transection injuries. *Journal of Neurotrauma* 18, 187–201.

Haskell, M.J., Vilariño, M., Savina, M., Atamna, J. and Picard, M. (2001) Do broiler chicks have a cognitive representation of food quality? Appetitive, behavioural and ingestive responses to a change in diet quality. *Applied Animal Behaviour Science* 72, 63–77.

Heeren, D.J. and Cools, A.R. (2000) Classifying postures of freely moving rodents with the help of Fourier descriptors and a neural network. *Behavior Research Methods, Instruments, and Computers* 32, 56–62.

Houx, B.B., Buma, M.O.S. and Spruijt, B.M. (2000) Non-invasive temperature tracking with automated infrared thermography: measuring inside out. In: *Proceedings of Measuring Behavior 2000, 3rd International Conference on Methods and Techniques in Behavioral Research, Nijmegen, 15–18 August 2000*, pp. 151–152.

Jahns, G., Kowalczyk, W. and Walter, K. (1997) An application of sound processing techniques for determining condition of cows. In: *Proceedings of the 4th International Workshop on Systems, Signals and Image Processing, IWSSP '97, Poznan, Poland*, pp. 105–109.

Kestin, S.C., Knowles, T.G., Tinch, A.E. and Gregory, N.G. (1992) Prevalence of leg weakness in broiler chickens and its relationship with genotype. *Veterinary Record* 131, 190–194.

Koene, P., van Ruiten, S. and Bokkers, E.A.M. (1999) Increasing behavioural possibilities of broilers by giving extra furniture and a slimmer body: the effects of perches and feed restriction. In: *Proceedings of the 33rd International Congress of the ISAE, Lillehammer, 17–21 August 1999*, p. 136.

Koene, P., Zimmerman, P., Bokkers, E. and Rodenburg, B. (2001) Vocalisation due to frustration in layer and broiler chickens. In: *Proceedings of the 6th European Symposium on Poultry Welfare, Zollikofen, 1–4 September 2001*, pp. 95–100.

Lochem, P.B.A. van, Buma, M.O.S., Rousseau, J.B.I. and Noldus, L.P.J.J. (1998) Automatic recognition of behavioral patterns of rats using video imaging and statistical classification. In: *Proceedings of Measuring Behavior '98, 2nd International Conference on Methods and Techniques in Behavioral Research, Groningen, 18–21 August 1998*, pp. 203–204.

Magnusson, M.S. (2000) Discovering hidden patterns in behavior: T-patterns and their detection. *Behavior Research Methods, Instruments, and Computers* 32, 93–110.

Martaresche, M., Le Fur, C., Magnusson, M., Faure, J.M. and Picard, M. (2000) Time structure of behavioral patterns related to feed pecking in chicks. *Physiology and Behavior* 70, 443–451.

McGary, S., Estevez, I. and Russek-Cohen, E. (2003) Reproductive and aggressive behavior in male broiler breeders with varying fertility levels. *Applied Animal Behaviour Science* 82, 29–44.

Meyhöfer, R. (2001) Intraguild predation by aphidophagous predators on parasitised aphids: the use of multiple video cameras. *Entomologia Experimentalis et Applicata* 100, 77–87.

Mukhina, T.V., Bachurin, S.O., Lermontova, N.N. and Zefirov, N.S. (2001) Versatile computerized system for tracking and analysis of water maze tests. *Behavior Research Methods, Instruments and Computers* 33, 371–380.

Noldus, L.P.J.J. (1991) The Observer: a software system for collection and analysis of observational data. *Behavior Research Methods, Instruments, and Computers* 23, 415–429.

Noldus, L.P.J.J., Trienes, R.J.H., Hendriksen, A.H., Jansen, H. and Jansen, R.G. (2000) The Observer Video-Pro: new software for the collection, management, and presentation of time-structured data from videotapes and digital media files. *Behavior Research Methods, Instruments and Computers* 32, 197–206.

Noldus, L.P.J.J., Spink, A.J. and Tegelenbosch, R.A.J. (2001) EthoVision: a versatile video tracking system for automation of behavioral experiments. *Behavior Research Methods, Instruments and Computers* 33, 392–414.

Picard, M., Plouzeau, M. and Faure, J.M. (1999) A behavioural approach to feeding broilers. *Annales de Zootechnie* 48, 233–245.

Richard, P., Vilariño, M., Faure, J.M., León, A. and Picard, M. (1997) Etude du comportement du poulet de chair dans un élevage intensif tropical au Venezuela. *Revue d'Élevage et de Médecine Vétérinaire des Pays Tropicaux* 50, 65–74.

Rodenburg, T.B. and Koene, P. (2003) Comparison of individual and social feather pecking tests in two lines of laying hens at ten different ages. *Applied Animal Behaviour Science* 81, 133–148.

Rodenburg, T.B., Zimmerman, P.H. and Koene, P. (2002) Reaction to frustration in high and low feather pecking laying hens. *Behavioural Processes* 59, 121–129.

Rousseau, J.B.I., Spruijt, B.M. and Gispen, W.H. (1996) Automated observation of multiple individually identified rats. In: *Proceedings of Measuring Behavior '96, International Workshop on Methods and Techniques in Behavioral Research, Utrecht, 16–18 October 1996*, pp. 86–87.

Rousseau, J.B.I., van Lochem, P.B.A., Gispen, W.H. and Spruijt, B.M. (2000) Classification of rat behavior with an image-processing method and a neural network. *Behavior Research Methods, Instruments and Computers* 32, 63–71.

Sams-Dodd, F. (1995) Automation of the social interaction test by a video tracking system: behavioural effects of repeated phencyclidine treatment. *Journal of Neuroscience Methods* 59, 157–167.

Savenije, B., Lambooij, B., Gerritzen, M., Venema, K. and Korf, J. (2003) Practical *in vivo* monitoring of glucose and lactate as indicators of stress [abstract]. In: Weeks, C. and Butterworth, A. (eds) *Measuring and Auditing Broiler Welfare, Bristol, 30–31 May 2003*, p. 43.

Savory, C.J., McGovern, R.E. and Sandilands, V. (2003) Objective assessment of broiler walking style using a force plate [abstract]. In: Weeks, C. and Butterworth, A. (eds) *Measuring and Auditing Broiler Welfare, Bristol, 30–31 May 2003*, p. 42.

Schütz, K., Kerje, S., Carlborg, Ö., Jacobsson, L., Anderson, L. and Jensen, P. (2002) QTL analysis of a red junglefowl × White Leghorn intercross reveals trade-off in resource allocation between behavior and production traits. *Behavior Genetics* 32, 423–433.

Sergeant, D.M., Boyle, R.D. and Forbes, J.M. (1998) Computer visual tracking of poultry. *Computers and Electronics in Agriculture* 21, 1–18.

Sgoifo, A., Kole, M.H., Buwalka, B., de Boer, S.F. and Koolhaas, J.M. (1998) Video tracking of social behaviors and telemetered ECG measurements in colony housed rats. In: *Proceedings of Measuring Behavior '98, 2nd International Conference on Methods and Techniques in Behavioral Research, Groningen, 18–21 August 1998*, pp. 258–259.

Siegel, P.B. (1989) The genetic–behaviour interface and well-being of poultry. *British Poultry Science* 30, 3–13.

Spink, A.J., Tegelenbosch, R.A.J., Buma, M.O.S. and Noldus, L.P.J.J. (2001) The EthoVision video tracking system – a tool for behavioral phenotyping of transgenic mice. *Physiology and Behavior* 73, 731–744.

Spooner, R.I.W., Thomson, A., Hall, J., Morris, R.G.M. and Salter, S.H. (1994) The Atlantis platform: a new design and further developments of Buresova's on-demand platform for the water maze. *Learning and Memory* 1, 203–211.

Spruijt, B.M., Hol, T. and Rousseau, J.B.I. (1992) Approach, avoidance, and contact behavior of individually recognized animals automatically quantified with an imaging technique. *Physiology and Behavior* 51, 747–752.

Spruijt, B.M., Buma, M.O.S., van Lochem, P.B.A. and Rousseau, J.B.I. (1998) Automatic behavior recognition: what do we want to recognize and how do we measure it? In: *Proceedings of Measuring Behavior '98, 2nd International Conference on Methods and Techniques in Behavioral Research, Groningen, 18–21 August 1998*, pp. 264–266.

Steiner, I., Bürgi, C., Werffeli, S., Dell'Omo, G., Valenti, P., Tröster, G. and Lipp, H.P. (2000) A GPS logger and software for analysis of homing in pigeons and small mammals. *Physiology and Behavior* 71, 589–596.

Šustr, P., Spinka, M., Clouthier, S. and Newberry, R.C. (2001) Computer-aided method for calculating animal configurations during social interactions from two-dimensional coordinates of color-marked body parts. *Behavior Research Methods, Instruments, and Computers* 33, 364–370.

Twining, C.J., Taylor, C.J. and Courtney, P. (2001) Characterising behavioural phenotypes using automated image analysis.

Behavior Research Methods, Instruments, and Computers 33, 381–391.

Vilariño, M., Leon, A., Faure, J.M. and Picard, M. (1998) Ability of broiler chicks to detect hidden food. *Archiv für Geflügelkunde* 62, 156–162.

Vries, H. de, Netto, W. and Hanegraaf, P.L.H. (1993) MatMan: a program for the analysis of sociometric matrices and behavioural transition matrices. *Behaviour* 125, 157–175.

Yo, T., Vilariño, M., Faure, J.M. and Picard, M. (1997a) Feed pecking in young chickens: new techniques of evaluation. *Physiology and Behavior* 61, 803–810.

Yo, T., Siegel, P.B., Guerin, H. and Picard, M. (1997b) Self-selection of dietary protein and energy by broilers grown under a tropical climate: effect of feed particle size on the feed choice. *Poultry Science* 76, 1467–1473.

Zimmerman, P.H. (1999) 'Say what?' Vocalisation as an indicator of welfare in the domestic laying hen. PhD thesis, Wageningen University, Wageningen, The Netherlands.

Zimmerman, P.H. and Koene, P. (1998) The effect of frustrative nonreward and behaviour in the laying hen, *Gallus gallus domesticus*. *Behavioural Processes* 44, 73–79.

Zimmerman, P.H., Koene, P. and Van Hooff, J.A.R.A.M. (2000a) Thwarting of behaviour in different contexts and the gakel-call in the laying hen. *Applied Animal Behaviour Science* 69, 255–264.

Zimmerman, P.H., Koene, P. and Van Hooff, J.A.R.A.M. (2000b) The vocal expression of feeding motivation and frustration in the domestic laying hen, *Gallus gallus domesticus*. *Applied Animal Behaviour Science* 69, 265–273.

4 Poster Papers

Poster 1

Reduced Welfare in Broiler Chickens may be Associated with Increased Risk of Colonization with *Campylobacter* Species

S. Haslam, S. Kestin and J. Corry

School of Veterinary Science, University of Bristol, Langford House, Langford, Bristol BS40 5DU, UK

Introduction

We hypothesized that colonization of birds by the human food-borne poisoning pathogens *Campylobacter* spp. might be affected by their state of welfare. Thus, as part of a study to evaluate the validity of a composite index (unitary welfare index, UWI), we measured numbers of *Campylobacter* spp. in caecal contents and compared them with flock UWI scores.

Method

We visited ten broiler flocks and assessed a random sample of birds for walking ability using the method described by Kestin *et al.* (1992). We recorded the mortality for the flock, the floor area of the house and the measures taken to enrich the environment of the birds. We killed 50 birds and took samples of caecal contents, which were examined for numbers of colony-forming units per gram of thermophilic *Campylobacter* spp. by plating decimal dilutions on to modified cefaperazone charcoal deoxycholate agar, and incubating at 42°C for 48 h in a microaerobic atmosphere. We also obtained standardized data from the slaughter plant for footpad dermatitis and bird weight. From the measures taken, we calculated the UWI score for each flock. We analysed the data using bivariate correlation, testing for significance with the Spearman rank correlation coefficient.

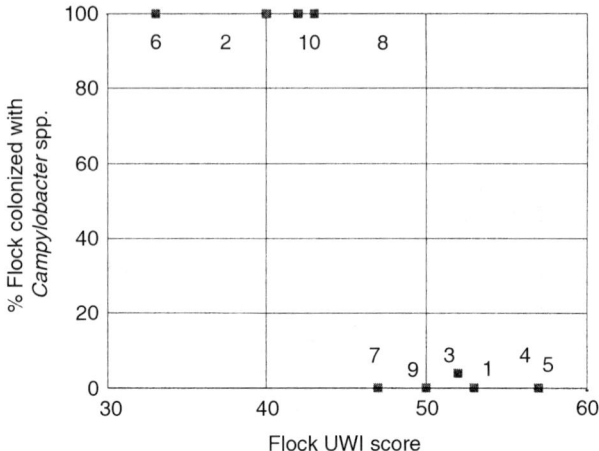

Fig. 1. A scatter plot of percentage of birds colonized with *Campylobacter* spp. and UWI score.

Results

There was a negative correlation between bird welfare and colonization with *Campylobacter* (Fig. 1). The Spearman rank correlation coefficient for *Campylobacter* colonization with UWI score was –0.83 ($P < 0.01$).

Discussion and Conclusions

This preliminary study in broiler chickens strongly suggests that good welfare reduces and poor welfare increases the probability of colonization by *Campylobacter* spp. However, further studies are necessary to confirm this.

Reference

Kestin, S.C., Knowles, T.G., Tinch, A.E. and Gregory, N.G. (1992) Prevalence of leg weakness in broiler chickens and its relationship with genotype. *Veterinary Record* 131, 190–194.

Poster 2

Estimating the Number of Broilers to Sample to Determine the Prevalence of Lameness

S.C. KESTIN AND T.G. KNOWLES

School of Veterinary Science, University of Bristol, Langford, Bristol BS40 5DU, UK

Introduction

In a study of leg disorders in broilers we wished to establish the prevalence of lameness and other disorders in several flocks. We were using the method of Kestin *et al.* (1992), which categorizes lameness into six gait score (GS) bands. We needed to know how many birds to assess so that: (i) we could have confidence in our estimates of prevalence; and (ii) we could identify important differences between flocks.

Method

First, we conducted sample size calculations according to the approach described in Altman *et al.* (2000) and employing Wilson's (1927) method. In this case, the flocks were assumed to have only two categories of lameness: GS ≤ 2 and GS > 2.

Secondly, we obtained information relating to the distribution of gait scores for different commercial flocks from the literature (Kestin *et al.*, 1992, 1999; Sanotra *et al.*, 2001). Using these data, we carried out a series of pairwise comparisons of all flocks, estimating the sample size required to identify a significant difference in the distribution of GS values between the flocks at the 5% level and with 80% power. Sample size was calculated for a two-group χ^2 test comparing proportions in six categories using NQUERY ADVISOR software version 4.0.

Results

Estimating the proportion of lame birds

Figure 1 shows how, as sample size increases from 50 to 500, the 95% confidence interval for the proportion reduces. It also shows how, as the proportion of lame birds changes from 0.02 to 0.40, the confidence interval increases for a set sample size.

Identifying a difference in GS distribution between pairs of flocks

Figure 2 shows, for each of the paired comparisons tested, the sample size required to identify the difference in the distributions of GS values between the flocks (note: because of the dimensional limitations of paper we have used the difference in mean gait score as an indicator of the different distributions).

Discussion and Conclusions

These modelling exercises clearly demonstrate how important it is to sample enough birds from each flock if estimates of the prevalence of lameness are to be reliable and important differences between flocks are to be identified.

Figure 1 indicates that if the expected prevalence of lameness (birds with GS > 2) in a flock is 0.20 (fairly low according to published information) and we require to know the true prevalence to ± 0.05 (i.e. 15–25% lame birds), it will be necessary

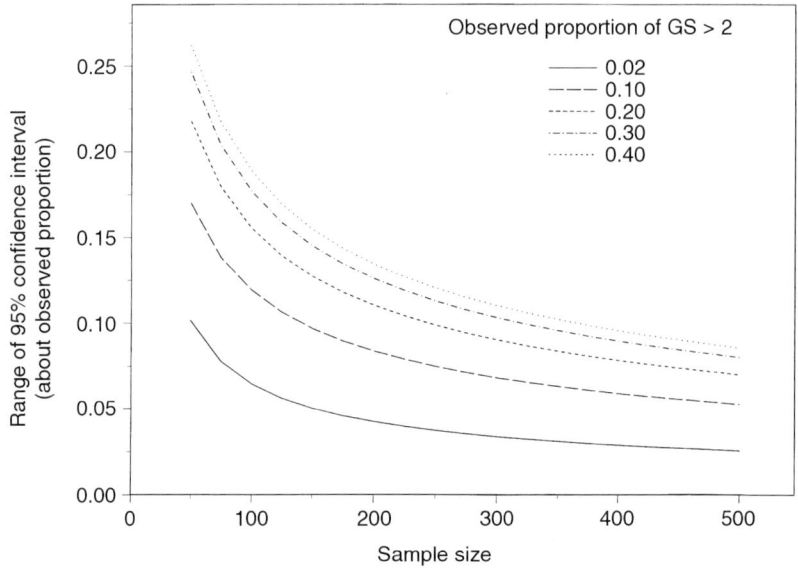

Fig. 1. Range of 95% confidence interval as affected by sample size and proportion of lame birds.

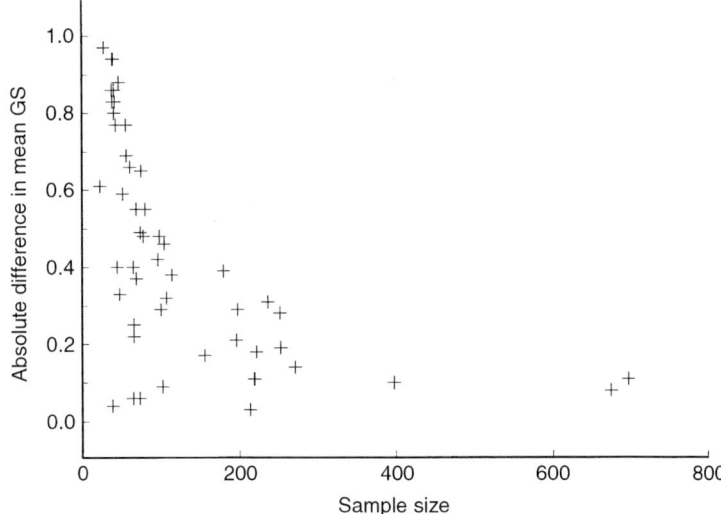

Fig. 2. Sample size required to identify the difference in distributions of mean GS values between flocks (based on paired comparisons).

to examine at least 250 birds. However, if, when measured, the true prevalence of lameness in the flock is 0.40, Fig. 1 indicates that to identify this with the same confidence interval (i.e. 35–45% lame birds) we will need to examine 370 birds. Similarly, with lower prevalences and the same sample size the confidence interval will be smaller (i.e. the precision of the estimate will be improved).

Using the data in Fig. 2, it can be calculated that a sample size of 250 birds per flock would have led to 85% of the paired comparisons of GS distribution being identifed as statistically significantly different. Similarly, sample sizes of 200 and 300 would have identified 75 and 93%, respectively, of the comparisons as showing a significant difference.

The results of these modelling exercises, together with the expected prevalence in lameness from previous studies, indicate that it will be necessary to sample at least 250 birds in each flock. These results apply to whatever method is used to assess lameness.

It follows that it is important that, for any method to be practicable for examining commercial flocks, lameness assessment methods should be able to assess birds relatively rapidly.

References

Altman, D.G., Machin, D., Bryant, T.N. and Gardner, M.J. (2000) *Statistics with Confidence*, 2nd edn. BMJ Books, London.

Kestin, S.C., Knowles, T.G., Tinch, A.E. and Gregory, N.G. (1992) The prevalence of leg weakness in broiler chickens assessed by gait scoring and its relationship to genotype. *Veterinary Record* 131, 190–194.

Kestin, S.C., Su, G. and Sørensen, P. (1999) Different commercial broiler crosses have different susceptibilities to leg weakness. *Poultry Science* 78, 1085–1090.

Sanotra, G.S., Lund, J.D., Ersboll, A.K., Petersen, J.S. and Vestergaard, K.S. (2001) Monitoring leg problems in broilers: a survey of commercial broiler production in Denmark. *World's Poultry Science Journal* 57, 55–69.

Wilson, E.B. (1927) Probable inference, the law of succcession, and statistical inference. *Journal of the American Statistical Association* 22, 209–212.

Poster 3

Practical *in vivo* Monitoring of Glucose and Lactate as Indicators of Stress

Bart Savenije[1], Bert Lambooij[2], Marien Gerritzen[2], Kor Venema[3] and Jakob Korf[3]

[1]CCL Research, PO Box 107, 5460 AC Veghel, The Netherlands; [2]ID-Lelystad, PO Box 65, 8200 AB Lelystad, The Netherlands; [3]Department of Biological Psychiatry, University Psychiatric Clinic, PO Box 30001, 9700 RB Groningen, The Netherlands

Metabolic parameters may be used as indicators of welfare. For example, acute stress mobilizes glucose and prolonged feed deprivation exhausts the readily available energy for coping. Monitoring metabolites may clarify the effects of treatments to which we expose broilers. Recently we developed a system based on ultrafiltration for the remote, continuous and nearly stress-free sampling of small amounts (50 nl/min) of subcutaneous fluid. Samples were analysed off-line for glucose and lactate levels, generating one measurement for every 2 minutes of sampling time.

The sampling device is small enough to be carried by a bird under the wing without obstructing normal behaviour. The animal needs to be restrained only briefly for the device to be attached and removed. The bird can be replaced in its home pen, even if group-housed, or undergo any regular treatments. Little pecking at the device was observed. Subcutaneous rather than intravenous sampling was studied to determine its advantages in ease and speed of application and the risk of haemorrhage or infection. Subcutaneous measurements were validated with intravenous measurements: while levels differed, the patterns were identical.

Although invasive sampling may be seen as compromising welfare ethics, methods for judging the effects of practical treatments over time are scarce, but much needed. For example, it is difficult to assess differences in the relative effects of preslaughter treatments under practical conditions because of the nature of these treatments. Also, point measurements have been shown to be of limited value. The experimental use of invasive sampling with a limited impact on welfare to generate relevant data patterns to assess practical treatments may be justified.

Poster 4

Objective Assessment of Broiler Walking Style Using a Force Plate

C.J. Savory[1], R. McGovern[2] and V. Sandilands[1]

[1]Avian Science Research Centre, Scottish Agricultural College, Ayr KA6 HW, UK; [2]Engineering Resources Group, Scottish Agricultural College, Aberdeen AB21 TR, UK

It is necessary to assess the incidence of lameness in commercial broiler flocks regularly as part of continued effort to minimize this problem. The aim of this DEFRA-funded work was to design a method for objective assessment of walking style, based on recording forces applied by each leg as a bird walks along a force-measuring plate. Methods of motivating birds to walk on such a plate were also tested. Broilers that were deemed to be either lame ($n = 29$) or not lame ($n = 33$) were tested at the Scottish Agricultural College and two commercial farms when they were 4–6 weeks old. They were placed one at a time at the head of the 1.6 m long force plate and induced to walk along it by blowing compressed air, clapping sticks and/or rustling a plastic bag, all behind the bird. Four load cells incorporated into the plate recorded forces in three dimensions from (overall mean) 8.9 footsteps per run and 5.4 runs per bird. Associated computer software was developed to automatically identify and analyse the data obtained for each foot placement. Mean values were calculated for each of 24 quantitative descriptors of walking style and for each of a subject bird's feet. These were analysed by analysis of variance, principal components analysis and discriminant analysis. The most useful information came from the last approach, which revealed that just two descriptors of walking style (step length and standard deviation of vertical force, standardized for body weight) could correctly classify as either lame or not lame 90% of the birds that walked along the plate in an acceptable manner. However, when we included birds that refused to walk on the plate and ones that produced data that were not analysable, the success rate dropped to about 50%.

Poster 5

Evaluation of Fluctuating Asymmetry as an Indicator of Broiler Welfare

A. Van Nuffel[1], W. Talloen[2], B. Sonck[1], L. Lens[3] and F. Tuyttens[1]

[1]Department of Mechanisation, Labour, Buildings, Animal Welfare and Environmental Protection, Agricultural Research Centre, Burgemeesters Van Gansberghelaan 115, 9820 Merelbeke, Belgium; [2]Terrestrial Ecology Unit, Laboratory of Animal Ecology, Department of Biology, University of Antwerp, Universiteitsplein, 1 2610 Wilrijk, Antwerp, Belgium; [3]Department of Biology, Ghent University, K. Ledeganckstraat 35, 9000 Ghent, Belgium

Many morphological traits of animals are genetically programmed to be bilaterally symmetrical. Under optimal circumstances, these traits are expected to develop identically on the left and right sides of the body. The extent to which an animal succeeds in undergoing stable development of the phenotype can be measured in terms of fluctuating asymmetry. This is a measure of the small, random deviations of the perfect symmetrical development of bilateral traits. Because the degree of fluctuating asymmetry appears to be a good indicator of a developing animal's ability to cope with environmental stressors, it may be useful in measuring animal welfare. However, a well-founded basis for fluctuating asymmetry measurements is still lacking for most species. Many fundamental questions remain unanswered, such as: (i) which and how many morphological structures ought to be measured, (ii) the baseline degree of variation of fluctuating asymmetry in a population, and (iii) the influences of environmental factors and genetic selection on the degree of fluctuating asymmetry.

The general objective of our research is to evaluate the extent to which measurements of fluctuating asymmetry can be used in order to measure and monitor farm animal welfare objectively and efficiently. As a first step towards this goal, it is crucial to develop a measuring protocol that specifies which combination of morphological traits provides the best estimate of the degree of fluctuating asymmetry of a dead or living bird, and how these traits can be measured precisely and efficiently. We will also investigate the stage of development at which these traits are most sensitive to stress. Next, these measurements of fluctuating asymmetry will be examined as valid and efficacious indicators of broiler welfare by: (i) investigating the effect of known stressors on fluctuating asymmetry in an experimental setting, (ii) comparing on-farm assessments of broiler welfare with fluctuating asymmetry, and (iii) comparing the degree of fluctuating asymmetry in individual birds with other indicators of welfare, such as parasite load, tonic immobility and gait score.

Poster 6

Effects of Growth Rate and Dietary Calcium and Phosphorus Content on Performance, Bone Mineralization and Gait Score of Broilers

E. VENÄLÄINEN, J. VALAJA AND T. JALAVA

Animal Production Research, MTT Agrifood Research Finland, FIN-31600 Jokioinen, Finland

A 37-day growth experiment was carried out with 2900 Ross 208 broilers. The experiment was conducted as a continuous design with a $2 \times 2 \times 4$ factorial arrangement of treatments. Treatments consisted of sex (male and female), two growth rates, achieved by two levels of energy content (11 and 12 MJ ME/kg), and four levels of calcium and phosphorus. The calculated calcium:available phosphorus ratio was 2.0 for all experimental diets and was adjusted to account for differences in dietary energy content. Starter and grower diets containing 12 MJ ME/kg were calculated to contain 0.4, 0.45, 0.5 or 0.55% and 0.35, 0.4, 0.45 or 0.5% available phosphorus respectively.

The difference in body weight between dietary energy groups was on average 144 g at the end of the study ($P < 0.001$). Tibia ash, calcium and phosphorus content of slower-growing broilers was higher ($P < 0.01$) than for faster-growing broilers. Tibia ash, calcium and phosphorus content and tibia breaking strength of males was higher ($P < 0.05$) than for female birds. Growth rate and dietary level of calcium and phosphorus had no effect on tibia breaking strength. Tibia ash, calcium and phosphorus content increased curvilinearly in response to increases in dietary calcium and phosphorus content. Dietary calcium and phosphorus content had no effect on feed intake or feed conversion ratio.

Higher growth rate appeared to increase gait score of broilers at 23 days of age ($P < 0.001$). At 37 days of age, gait scores of males were higher than females ($P < 0.001$). Dietary calcium and phosphorus content had no effect on broiler gait score.

©CAB International 2004. *Measuring and Auditing Broiler Welfare*
(eds C. Weeks and A. Butterworth)

Index